Lecture Notes in Mathematics

Edited by A. Dold and B. Eckmann

T0210792

1036

Combinatorial Mathematics X

Proceedings of the Conference
held in Adelaide, Australia, August 23–27, 1982

Edited by L. R. A. Casse

Springer-Verlag
Berlin Heidelberg New York Tokyo 1983

Editor

Louis Reynolds Antoine Casse
The University of Adelaide
G.P.O., Box 498, Adelaide, South Australia 5001, Australia

AMS Subject Classifications (1980): 05A05, 05A15, 05B05, 05B10, 05B15, 05B20, 05B25, 05B30, 05B35, 05B40, 05B45, 05B50, 05C05, 05C25, 05C30, 05C60, 05C65, 05C75, 12C20, 15A23, 15A33, 20B 22, 20B25, 51E05, 51E15, 51E25, 62K10, 68E10, 90B35, 94B05, 94B70

ISBN 3-540-12708-9 Springer-Verlag Berlin Heidelberg New York Tokyo
ISBN 0-387-12708-9 Springer-Verlag New York Heidelberg Berlin Tokyo

Printing and binding: Beltz Offsetdruck, Hemsbach/Bergstr.
2146/3140-543210

PREFACE

The Tenth Australian Conference on Combinatorial Mathematics was held at the University of Adelaide from 23rd to 27th August 1982.

The Conference was fortunate enough to hear addresses given by distinguished combinatorialists: C.C. Chen (Singapore), J.W.P. Hirschfeld (U.K.), D.A. Holton (Australia), A.D. Keedwell (U.K.), C. Lindner (U.S.A.), N.J. Pullman (Canada), D. Stinson (Canada), and J.A. Thas (Belgium). This volume contains the text of seven of these invited addresses and of twenty-three contributed talks. Manuscripts of the remaining contributed talks given at the conference are to be published elsewhere.

Many people helped with the organisation of the conference and with the publication of this volume, and we are grateful to all of them. We thank all those who chaired sessions and refereed papers. Our special thanks go to the members of the "Geometry Seminar" and of the departments of Mathematics at the University of Adelaide.

We particularly acknowledge the generous support of the following

Australian Mathematical Society

A.N.Z. Bank

Trans Australia Airlines.

The University of Adelaide, the Student Union, and the University of Adelaide Club allowed us the use of their facilities. Much financial assistance was provided by the Departments of Applied Mathematics, Computer Science, Pure Mathematics, and Statistics.

Finally we thank Ms. Henderson, Mrs. Renshaw and Mrs. Halsey for their kindness and typing.

L.R.A. Casse

PARTICIPANTS

Mr. R. Aldred	Department of Mathematics, University of Melbourne, Parkville, Vic., 3052.
Dr. J. Arkinstall	Mathematics and Computing Department, South Australian Institute of Technology, Whyalla, S.A., 5600.
Prof. G. Berzsenyi	Department of Mathematics, Lamar University, Beaumont, Texas, U.S.A.
Mr. S. Bourn	Pure Mathematics Department, University of Adelaide, Adelaide, S.A., 5001.
Dr. D.R. Breach	Department of Mathematics, University of Canterbury, Christchurch, New Zealand.
Dr. R. Buttsworth	Department of Mathematics, University of Queensland, St. Lucia, Qld., 4067.
Dr. L. Caccetta	School of Mathematics and Computing, Western Australian Institute of Technology, South Bentley, W.A., 6102.
Dr. R. Casse	Department of Pure Mathematics, University of Adelaide, Adelaide, S.A., 5001.
Prof. C.C. Chen	Department of Mathematics, National University of Singapore, Bukit Timah Road, Singapore, 1025.
Dr. R.J. Clarke	Department of Pure Mathematics, University of Adelaide, Adelaide, S.A., 5001.
Prof. J. Colbourn	Department of Computational Science, University of Saskatchewan, Saskatchewan, Saskatoon, S7N 0W0, Canada.
Dr. M.J. Colbourn	Department of Computational Science, University of Saskatchewan, Saskatchewan, Saskatoon, S7N 0W0, Canada.
Dr. W.H. Cornish	School of Mathematical Sciences, The Flinders University of S.A., Bedford Park, S.A., 5042.

Dr. E.A. Cousins	Department of Applied Mathematics, University of Adelaide, Adelaide, S.A., 5001.
Mr. E. Dawson	Department of Mathematics, Queensland Institute of Technology, P.O. Box 2434, Brisbane, Qld., 4001.
Mr. W. De Launay	Department of Applied Mathematics, University of Sydney, Sydney, N.S.W., 2006.
Dr. P. Eades	Department of Computing Science, University of Queensland, St. Lucia, Qld., 4067.
Mr. M. Ellingham	Department of Mathematics, University of Melbourne, Parkville, Vic., 3052.
Dr. D.G. Glynn	Department of Pure Mathematics, University of Adelaide, Adelaide, S.A., 5001.
Dr. J. Hammer	Department of Pure Mathematics, University of Sydney, Sydney, N.S.W., 2006.
Dr. W. Henderson	Department of Applied Mathematics, University of Adelaide, Adelaide, S.A., 5001.
Dr. J. Hirschfeld	School of Mathematics and Physical Sciences, University of Sussex, Falmer, Brighton, BN1 9QH, England.
Dr. D.A. Holton	Department of Mathematics, University of Melbourne, Parkville, Vic., 3052.
Prof. A.F. Horadam	Department of Mathematics, University of New England, Armidale, N.S.W., 2351.
Mr. R. Johnston	Department of Applied Mathematics, La Trobe University, Bundoora, Vic., 3083.
Dr. A.D. Keedwell	Department of Mathematics, University of Surrey, Guildford, Surrey, GU2 5XH, England.
Mr. R.W. Kennington	Department of Applied Mathematics, University of Adelaide, Adelaide, S.A., 5001.

Mr. S. Kettle	Department of Mathematics, Monash University, Clayton, Vic., 3168.
Prof. C. Lindner	Mathematics Department, Auburn University, Auburn, Alabama, 36849, U.S.A.
Dr. K.L. McAvaney	Division of Computing and Mathematics, Deakin University, P.O. Box 125, Belmont, Vic., 3217.
Dr. R.G.J. Mills	South Australian Institute of Technology, P.O. Box 1, Ingle Farm, S.A., 5098.
Mr. P. O'Halloran	School of Information Sciences, Canberra College of Advanced Education, P.O. Box 1, Belconnen, A.C.T., 2600.
Ms. C. O'Keefe	Department of Pure Mathematics, University of Adelaide, Adelaide, S.A., 5001.
Dr. D. Parrott	Department of Pure Mathematics, University of Adelaide, Adelaide, S.A., 5001.
Dr. C. Pearce	Department of Applied Mathematics, University of Adelaide, Adelaide, S.A., 5001.
Mr. T. Pentilla	Department of Pure Mathematics, University of Adelaide, Adelaide, S.A., 5001.
Dr. B.B. Phadke	School of Mathematical Sciences, The Flinders University of S.A., Bedford Park, S.A., 5042.
Dr. E.J. Pitman	Department of Pure Mathematics, University of Adelaide, Adelaide, S.A., 5001.
Dr. R. Potter	75 Alexander Drive, River Plaza, New Jersey, 07701, U.S.A.
Dr. C. Praeger	Department of Mathematics, University of Western Australia, Nedlands, W.A., 6009.
Prof. N.J. Pullman	Queens University, Kingston, Ontario, K7L 3N6, Canada.

Dr. A.J. Rahilly	Gippsland Institute of Advanced Education, Switchback Road, Churchill, Vic., 3842.
Dr. C. Rodger	Mathematics Department, Auburn University, Auburn, Alabama, 36849, U.S.A.
Dr. D. Rogers	Mathematics Department, The University, Reading, RG6 2AX, England.
Dr. F. Salzborn	Department of Applied Mathematics, University of Adelaide, Adelaide, S.A., 5001.
Dr. J. Seberry	Department of Applied Mathematics, University of Sydney, Sydney, N.S.W., 2006.
Dr. B. Sherman	Department of Education, University of Adelaide, Adelaide, S.A., 5001.
Mr. R.J. Simpson	Department of Pure Mathematics, University of Adelaide, Adelaide, S.A., 5001.
Dr. D. Skillicorn	Department of Computing Information Sciences, Queens University, Kingston, Ontario, K7L 3N6, Canada.
Mr. D. Skilton	Department of Mathematics, University of Newcastle, Shortland, N.S.W., 2303.
Mr. B. Smetaniuk	Department of Applied Mathematics, University of Sydney, Sydney, N.S.W., 2006.
Dr. D.R. Stinson	University of Manitoba, Winnipeg, Manitoba, R3T 2N2, Canada.
Dr. A.P. Street	Department of Mathematics, University of Queensland, St. Lucia, Qld., 4067.
Mrs. M. Sved	Department of Pure Mathematics, University of Adelaide, Adelaide, S.A., 5001.
Mrs. E. Szekeres	School of Mathematics, University of New South Wales, P.O. Box 1, Kensington, N.S.W., 2033.

Prof. G. Szekeres	School of Mathematics, University of New South Wales, P.O. Box 1, Kensington, N.S.W., 2033.
Prof. J.A. Thas	Seminar of Geometry and Combinatorics, University of Ghent, Krijgslaan 271, 9000 - Ghent, Belgium.
Mr. R. Turner	Department of Mathematics, University of Melbourne, Parkville, Vic., 3052.
Dr. W. Venables	Department of Statistics, University of Adelaide, Adelaide, S.A., 5001.
Prof. W.D. Wallis	Department of Mathematics, University of Newcastle, Shortland, N.S.W., 2308.
Dr. A. Werner	P.O. Box 252, Elizabeth, S.A., 5112.
Dr. P. Wild	Department of Pure Mathematics, University of Adelaide, Adelaide, S.A. 5001.
Dr. N. Wormald	Department of Mathematics, University of Newcastle, Shortland, N.S.W., 2308.

TABLE OF CONTENTS

In the case of co-authored papers, an asterisk(*) indicates the author who spoke at the conference

INVITED PAPERS

C.C. Chen* and N. Quimpo:
Hamiltonian Cayley graphs of order pq . 1

J.W.P. Hirschfeld:
The Weil conjectures in finite geometry. 6

D.A. Holton:
Cycles in graphs. 24

A.D. Keedwell:
Sequenceable groups, generalized complete mappings, neofields, and block designs. 49

N.J. Pullman:
Clique coverings of graphs - A survey. 72

D. Stinson:
Room squares and subsquares. 86

J.A. Thas:
Geometries in finite projective spaces: recent results. 96

CONTRIBUTED PAPERS

S. Bourn:
A canonical form for incidence matrices of finite projective planes and their associated Latin squares and planar ternary rings. 111

L. Caccetta* and N. Pullman:
On clique covering numbers of cubic graphs. 121

R.J. Clarke:
Modelling competitions by poset multiplication. 128

C.J. Colbourn and M.J. Colbourn:
Decomposition of block designs: computational issues. 141

W.H. Cornish:
A combinatorial problem and the generalized cosh. 147

W. DeLauney:
Generalised Hadamard matrices whose rows and columns form a group. 154

M.N. Ellingham:
The asymptotic connectivity of labelled coloured regular bipartite graphs. 177

H.M. Gastineau-Hills:
Kronecker products of systems of orthogonal designs. 189

H.M. Gastineau-Hills and J. Hammer*:
Kronecker products of systems of higher dimensional orthogonal designs. 206

D. Glynn:
Two new sequences of ovals in finite Desarguesian planes of even order. 217

W. Henderson, R.W. Kennington and C.E.M. Pearce:
Stochastic processes and combinatoric identities. 230

S.G. Kettle:
Families enumerated by the Schroder-Etherington sequence and a renewal array it generates. 244

S.G. Kettle:
Classifying and enumerating some freely generated families of objects. 275

K.L. McAvaney:
Composite graphs with edge stability index one. 305

J. Pitman and P. Leske:
A number-theoretical note on Cornish's paper. 316

C.E. Praeger* and P. Schultz:
On the automorphisms of rooted trees with height distributions. 319

A. Rahilly and D. Searby:
On partially transitive planes of Hughes type (6,m). 335

C.A. Rodger:
Embedding incomplete idempotent Latin squares. 355

B. Smetaniuk:
The completion of partial f-squares. 367

M. Sved:
Baer subspaces in the n-dimensional projective space. 375

G. Szekeres:
Distribution of labelled trees by diameter. 392

W.D. Wallis* and L. Zhu:
Orthogonal Latin squares with small subsquares. 398

P.R. Wild* and L.R.A. Casse:
K-sets of (n-1)-dimensional subspaces of PG(3n-1,q). 410

N.C. Wormald:
Subtrees of large tournaments. 417

THE FOLLOWING TALKS WERE ALSO GIVEN AT THE CONFERENCE

Invited Talk:

C. Lindner:
How many triples can a pair of Steiner triple systems have in common?

Contributed Papers:

R.E.L. Aldred:
C(m,n) properties of graphs.

D.R. Breach* and A.R. Thompson:
A census of 3-(12,6,4) designs.

R.N. Buttsworth:
Polynomial representation of generalised Steiner systems.

L.R.A. Casse* and D.G. Glynn:
Recent results on $(q+1)_r$-arcs, $q=2^h$.

P. Eades:
Some minimal change algorithms.

J.A. Hoskins, C.E. Praeger and A. Penfold Street:
Twills with bounded float length.

D. Rogers* and D.J. Crampin:
Harmonious windmills and other graphs.

D. Rogers:
Some rewards on additive and complete permutations.

J. Seberry:
The skew-weighing matrix conjecture.

J. Seberry:
Bhaskar Rao designs of block size three over groups of order divisible by three.

D.B. Skillicorn:
Computer network design using combinatorial methods.

D.K. Skilton:
Infinite graphs and transection-free chain decompositions.

HAMILTONIAN CAYLEY GRAPHS OF ORDER PQ

C.C. CHEN and N. QUIMPO

Every abelian Cayley graph is edge-hamiltonian. Every Cayley graph of order p,q where p,q are primes is also edge-hamiltonian.

1. INTRODUCTION

Let G be a (finite) group generated by X. We shall denote by $G(X)$ the graph whose vertices are elements of G and any two vertices a,b are adjacent (written $a \sim b$) if and only if $a^{-1}b \in X \cup X^{-1}$, where $X^{-1} = \{x^{-1} : x \in X\}$. The graph $G(X)$ thus obtained is called a *Cayley graph*. If G is an abelian group, then it is called an *abelian Cayley graph*. Every Cayley graph is connected and vertex-transitive. The well-known Lovasz conjecture says that every finite connected vertex-transitive graph has a hamiltonian path. (See Problem 20 in [1]). It is therefore natural to ask if every Cayley graph has a hamiltonian path or cycle. In [2], it is shown that every abelian Cayley graph is hamiltonian. For non-abelian Cayley graphs, the problem seems a lot more difficult. The main object of this paper is to prove that every abelian Cayley graph is edge-hamiltonian (i.e. every edge lies in a hamiltonian cycle), and that every Cayley graph of order pq, where p,q are primes, is also edge-hamiltonian.

2. ABELIAN CAYLEY GRAPHS

We first establish the following lemmas.

Lemma 1. *Let* $G(X)$ *be an abelian Cayley graph. Then for each* $x \in X \cup X^{-1}$, $G(X)$ *contains a hamiltonian cycle through the edge* $\langle e,x\rangle$ *where* e *denotes the identity of the group* G.

Proof. We list elements of $X \cup X^{-1}$ in a sequence $x_1, x_2, \dots, x_n$ with $x_1 = x$ and $n = |X \cup X^{-1}|$. For each $i=1,2,\dots,n$, let $X_i = \{x_1, \dots, x_i\}$ and $G_i = [X_i]$, the subgroup generated by X_i.

Claim. *Each* $G_i(X_i)$ *contains a hamiltonian cycle through the edge* $\langle e,x\rangle$.

Indeed, as $x = x_1 \in X_i$, each $G_i(X_i)$ contains $\langle e,x\rangle$ as an edge. We shall prove our claim by induction on i. For $i=1$, $G_1(X_1)$ is itself a cycle through $\langle e,x\rangle$. Assume that the claim is valid for

$i < k$ and consider the case $i = k > 1$. If $x_k \in G_{k-1}$, then $G_k(X_k) = G_{k-1}(X_{k-1})$. Hence by induction hypothesis, $G_k(X_k)$ contains a hamiltonian cycle through $<e,x>$. On the other hand, if $x_k \notin G_{k-1}$, then G_k is the disjoint union $G_{k-1} \dot{\cup} x_k G_{k-1} \dot{\cup} \ldots \dot{\cup} x_k^{r-1} G_{k-1}$ of cosets where r is the smallest positive integer such that $x_k^r \in G_{k-1}$. By induction hypothesis, the subgraph G_{k-1} of $G_k(X_k)$ contains a hamiltonian cycle C through $<e,x>$ where $|C| = m = |G_{k-1}|$. As each other subgraph $x_k^j G_{k-1}$ is isomorphic to G_{k-1}, it is easy to see that $G_k(X_k)$ contains a spanning subgraph isomorphic to $C_m \times P_n$ with $<e,x>$ as an edge, where C_m and P_n denote the cycle and the path of order m and n respectively. As $C_m \times P_n$ is edge-hamiltonian, it follows that $G_k(X_k)$ contains a hamiltonian cycle through $<e,x>$, establishing our claim.

Lemma 1 now follows immediately from our claim as $G_n(X_n) = G(X)$.

Lemma 2. *Let* $G(X)$ *be any Cayley graph. If for each* $x \in X$, $G(X)$ *contains a hamiltonian cycle through* $<e,x>$, *then* $G(X)$ *is edge-hamiltonian.*

Proof. Let $<x,y>$ be any edge of $G(X)$. Then either $x^{-1}y \in X$ or $y^{-1}x \in X$, say the former. By hypothesis, $G(X)$ contains a hamiltonian cycle C through the edge $<e,x^{-1}y>$. Consider then the mapping $f:G \to G$ defined by $f(u) = xu$ for each $u \in G$. Then f is an automorphism of $G(X)$. Hence $f(C)$ is also a hamiltonian cycle of $G(X)$. Moreover, as $<e,x^{-1}y>$ lies in C, $<f(e),f(x^{-1}y)> = <x,y>$ must lie in $f(C)$. Hence $G(X)$ is edge-hamiltonian, as required.

Combining Lemmas 1 and 2, we have:

Theorem 3. *Every abelian Cayley graph is edge-hamiltonian.*

3. CAYLEY GRAPHS OF ORDER pq

From now on, we consider only non-abelian groups $G(X)$ of order pq where p,q are primes. As every group of order p^2 is abelian, by Theorem 1, to show that $G(X)$ is edge-hamiltonian, we may assume that $p < q$ in the sequel. The following lemmas will be useful later.

Lemma 4. G *contains a unique subgroup* B *of order* q *and* B *is normal.*

Proof. By the third Sylow's Theorem, G contains $1+kq$ subgroups of order q and $1+kq$ divides p. This forces $1+kq = 1$. Hence there exists exactly one subgroup of order q. It follows immediately from the uniqueness that this subgroup is normal.

Lemma 5. *Let* a,b *be elements of* G *of order* p *and* q *respectively. Then, we have:*

(i) $G = [a,b]$ *(i.e.* G *is generated by* a,b*);*

(ii) $ba = ab^r$ *for some integer* r *with* $r \not\equiv 1 (\mathrm{mod}\ q)$ *and* $r^p \equiv 1 (\mathrm{mod}\ q)$*;*

(iii) $b^n a^m = a^m b^{nr^m}$ *for any integers* m,n.

Proof. Let $H = [a,b]$. Then p,q divide $|H|$ and so $|H| \geq pq = |G|$. Hence $H = G$ proving (i).

Next by Lemma 4, the subgroup $[b]$ generated by b is normal, and so $a^{-1}ba = b^r$ for some $r=1,2,\ldots,q-1$. Evidently $r \neq 1$ or else $ab = ba$ which contradicts the fact G is non-abelian. Hence $r \not\equiv 1(\mathrm{mod}\ q)$. Now, $(a^{-1}ba)^r = (b^r)^r$ and so $b^{r^2} = a^{-1}b^r a = a^{-2}ba^2$. By induction, $b^{r^n} = a^{-n}ba^n$ for all positive integers n. In particular, $b^{r^p} = a^{-p}ba^p = b$ which implies that $r^p \equiv 1(\mathrm{mod}\ q)$. This establishes (ii).

From the proof of (ii), we see that $b^n a = ab^{r^n}$ for any integer n. Using this, (iii) can easily be established by induction.

Remark. *For convenience, for any path* $P = \langle a_1, a_2, \ldots, a_n \rangle$, *we shall denote by* P^- *the initial vertex* a_1 *of* P *and by* P^+ *the terminal vertex* a_n *of* P. *Also, for any two paths* P *and* Q, *if* $P^+ \sim Q^-$, *then we simply write:* $P \sim Q$.

Theorem 6. *The Cayley graph* $G(X)$ *is hamiltonian.*

Proof. By Lemma 4, X must contain an element a of order p. We then have the following cases to consider.

Case 1. X *contains an element* b *of order* q.

In this case, we have $[a,b] = G$ by Lemma 4. Thus, we may assume without loss of generality that $X = \{a,b\}$. Let $B = [b]$. Then G can be decomposed into p cosets $B, aB, \ldots, a^{p-1}B$. Now let $j_1, j_2, \ldots, j_{p-1}$ be integers such that $b^{q-1}a = ab^{j_1}$, $b^{j_1-1}a = ab^{j_2}$, $b^{j_2-1}a = ab^{j_3}$, and in general $b^{j_t-1}a = ab^{j_{t+1}}$ for $t=1,2,\ldots,p-1$. Also, let P_0 be the path $\langle e, b, b^2, \ldots, b^{q-1} \rangle$ in B, and P_i be the path $\langle a^i b^{j_i}, a^i b^{j_i+1}, \ldots, a^i b^{j_i+q-1} \rangle$ in $a^i B$ for $i=1,2,\ldots,p-1$. Note that $(P_0^+)^{-1}(P_1^-) = b^{-(q-1)}ab^{j_1} = b^{-(q-1)}b^{(q-1)}a = a$. Hence $P_0 \sim P_1$. Similarly, $P_i \sim P_{i+1}$ for all $i=0,1,\ldots,p-2$. This gives us a hamiltonian path $P = P_0 P_1 \ldots P_{p-1}$ in $G(X)$. Finally, $P_0^- = e$ and $P_{p-1}^+ = a^{p-1}b^{j_{p-1}-1}$. However, by Lemma 4, $j_1 \equiv (q-1)r(\mathrm{mod}\ q)$; $j_2 \equiv (j_1-1)r = j_1 r - r \equiv qr^2 - r^2 - r(\mathrm{mod}\ q)$; and eventually $j_{p-1} \equiv qr^{p-1} - r^{p-1} - r^{p-2} - \ldots - r(\mathrm{mod}\ q)$. By Lemma 4 again,

$r^p \equiv 1 \pmod q$ which implies that $(r-1)(r^{p-1}+\dots+1) \equiv 0 \pmod q$. As $r \not\equiv 1 \pmod q$, we must therefore have $r^{p-1} + \dots + r + 1 \equiv 0 \pmod q$. From this, we conclude that $j_{p-1} \equiv 1 \pmod q$. Hence $P^+_{p-1} = a^{-1} \sim e = P^-_0$. This shows that P is a hamiltonian cycle of $G(X)$, settling Case 1.

<u>Case 2</u>. *All elements of* X *are of order* p.

In this case, apart from a, X must contain another element c of order p not in $[a]$. Let B be the unique normal subgroup of G of order q. Then c must be in a^iB for some i. That is $c = a^ib$ for some $i=1,2,\dots,p-1$ and some element b of G of order q. As $b = a^{-i}c$, it is clear that $[a,c] = [a,b] = G$. Hence we may assume without loss of generality that $X = \{a,c\}$. Also, by Lemma 4, $a^{-1}ba = b^r$ for some integer r with $r \not\equiv 1 \pmod q$ and $r^p \equiv 1 \pmod q$.

If $p=2$, then $c = ab$. Then it is easy to see that $\langle e,a,b,ba,b^2,b^2a,\dots,b^{q-1},b^{q-1}a\rangle$ is a hamiltonian cycle of $G(X)$. Hence we may assume that $2 < p < q$. Then let P_t be the path $\langle b^t a^{p-1}, b^t a^{p-2},\dots,b^t a^{p-i}, b^{t+1}, b^{t+1}a,\dots,b^{t+1}a^{p-i-1}\rangle$ for any integer t. Let $t_0 = 0$, $t_1 = r+1$, $t_2 = 2r+2$, $\dots$, $t_{q-1} = (q-1)r + q - 1$. Then $(P^+_{t_0})^-(P^{-1}_{t_1}) = (ba^{p-i-1})^{-1}b^{r+1}a^{p-1} = a^{i+1}b^ra^{p-1} = a^{i+1}a^{p-1}b^{r^p}$ $= a^ib = c$. Hence $P_{t_0} \sim P_{t_1}$. Similarly $P_{t_n} \sim P_{t_{n+1}}$ for $n=0,1,\dots,q-1$. Let $P = P_{t_0}P_{t_1}\dots P_{t_{q-1}}$. For P to be a hamiltonian path, we need to show that $t_m \not\equiv t_n \pmod q$ for m different from n. Indeed, suppose to the contrary that $t_m \equiv t_n \pmod q$. Then $mr + m \equiv nr + n \pmod q$. Therefore, $(m-n)(r+1) \equiv 0 \pmod q$ which implies that $r \equiv -1 \pmod q$. Hence $r^p \equiv (-1)^p = -1 \pmod q$. But by Lemma 4, $r^p \equiv 1 \pmod q$ and so $1 \equiv -1 \pmod q$, which is impossible. Hence we conclude that P is a hamiltonian path in $G(X)$. Finally,

$$(P^+_{t_{q-1}})^{-1}(P^-_{t_0}) = (b^{t_{q-1}+1}a^{p-i-1})^{-1}a^{p-1} = a^{i+1}b^{-t_{q-1}-1}a^{p-1}$$
$$= a^ib^{r^{p-1}(-t_{q-1}-1)} = a^ib^{r^{p-1}(-(q-1)r-q-1+1)} = a^ib^{r^{p-1}(-qr-q+r)}$$
$$= a^ib^{r^{p-1}r} = a^ib^{r^p} = a^ib = c \in X.$$

This shows that P is a hamiltonian cycle of $G(X)$, completing the proof.

Now, for each a in X, there always exists b in X such that $[a,b] = G$. Let $X_1 = \{a,b\}$. From the proof of Theorem 6, we see that $G(X_1)$ always contains a hamiltonian cycle through the edge $\langle e,a\rangle$. Combining this fact and Lemma 2, we have:

<u>Theorem 7</u>. *Every Cayley graph of order* p,q *where* p,q *are primes is edge-hamiltonian.*

4. A FINAL REMARK

To end this paper, we would like to note that if G is abelian then it is shown in [3] that G(X) is hamilton-connected if and only if it is neither a cycle with more than three elements nor a bipartite graph. The following problem thus arises naturally:

Problem. *Characterize hamilton-connected Cayley graphs, and in particular those of order* pq, *where* p,q *are primes.*

REFERENCES

[1] J.A. Bondy and U.S.R. Murty, *Graph Theory with Applications*, American Elsevier, New York, 1976.

[2] C.C. Chen and N. Quimpo, On some classes of hamiltonian graphs, *Southeast Asian Bull. Maths.*, Special Issue (1976), 252-258.

[3] C.C. Chen and N. Quimpo, On strongly hamiltonian abelian group graph, *Combinatorial Maths. VIII*, (Proceedings, Geelong, Australia 1980), 23-24, (Springer-Verlag, Berlin).

THE WEIL CONJECTURES IN FINITE GEOMETRY

J.W.P. HIRSCHFELD

In the first section the Weil conjectures for non-singular primals are stated and several examples are given. Particularities for curves are described in section two. The remaining sections are devoted to elliptic cubic curves. In particular, the number of points that a cubic can have is precisely given, as well as the number of inequivalent curves with a fixed number of points.

1. THE HASSE-WEIL-DWORK-DELIGNE THEOREM

After first being posited in 1949, the final part of these conjectures was proved in 1974. The method of proof requires very deep mathematics; but it is possible to understand what the result actually says without understanding the details of the proof. Although the theorem will not be given in the most general case, it will be described in the strongest and most accessible case.

Let $K = GF(q)$, the Galois field of q elements, and let $\bar{K}$ be the algebraic closure of K. Let F be a form (homogeneous polynomial) of degree m in $K[X_0,X_1,\ldots,X_n]$, let $PG(n,q)$ be projective space of n dimensions over K, and let $P(X) = P(x_0,\ldots,x_n)$ be the point of $PG(n,q)$ with coordinate vector $X = (x_0,x_1,\ldots,x_n)$. Then

$$\mathcal{F} = V_{n,q}(F) = \{P(X) \in PG(n,q) \mid F(X) = 0\}.$$

When there is no ambiguity, we write

$$\mathcal{F} = V_{n,q}(F) = V(F).$$

We impose two conditions on $\mathcal{F}$.

(i) $\mathcal{F}$ is a *non-singular*; that is there does not exist $X = (x_0,\ldots,x_n)$ in $\bar{K}^{n+1}$ such that

$$F = \frac{\partial F}{\partial X_0} = \ldots = \frac{\partial F}{\partial X_n} = 0 \text{ at } X.$$

(ii) $\mathcal{F}$ is *absolutely irreducible*; that is, there do not exist G, H in $\bar{K}[X_0,\ldots,X_n]\backslash\bar{K}$ with $F = GH$.

Some points should be noted.

(1) F is a form, so that we are only considering projective varieties $\mathcal{F}$; otherwise, in counting arguments, some "points at infinity" would be lost.

(2) Varieties defined over K may have singular points over $\bar{K}\backslash K$. For example, if $K = GF(q)$ with $q \equiv -1 \pmod 4$, then $\mathcal{F} = V(F)$ where

$$F = (X_1^2 + X_2^2)^2 + (X_1^2 - X_2^2)X_0^2 + X_0^4$$

has the singular points $P(0,\pm i,1)$, $i^2 = -1$; the singular points lie over the quadratic extension of K, but not over K. Over the reals, F is known as a "bicircular" quartic.

(3) (i) implies (ii), since if F is reducible over $\bar{K}$ to $F = G \cup H$, then the points over $\bar{K}$ of $G \cap H$ will be singular points of F. Later, condition (i) will be dropped for part of the discussion on curves.

Now, still with F a form in $n + 1$ variables of degree m over GF(q) satisfying (i) and (ii), let

$$F_i = V_{n,q^i}(F), \quad N_i = |F_i|.$$

We wish to find N_i, the number of points defined by F over $GF(q^i)$, for all i together.

The *zeta function* of F is

$$\zeta(F) = \zeta(F;T;q) = \exp(\Sigma N_i T^i/i).$$

Theorem 1.

(i) $\zeta(F) = f(T)^{(-1)^n}/\{(1-T)(1-qT)\dots(1-q^{n-1}T)\}$.

(ii) $f(T) = (1-\alpha_1 T)\dots(1-\alpha_r T) \in 1 + T\mathbb{Z}[T]$ *and has the properties:*

(a) $r = \left(\frac{m-1}{m}\right)\{(m-1)^n - (-1)^n\}$;

(b) *if* α_i *in* $\mathbb{C}$ *is an inverse root of* f, *then so is* q^{n-1}/α_i;

(c) $|\alpha_i| = q^{(n-1)/2}$. □

Remarks. (1) The theorem was proved for elliptic curves by Hasse (1934), for curves of arbitrary genus by Weil (1948) and for primals of arbitrary dimension by Dwork (1960) as far as (ii)(b); the final part (ii)(c) for primals was proved by Deligne (1974). In the paper in which the conjectures were formulated, Weil (1949) proved the whole theorem for a particular class of primals.

In fact, the results of Dwork and Deligne are valid for more general varieties than primals.

(2) For an exposition of Dwork's proof, see Koblitz [9]; for Deligne's proof, see Mazur [10] and Katz [8]. For an elementary treatment of Weil's theorem for curves, see Schmidt [11].

(3) Part (ii)(c) is known as the "Riemann hypothesis for function fields over finite fields". For varieties, under certain restrictions, of dimension d, this becomes

$$|\alpha_i| = q^{d/2}.$$

Here we are dealing just with primals (hypersurfaces) and so $d = n-1$.

Corollary 1.

$$N_i = 1 + q^i + q^{2i} + \dots + q^{(n-1)i} + (-1)^{n+1}(\alpha_1^i + \dots + \alpha_r^i).$$

Proof. This follows immediately by taking logarithms of both sides in (i) and expanding formally. □

An alternative form for f(T) with only [½r] constants to be determined is the following.

Corollary 2. *Let* $\beta_j = \alpha_j + q^{n-1}/\alpha_j$, $j = 1,2,\ldots,s$.

(i) *If* $r = 2s$,

$$f(T) = (1-\beta_1 T + q^{n-1}T^2)\ldots(1-\beta_s T + q^{n-1}T^2).$$

(ii) *If* $r = 2s+1$,

$$f(T) = (1-\beta_1 T + q^{n-1}T^2)\ldots(1-\beta_s T + q^{n-1}T^2)(1 \pm q^{(n-1)/2}T). \quad \square$$

The estimate for N_i that follows is important for applications. It also expresses the idea that the number of points on a primal is fairly close to the number of points in a prime (hyperplane).

Corollary 3.

$$|N_i - (1+q^i+q^{2i}+\ldots+q^{(n-1)i})| \le rq^{(n-1)i/2}.$$

Proof. By Corollary 1,

$$|N_i - (1+q^i+q^{2i}+\ldots+q^{(n-1)i}|$$
$$= |\alpha_1^i + \ldots + \alpha_r^i| \le |\alpha_1^i| + \ldots + |\alpha_r^i| = rq^{(n-1)i/2}. \quad \square$$

In particular, if $f(T) = 1 + c_1T + \ldots + c_rT^r$, this corollary gives the following for N_1.

Corollary 4.

$$N_1 = 1+q+q^2+\ldots+q^{n-1}+(-1)^{n+1}(\alpha_1+\ldots+\alpha_r)$$
$$= 1+q+q^2+\ldots+q^{n-1}+(-1)^n c_1$$

and

$$|N_1 - (1+q+\ldots+q^{n-1})| \le rq^{(n-1)/2}. \quad \square$$

Corollary 5. *For plane curves of order* m,

$$|N_1 - (1+q)| \le (m-1)(m-2)\sqrt{q}. \quad \square$$

Corollary 6. *For surfaces of order* m,

$$|N_1 - (1+q+q^2)| \le (m-1)(m^2-3m+3)q. \quad \square$$

To get some feeling for the zeta function and the theorem, we obtain some familiar results.

Example 1. $\mathcal{F}$ is a prime, whence $m = 1$ and $r = 0$. So $f = 1$. Thus

$$\zeta(\mathcal{F}) = 1/\{(1-T)\ldots(1-q^{n-1}T)\},$$
$$\log\zeta(\mathcal{F}) = -\log(1-T) - \ldots - \log(1-q^{n-1}T)$$
$$= \Sigma N_i T^i/i.$$

Hence

$$N_1 = 1+q+\ldots+q^{n-1} = (q^n-1)/(q-1),$$

the number of points in $PG(n-1,q)$.

Example 2. F is a quadric, whence $m = 2$ and

$$r = \tfrac{1}{2}\{1^n - (-1)^n\} = \begin{cases} 0, & n \text{ even} \\ 1, & n \text{ odd} \end{cases}.$$

So, if n is even, $f = 1$ and, as in example 1,

$$\zeta(F) = 1/\{(1-T)(1-qT)\dots(1-q^{n-1}T)\},$$

$$N_1 = 1+q+\dots+q^{n-1}.$$

If n is odd, $f = 1 + c_1T$. By (ii)(b) of the theorem, since $-c_1$ is an inverse root of f, so is $q^{n-1}/(-c_1)$; hence $c_1^2 = q^{n-1}$ and $c_1 = \pm q^{(n-1)/2}$. Equivalently, by (ii)(c), since c_1 is an integer and $|-c_1| = q^{(n-1)/2}$, so again $c_1 = \pm q^{(n-1)/2}$. Thus

$$\zeta(F) = 1/\{(1-T)(1-qT)\dots(1-q^{n-1}T)(1 \pm q^{(n-1)/2}T)\},$$

$$N_1 = 1+q+\dots+q^{n-1} \pm q^{(n-1)/2},$$

corresponding to the cases of a hyperbolic and elliptic quadric. We also note that

$$N_2 = 1+q^2+\dots+q^{2(n-1)} + q^{n-1}$$

in both cases. This corresponds to the fact that, whether F_1 is hyperbolic or elliptic, F_2 is hyperbolic. This is also apparent from the canonical forms for a hyperbolic quadric H_n and an elliptic quadric E_n in $PG(n,q)$, n odd.

$$H_n = V(X_0X_1 + X_2X_3 + \dots + X_{n-1}X_n)$$

$$E_n = V(g(X_0,X_1) + X_2X_3 + \dots + X_{n-1}X_n),$$

where g is any binary, quadratic form irreducible over $GF(q)$.

For n even, the canonical form of the (parabolic) quadric is

$$P_n = V(X_0^2 + X_1X_2 + \dots + X_{n-1}X_n).$$

So, to combine all three cases,

$$N_1 = 1+q+\dots+q^{n-1} + (w-1)q^{(n-1)/2},$$

where $w = 2$, 1 or 0 according as F is hyperbolic, parabolic or elliptic.

The hyperbolic quadric satisfies the upper estimate in corollary 4 and the elliptic quadric the lower estimate.

Example 3. F is a Hermitian variety; that is, q is a square and F is a Hermitian form. If F is canonical, then

$$F = X_0\bar{X}_0 + X_1\bar{X}_1 + \dots + X_n\bar{X}_n$$

with $\bar{X}_i = X_i^{\sqrt{q}}$. From [7], p.102,

$$N_1 = [q^{(n+1)/2} + (-1)^n][q^{n/2} - (-1)^n]/(q-1).$$

With $m = \sqrt{q} + 1$,

$$r = \frac{\sqrt{q}}{\sqrt{q}+1}[(\sqrt{q})^n - (-1)^n]$$

and

$$1 + q + q^2 + \ldots + q^{n-1} + rq^{(n-1)/2}$$

$$= \frac{q^n - 1}{q - 1} + \frac{q^{n/2}}{\sqrt{q}+1}[q^{n/2} - (-1)^n]$$

$$= N_1 .$$

So the Hermitian variety provides an example of high order in which the upper limit in corollary 4 is attained.

Example 4. F is a cubic surface in PG(3,q) with 27 lines. Then, [6] for example,

$$N_i = q^{2i} + 7q^i + 1.$$

Hence

$$\zeta(F) = \{(1 - T)(1 - qT)^7(1 - q^2T)\}^{-1}.$$

Here $f(T) = (1 - qT)^6$ and again we see that the theorem is true in this case. The cubic surface F also satisfies the upper limit in corollary 6.

In particular, for the Hermitian surface over GF(4),

$$\zeta(X_0^3 + X_1^3 + X_2^3 + X_3^3;T;4) = \{(1 - T)(1 - 4T)^7(1 - 16T)\}^{-1}.$$

2. CURVES

So far the theorem has only been seen to work to verify known results. Now it will be used to deduce for curves results which are not immediately obvious.

Firstly, we restate theorem 1 for curves in a more general situation. It is no longer required that the curve be plane and it is also not required that the curve be non-singular. So F is merely an absolutely irreducible curve.

Let us recall that if F is a plane curve of order m with ordinary singularities over $\bar{K}$ of multiplicities $m_1,\ldots,m_t$, then the *genus*

$$g = \tfrac{1}{2}(m - 1)(m - 2) - \tfrac{1}{2}\Sigma_{i=1}^t m_i(m_i - 1).$$

If F is a plane curve with no singularities over $\bar{K}$, then

$$g = \tfrac{1}{2}(m - 1)(m - 2).$$

For example, the bicircular quartic $V((X_1^2 + X_2^2)^2 + (X_1^2 - X_2^2)X_0^2 + X_0^4)$ with two ordinary double points has genus

$$g = \tfrac{1}{2}(4 - 1)(4 - 2) - \tfrac{1}{2}\{2(2 - 1) + 2(2 - 1)\}$$

$$= 1.$$

COUNTING PRINCIPLE

The points on a curve F over K are counted in theorem 2 according to the number of points on a non-singular model of F over K. This means that

1) each simple point of F is counted once;

2) each multiple point P of F is counted according to the number of distinct tangents at P lying over K.

As an example, the twisted cubic

$$F' = \{P(t^3, st^2, s^2t, s^3) \mid s,t \in GF(q)\}$$

has exactly $q + 1$ simple points and no multiple points. It is a non-singular model of the curve

$$F = V((X_1^2 + X_2^2)X_0 - X_1^3).$$

The point P(1,0,0) is a node, an isolated double point or a cusp of F according as $X_1^2 + X_2^2$ is the product of two distinct linear factors, is irreducible or is the square of a linear factor; that is, according as $q \equiv 1 \pmod 4$, $q \equiv -1 \pmod 4$ or $q \equiv 0 \pmod 2$. Correspondingly, the double point counts 2, 0 or 1 time in the model.

	Number of simple points on F	Number of tangents at P(1,0,0) over GF(q)	Total number of model points	Actual number of points over GF(q)
$q \equiv 1 \pmod 4$	$q - 1$	2	$q + 1$	q
$q \equiv -1 \pmod 4$	$q + 1$	0	$q + 1$	$q + 2$
$q \equiv 0 \pmod 2$	q	1	$q + 1$	$q + 1$

The three cases may be respectively compared with the following three curves in the real Euclidean plane.

$y^2 - x^2 = x^3$

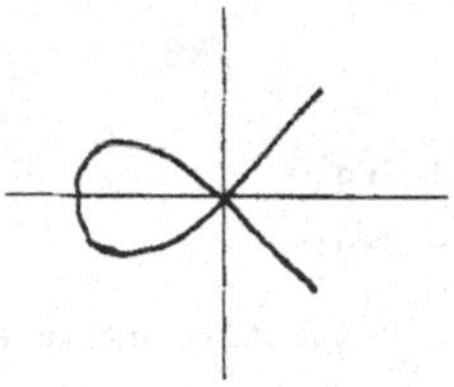

node

$y^2 + x^2 = x^3$

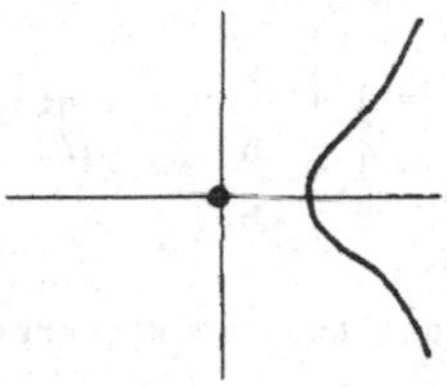

isolated double point

$y^2 = x^3$

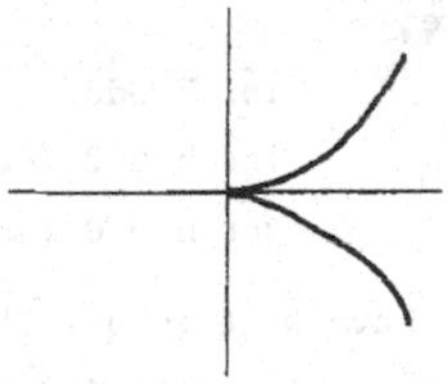

cusp

Theorem 2. *If F is an absolutely irreducible curve of genus g, then*

$$\begin{aligned}\zeta(F) &= \exp(\Sigma N_i T^i/i)\\ &= \frac{1 + c_1T + c_2T^2 + \dots + c_{2g-1}T^{2g-1} + q^gT^{2g}}{(1-T)(1-qT)}\\ &= \frac{(1-\beta_1T+qT^2)\dots(1-\beta_gT+qT^2)}{(1-T)(1-qT)}\end{aligned}$$

where each root α^{-1} of the numerator has $|\alpha| = \sqrt{q}$. □

Corollary. *If F has genus* $g = 1$, *such as a plane non-singular cubic curve, then*

$$\zeta(F) = \frac{1 + c_1T + qT^2}{(1-T)(1-qT)} .$$

Also

$$(\sqrt{q}-1)^2 \le N_1 \le (\sqrt{q}+1)^2,$$

$$N_2 = N_1[2(q+1) - N_1].$$

Proof. The zeta function is obtained by putting $g = 1$ into that of the theorem. Now,

$$\log\zeta(F) = \Sigma N_i T^i/i = (c_1T + qT^2) - \tfrac{1}{2}(c_1T + qT^2)^2 + \ldots$$
$$+ T + \tfrac{1}{2}T^2 + \ldots + qT + \tfrac{1}{2}(qT)^2 + \ldots .$$

So

$$N_1 = c_1 + 1 + q,\ N_2 = 2q - c_1^2 + 1 + q^2 = (1+q)^2 - c_1^2 .$$

Hence the final equality follows. The inequality is that of theorem 1, corollary 5, with $m = 3$. A curve of genus one is *elliptic*. □

Example 5. $q = 2$, $F = X_0^3 + X_1^3 + X_2^3$, $N_h = |V_{2,2^h}(F)|$.

Since $x^3 = x$ over GF(2), so F is a line. Hence

$$3 = N_1 = c_1 + 1 + q = c_1 + 3.$$

Thus

$$\zeta(F) = \zeta(F;T;2) = \frac{1 + 2T^2}{(1-T)(1-2T)}$$

and

$$\log\zeta(F) = \Sigma T^i/i + \Sigma(2T)^i/i + 2\Sigma(-1)^{j-1}(2T^2)^j/(2j).$$

Therefore,

for h odd, $N_h = 1 + 2^h = 1 + q$;
for $h \equiv 2 \pmod 4$, $N_h = 1 + 2^h + 2.2^{h/2} = 1 + q + 2\sqrt{q}$;
for $h \equiv 0 \pmod 4$, $N_h = 1 + 2^h - 2.2^{h/2} = 1 + q - 2\sqrt{q}$;

here, we have written $q = 2^h$. The last two cases therefore give respective examples of the upper and lower limits in the inequality of the corollary being achieved.

Example 6. F is a Hermitian curve over GF(4); for example $F = X_0^3 + X_1^3 + X_2^3$. Then

$$\zeta(F;T;4) = \frac{(1+2T)^2}{(1-T)(1-4T)} .$$

This can be obtained from example 5, since we require the coefficients N_h, h even, there given. Alternatively, directly from the corollary to theorem 2,

$$9 = N_1 = c_1 + 1 + q = c_1 + 5.$$

Example 7. $q = 3$, $F = X_0^4 + X_1^4 + X_2^4$. $V_{2,3}(F) = V_{2,3}(X_0^2 + X_1^2 + X_2^2)$, whence $N_1 = 4$. $V_{2,9}(F)$ is a Hermitian curve whence $N_2 = 28$. By direct calculation, $V_{2,27}(F)$ also has 28 points; i.e. $N_3 = 28$.

By the corollary to theorem 2,

$$\zeta(F) = \frac{(1-\beta_1 T+3T^2)(1-\beta_2 T+3T^2)(1-\beta_3 T+3T^2)}{(1-T)(1-3T)},$$

$$N_1 = -(\beta_1+\beta_2+\beta_3) + 1 + 3 = 4,$$

$$N_2 = 18 - (\beta_1^2+\beta_2^2+\beta_3^2) + 1 + 9 = 28,$$

$$N_3 = -3\beta_1\beta_2\beta_3 + 1 + 27 = 28.$$

Hence $\beta_1 = \beta_2 = \beta_3 = 0$ and

$$\zeta(F) = \frac{(1+3T^2)^3}{(1-T)(1-3T)}.$$

Thus,

for h odd, $N_h = 1 + 3^h = 1 + q$;

for $h \equiv 2 \pmod 4$, $N_h = 1 + 3^h + 6.3^{h/2} = 1 + q + 6\sqrt{q}$;

for $h \equiv 0 \pmod 4$, $N_h = 1 + 3^h - 6.3^{h/2} = 1 + q - 6\sqrt{q}$;

here $q = 3^h$.

Example 8. From the previous example, let F be a Hermitian curve over GF(9); with a particular F,

$$\zeta(X_0^4+X_1^4+X_2^4;T;9) = \frac{(1+3T)^6}{(1-T)(1-9T)}.$$

Exercise. For q square, let F be a Hermitian curve. Then, in canonical form, is

$$\zeta(X_0^{\sqrt{q}+1}+X_1^{\sqrt{q}+1}+X_2^{\sqrt{q}+1};T;q) = (1+\sqrt{q}T)^{q-\sqrt{q}}/\{(1-T)(1-qT)\}?$$

3. THE CLASSIFICATION OF CUBIC CURVES

Before further considering elliptic cubics, we will for completeness explain briefly the classification of singular cubics.

CURVES NOT ABSOLUTELY IRREDUCIBLE

Theorem 3. *In PG(2,q), there are eleven types of cubic curves reducible over $\bar{K}$, each of which is unique up to a projectivity of PG(2,q). The eleven types are given in the following diagrams with the number of points over GF(q) on each. The unbroken lines are over GF(q), dashed lines over GF(q^2)\GF(q), dotted lines over GF(q^3)\GF(q); the ovals are conics.* □

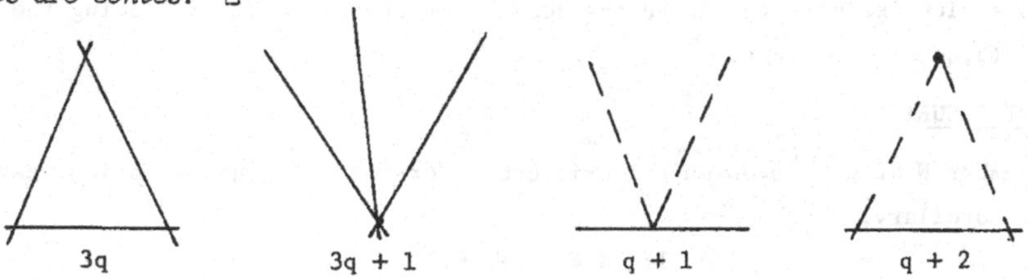

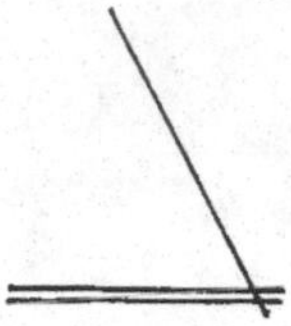

2q + 1

q + 1

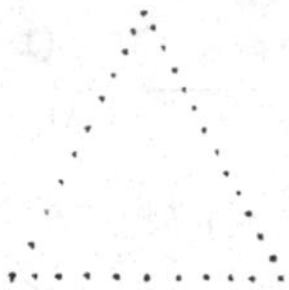

0

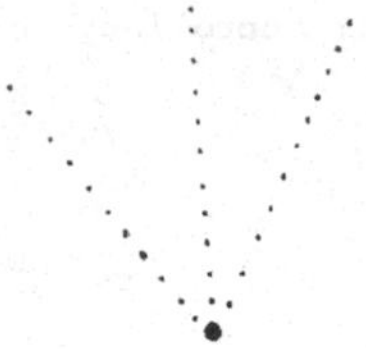

1

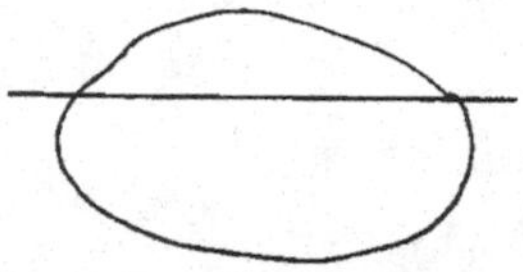

2q

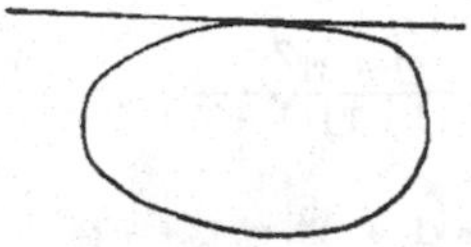

2q + 1

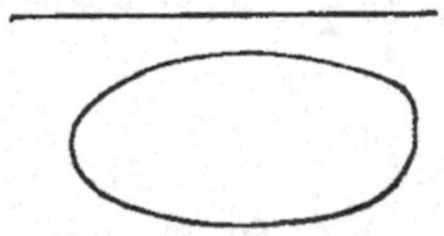

2q + 2

ABSOLUTELY IRREDUCIBLE SINGULAR CUBICS

Theorem 4. *For each* q, *there are exactly four, rational, projectively distinct cubics in* PG(2,q). *Each one is determined by the type of singularity and the number of inflexions according to the following table, where* $q \equiv c \pmod 3$.

Singularity	c	*Number of inflexions*	*Number of points*
	0	1	q
node	1	0 or 3	q
	-1	1	q
isolated	0	1	q + 2
double	1	1	q + 2
point	-1	0 or 3	q + 2
	0	0 or q	q + 1
cusp	1	1	q + 1
	-1	1	q + 1

Proof. For the details and canonical forms, see [7], §11.3. The number of points was explained in §2 in the example of the Counting Principle. □

It is a nice exercise to derive the details of theorem 4 by projecting the twisted cubic in PG(3,q) onto a plane.

4. ELLIPTIC CUBICS

The number N of points on an elliptic cubic curve in PG(2,q) satisfies, as in theorem 2, corollary,

$$(\sqrt{q}-1)^2 \le N \le (\sqrt{q}+1)^2.$$

To investigate cubics and the numbers of their points, we first extend example 5 and let

$$M_q = |V_{2,q}(X_0^3 + X_1^3 + X_2^3)| .$$

If $(q-1, 3) = 1$, then $M_q = q+1$, as the points of the curve are in bijective correspondence with PG(1,q).

For $(q-1, 3) = 3$ and $q < 100$, the values of M_q are given in the following table with $N_{min} = q+1-[2\sqrt{q}]$ and $N_{max} = q+1+[2\sqrt{q}]$.

q	4	7	13	16	19	25	31	37	43	49	61	64	67	73	79	97
M_q	9	9	9	9	27	36	36	27	36	63	63	81	63	81	63	117
N_{min}	1	3	7	9	12	16	21	26	31	36	47	49	52	57	63	79
N_{max}	9	13	21	25	28	36	43	50	57	64	77	81	84	91	97	117

For prime q, there exist unique integers $x \equiv -1 \pmod 3$ and $y \equiv 0 \pmod 3$ such that $4q = x^2 + 3y^2$; then $M_q = q+1-x$. For square q, the corollary to theorem 2 was used. We note that M_q is liberally scattered throughout the interval $[N_{min}, N_{max}]$. However, in each case, nine divides M_q for a classical reason.

The points of an arbitrary elliptic cubic F over GF(q) form an abelian group in the following way. Take any point O as the zero and let the tangent at O meet F again at O'. If P_1P_2 meets F again at R and OR meets F again at Q, then

$$P_1 + P_2 = Q .$$

If $P_1 = P_2$, then P_1P_2 is tangent to F at P_1. It follows that P is an inflexion when

$$3P = O' .$$

If O is an inflexion, then $O' = O$ and the inflexions form a subgroup.

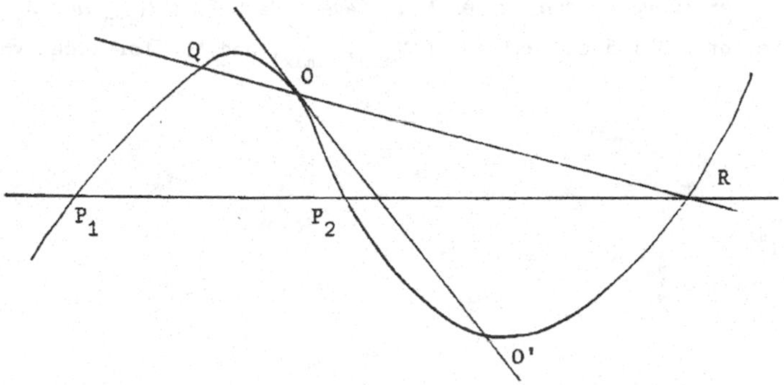

When $F = V(X_0^3 + X_1^3 + X_2^3)$ and $(q-1,3) = 3$, then F has nine inflexions

$$V(X_0X_1X_2) \cap V(X_0^3 + X_1^3 + X_2^3).$$

The possible numbers of inflexions that an arbitrary elliptic cubic F may have are the following:

$$q \equiv 0 \pmod 3): 0, 1, 3;$$
$$q \equiv -1 \pmod 3): 0, 1, 3;$$
$$q \equiv 1 \pmod 3): 0, 1, 3, 9.$$

When F has exactly one inflexion, then $|F| \equiv \pm 1 \pmod 3$; when F has 0 or 3 inflexions, $|F| \equiv 0 \pmod 3$; when F has 9 inflexions, then $|F| \equiv 0 \pmod 9$; see [2].

The number of points that an elliptic cubic can have is given by the following theorem.

Theorem 5. *For every integer* $N = q + 1 - t$ *with* $|t| \le 2\sqrt{q}$, *there exists an elliptic cubic in* $PG(2,q)$, $q = p^h$, *with exactly* N *points, providing one of the following holds:*

	Value of t	*Conditions on* p^h
(i)	$(t,p) = 1$	
(ii)	$t = 0$	h *odd or* $p \not\equiv 1 \pmod 4$
(iii)	$t = \pm\sqrt{q}$	h *even and* $p \not\equiv 1 \pmod 3$
(iv)	$t = \pm 2\sqrt{q}$	h *even*
(v)	$t = \pm\sqrt{(2q)}$	h *odd and* $p = 2$
(vi)	$t = \pm\sqrt{(3q)}$	h *odd and* $p = 3$

Proof. See Waterhouse [15], chapter 4 or Ughi [14]. □

This theorem says that the only values N does not take in the interval $[N_{min}, N_{max}]$ are of the form $q + 1 + kp$ with k an integer. Further, $N = q + 1 + kp$ for at most five values of k.

Corollary 1. *The number* N *assumes every value in the interval* $[N_{min}, N_{max}]$ *if and only if either (i)* $q = p$ *or (ii)* $q = p^2$ *with* $p = 2$, *or* $p = 3$ *or* $p \equiv 11 \pmod{12}$. □

Corollary 2. *Over every* q, *the number* N *achieves the values* N_{min} *and* N_{max}. □

Below, we give for $q \le 125$ the values of N_{min}, N_{max}, and the forbidden values of N.

q	N_{min}	N_{max}	Forbidden N
2	1	5	
3	1	7	
4	1	9	
5	2	10	
7	3	13	
8	4	14	7,11
9	4	16	
11	6	18	
13	7	21	
16	9	25	11,15,19,23
17	10	26	
19	12	28	
23	15	33	
25	16	36	26
27	18	38	
29	20	40	
31	21	43	
32	22	44	23,27,29,31,35,37,39,43
37	26	50	
41	30	54	
43	31	57	
47	35	61	
49	36	64	43,57
53	40	68	
59	45	75	
61	47	77	
64	49	81	51,53,55,59,61,63,67,69,71,75,77,79
67	52	84	
71	56	88	
73	57	91	
79	63	97	
81	64	100	67,70,76,79,85,88,94,97
83	66	102	
89	72	108	
97	79	117	
101	82	122	
103	84	124	
107	88	128	
109	90	130	
113	93	135	
121	100	144	
125	104	148	106,111,116,121,131,136,141,146

5. EQUIVALENCE CLASSES OF ELLIPTIC CUBICS

To an algebraic number theorist, given a non-singular cubic form over K, the isomorphism class of elliptic curves it defines is given by its function field defined over $\bar{K}$. So, in this sense, two elliptic curves are *isomorphic* if there exists a bijective polynomial map ϕ over $\bar{K}$ from one curve to the other with an inverse polynomial map over $\bar{K}$ such that ϕ preserves the zero of the group law. Effectively, this means that one considers non-singular plane cubics with at least one inflexion, two of which are isomorphic if there is a projectivity of PG(2,q) transforming one curve to the other; see, for example, Tate [13].

With this definition, the number of isomorphism classes of elliptic cubics is denoted by A_q and the number of isomorphism classes with exactly $q + 1 - t$ points by $A_q(t)$ or, if there is no ambiguity, simply by $A(t)$.

It should be noted that the collineation $P(x_0,x_1,x_2) \to P(x_0^p,x_1^p,x_2^p)$ of PG(2,q) is not an isomorphism in this sense.

Another point of view is to concentrate on projective equivalence. Two non-singular cubics in PG(2,q) are *projectively equivalent* if there exists a projectivity of the plane transforming one to the other.

To see that isomorphism and projective equivalence are different relations, it suffices to note that over any field there exists an elliptic cubic with no inflexions. For example, in PG(2,7), the curve $V(X_0^3 + 2X_1^3 + 3X_2^3)$ has any inflexions on $V(X_0X_1X_2)$. As, over GF(7), $x^3 = 0$, 1 or -1, the curve has no inflexions.

The number of projective equivalence classes of elliptic cubics is denoted by P_q and the number of classes with exactly $q + 1 - t$ points by $P_q(t)$ or, if unambiguous, by $P(t)$.

In [7], chapter 11, a projective classification of elliptic cubics is given. The numbers n_9, n_3, n_1 and n_0 denote the respective numbers of projectively distinct elliptic cubics in PG(2,q) with 9, 3, 1 and 0 inflexions. So, with the above definitions,

$$A_q = n_9 + n_3 + n_1,$$
$$P_q = n_9 + n_3 + n_1 + n_0.$$

These numbers are given in the following table, taken from [7], §11.10. Here $q \equiv m \pmod{12}$.

m	3	9	2,8	4	1	7	5	11
n_9	0	0	0	$\frac{q+8}{12}$	$\frac{q+11}{12}$	$\frac{q+5}{12}$	0	0
n_3	$q-1$	$q-1$	$q-1$	$\frac{2q+4}{3}$	$\frac{2q+4}{3}$	$\frac{2q+4}{3}$	$q-1$	$q-1$
n_1	$q+3$	$q+5$	$q+2$	$\frac{5q+12}{4}$	$\frac{5q+15}{4}$	$\frac{5q+9}{4}$	$q+3$	$q+1$
n_0	$q-1$	$q-1$	$q-1$	$q+1$	$q+1$	$q+1$	$q-1$	$q-1$
A_q	$2q+2$	$2q+4$	$2q+1$	$2q+5$	$2q+6$	$2q+4$	$2q+2$	$2q$
P_q	$3q+1$	$3q+3$	$3q$	$3q+6$	$3q+7$	$3q+5$	$3q+1$	$3q-1$

Alternative formulas for A_q and P_q are given by

$$A_q = 2q + 3 + \left(\frac{-4}{q}\right) + 2\left(\frac{-3}{q}\right),$$

$$P_q = 3q + 2 + \left(\frac{-4}{q}\right) + \left(\frac{-3}{q}\right)^2 + 3\left(\frac{-3}{q}\right).$$

Here $\left(\frac{b}{c}\right)$ is the usual Legendre-Jacobi symbol; effectively,

$$\left(\frac{-4}{c}\right) = \begin{cases} 1 & \text{if} \quad c \equiv 1 \pmod 4 \\ 0 & \text{if} \quad c \equiv 0 \pmod 2 \\ -1 & \text{if} \quad c \equiv -1 \pmod 4, \end{cases}$$

$$\left(\frac{-3}{c}\right) = \begin{cases} 1 & \text{if} \quad c \equiv 1 \pmod 3 \\ 0 & \text{if} \quad c \equiv 0 \pmod 3 \\ -1 & \text{if} \quad c \equiv -1 \pmod 3. \end{cases}$$

Now, the value of $A(t)$ will be given. To do this, it is necessary to recall the notion of the *class number* of an integral quadratic form.

Let $E = \{f = aX^2 + bXY + cY^2 \mid a, b, c \in \mathbb{Z},\ a > 0\}$. Consider $G = SL(2, \mathbb{Z})$ acting on E. For σ in G with $\sigma = \begin{pmatrix} A & B \\ C & D \end{pmatrix}$ and $AD - BC = 1$, let

$$f^{\sigma} = a(AX + BY)^2 + b(AX + BY)(CX + DY) + c(CX + DY)^2 .$$

With $\Delta(f) = b^2 - 4ac$, also $\Delta(f^{\sigma}) = \Delta(f)$. So all quadratic forms in the same orbit have the same discriminant.

The *class number* $H(\Delta)$

= the number of orbits of G acting on E whose representatives have discriminant Δ

= $|\{(a,b,c) \in \mathbb{Z}^3 \mid b^2 - 4ac = \Delta,\ a > 0, \text{ satisfying } (C_1) \text{ or } (C_2)\}|$.

(C_1) : $c > a$ and $-a < b \le a$;

(C_2) : $c = a$ and $0 \le b \le a$.

For a discussion of the equivalence of these two definitions of class number, see [1], chapter 6. When $\Delta < 0$, then $H(\Delta)$ is finite since $a \le \sqrt{(-\Delta/3)}$. Also $\Delta \equiv 0$ or $1 \pmod 4$.

For example, let us calculate $H(-16)$ using the second definition. We have $0 < a \le 2$, $c = (16 + b^2)/(4a)$. So $(a,b,c) = (1,0,4)$ or $(2,0,2)$. Hence $H(-16) = 2$.

For $0 < -\Delta \le 100$, $H(-\Delta)$ is given in the following table.

$-\Delta$	$H(\Delta)$	$-\Delta$	$H(\Delta)$	$-\Delta$	$H(\Delta)$
3	1	36	3	71	7
4	1	39	4	72	3
7	1	40	2	75	3
8	1	43	1	76	4
11	1	44	4	79	5
12	2	47	5	80	6
15	2	48	4	83	3
16	2	51	2	84	4
19	1	52	2	87	6
20	2	55	4	88	2
23	3	56	4	91	2
24	2	59	3	92	6
27	2	60	4	95	8
28	2	63	5	96	6
31	3	64	4	99	3
32	3	67	1	100	3
35	2	68	4		

Theorem 6. *The number* A(t) *of isomorphism classes of elliptic curves over* GF(q), $q = p^h$, *with* $q+1-t$ *points,* $|t| \le 2\sqrt{q}$, *is given by the following values. In all other cases,* A(t) = 0.

		t	A(t)
(i)		$(t,p) = 1$	$H(t^2 - 4q)$
(ii)	h *odd*		
	(a)	0	$H(-4p)$
	(b) $p = 2$	$\pm\sqrt{(2q)}$	1
	(c) $p = 3$	$\pm\sqrt{(3q)}$	1
(iii)	h *even*		
	(a)	0	$1 - \left(\frac{-4}{p}\right)$
	(b)	$\pm\sqrt{q}$	$1 - \left(\frac{-3}{p}\right)$
	(c)	$\pm 2\sqrt{q}$	$\dfrac{p+6-4\left(\frac{-3}{p}\right) - 3\left(\frac{-4}{p}\right)}{12}$

Proof. This comes from Schoof [12] . □

The values for P(t) can be deduced from those for A(t) and are given in the next theorem. The complication is due to the case that N is divisible by nine.

Firstly, we note two lemmas applicable when $q \equiv 1 \pmod 3$:

(a) For $p \equiv 1 \pmod 3$, there exists a unique solution t_0 for

$$(t,p) = 1, \quad t^2 - 4q = -3 \times \text{square}, \quad t \equiv q+1 \pmod 9;$$

(b) For $p \equiv 1 \pmod 4$, there exists at most one solution t_1 for

$$(t,p) = 1, \quad t^2 - 4q = -4 \times \text{square}, \quad t \equiv q+1 \pmod 9.$$

Theorem 7. *The number* P(t) *of projectivity classes of elliptic cubics over* GF(q), $q = p^h$, *with* $q+1-t$ *points,* $|t| \le 2\sqrt{q}$ *is given by the following description:*

(i) $P(t) = A(t)$ *when* $t \not\equiv q+1 \pmod 3$;

(ii) $P(t) = 2A(t)$ *when* $\begin{cases} t \equiv q+1 \pmod 3 \\ t \not\equiv q+1 \pmod 9; \end{cases}$

(iii) $P(t) = 2A(t)$ *when* $\begin{cases} t \equiv q+1 \pmod 9 \\ q \not\equiv 1 \pmod 3; \end{cases}$

(iv) $P(t) = 2H(t^2-4q) + 3H\left(\frac{t^2-4q}{9}\right)$ *when* $\begin{cases} t \equiv q+1 \pmod 9 \\ (t,p) = 1 \\ t \ne t_0, t_1 \\ q \equiv 1 \pmod 3; \end{cases}$

(v) $P(t) = 2H(t^2 - 4q) + 3H\left(\frac{t^2 - 4q}{9}\right) - 2$

$$\textit{when} \begin{cases} t = t_0 \text{ or } t_1 \\ q \equiv 1 \pmod 3; \end{cases}$$

(vi) $P(t) = \dfrac{5p + 6 - 8\left(\frac{-3}{p}\right) - 3\left(\frac{-4}{p}\right)}{12}$

$$\textit{when} \begin{cases} t = 2\sqrt{q} \\ \sqrt{q} \equiv 1 \pmod 3 \end{cases}$$

$$\textit{or} \begin{cases} t = -2\sqrt{q} \\ \sqrt{q} \equiv -1 \pmod 3). \end{cases}$$

Proof. This again comes from Schoof [12]. □

For any q, a check on the values A(t) and P(t) is provided by the formulas:

$$A_q = \Sigma A(t),$$
$$P_q = \Sigma P(t).$$

Example 9.

q = 4

t	N	$t^2 - 4q$	A(t)	P(t)	Theorem 6	Theorem 7
4	1		1	1	(iii)(c)	(i)
3	2	-7	1	1	(i)	(i)
2	3		2	4	(iii)(b)	(ii)
1	4	-15	2	2	(i)	(i)
0	5		1	1	(iii)(a)	(i)
-1	6	-15	2	4	(i)	(ii)
-2	7		2	2	(iii)(b)	(i)
-3	8	-7	1	1	(i)	(i)
-4	9		1	2	(iii)(c)	(vi)
			13	18		

Example 10.

q = 8

t	N	$t^2 - 4q$	A(t)	P(t)	Theorem 6	Theorem 7
5	4	-7	1	1	(i)	(i)
4	5		1	1	(ii)(b)	(i)
3	6	-23	3	6	(i)	(ii)
2	7		0	0		(i)
1	8	-31	3	3	(i)	(i)
0	9		1	2	(ii)(a)	(iii)
-1	10	-31	3	3	(i)	(i)
-2	11		0	0		(i)
-3	12	-23	3	6	(i)	(ii)
-4	13		1	1	(ii)(b)	(i)
-5	14	-7	1	1	(i)	(i)
			17	24		

Example 11.

$q = 13$

t	N	$t^2 - 4q$	A(t)	P(t)	Theorem 6	Theorem 7
7	7	-3	1	1	(i)	(i)
6	8	-16	2	2	(i)	(i)
5	9	-27	2	5	(i)	(v), t_0
4	10	-36	3	3	(i)	(i)
3	11	-43	1	1	(i)	(i)
2	12	-48	4	8	(i)	(ii)
1	13	-51	2	2	(i)	(i)
0	14		2	2	(ii)(a)	(i)
-1	15	-51	2	4	(i)	(ii)
-2	16	-48	4	4	(i)	(i)
-3	17	-43	1	1	(i)	(i)
-4	18	-36	3	7	(i)	(v), t_1
-5	19	-27	2	2	(i)	(i)
-6	20	-16	2	2	(i)	(i)
-7	21	-3	1	2	(i)	(ii)
			32	46		

The values of A(t) and P(t) in examples 9 - 11 all agree with those in the Grand Table in [2]. The totals A_q and P_q agree with the previous formulas.

ACKNOWLEDGEMENTS

I am most grateful to R.J. Schoof (University of Leiden) for supplying an alternative proof of theorem 5, for the proofs of theorems 6 and 7, and for many helpful discussions.

REFERENCES

[1] H. Davenport, *The Higher Arithmetic.* 4th edition (Hutchinson, 1970).

[2] R. De Groote and J.W.P. Hirschfeld, The number of points on an elliptic cubic curve over a finite field, *Europ. J. Combin.* 1 (1980), 327-333.

[3] P. Deligne, La conjecture de Weil, I, *Inst. Hautes Études Sci. Publ. Math.* 43 (1974), 273-307.

[4] B.M. Dwork, On the rationality of the zeta function of an algebraic variety, *Amer. J. Math.* 82 (1960), 631-648.

[5] H. Hasse, Abstrakte Begründung der komplexen Multiplikation und Riemannsche Vermutung in Funktionkörpern, *Abh. Math. Sem. Univ. Hamburg* 10 (1934), 325-348.

[6] J.W.P. Hirschfeld, Classical configurations over finite fields: I. The double-six and the cubic surface with 27 lines, *Rend. Mat. e Appl.* 26 (1967), 115-152.

[7] J.W.P. Hirschfeld, *Projective Geometries over Finite Fields.* (Oxford University Press, 1979).

[8] N.M. Katz, An overview of Deligne's proof of the Riemann hypothesis for varieties over finite fields, *Proc. Sympos. Pure Math.* 28 (1976), 275-306.

[9] N. Koblitz, *p-adic Numbers, p-adic Analysis, and Zeta Functions*. (Springer, 1977).

[10] B. Mazur, Eigenvalues of Frobenius acting on algebraic varieties over finite fields, *Proc. Sympos. Pure Math.* 29 (1975), 231-261.

[11] W.M. Schmidt, Equations over Finite Fields, an Elementary Approach. (Lecture Notes in Mathematics 536, Springer, 1976).

[12] R.J. Schoof, Unpublished manuscript.

[13] J.T. Tate, The arithmetic of elliptic curves, *Invent. Math.* 23 (1974), 179-206.

[14] E. Ughi, On the number of points of elliptic curves over a finite field and a problem of B. Segre, *Europ. J. Combin.*, to appear.

[15] W.C. Waterhouse, Abelian varieties over finite fields, *Ann. Sci. École Norm. Sup.* 2 (1969), 521-560.

[16] A Weil, *Sur les Courbes Algébriques et les Variétés qui s'en Déduisent.* (Hermann, 1948).

[17] A. Weil, Numbers of solutions of equations in finite fields, *Bull. Amer. Math. Soc.* 55 (1949), 497-508.

FURTHER ACKNOWLEDGEMENTS

The production of this paper owes a great deal to the University of Sydney for its support and to Miss Cathy Kicinski for her excellent typing.

CYCLES IN GRAPHS

D.A. HOLTON

Let M, N be any disjoint subsets of the vertex set of the graph G with $|M| = m$ and $|N| = n$. We say that $G \in C(m, n)$ if there is a cycle K in G such that $M \subseteq VK$ and $N \cap VK = \phi$.

If G is k-connected, then it is an old result of Dirac that $G \in C(k, 0)$. It is easy to produce k-connected graphs which are not $C(k + 1, 0)$. Hence the best we can hope of an arbitrary k-connected graph is that it is $C(k, 0)$. However if we restrict our attention to k-connected regular graphs we can improve on $C(k, 0)$. Indeed two recent papers have shown that 3-connected cubic graphs are $C(9, 0)$ but not $C(10, 0)$. In addition the 3-connected cubic graphs which are $C(9, 0)$ but not $C(10, 0)$ have also been characterised. Some interesting open questions exist for k-connected regular graphs in general.

Further results regarding the relation between graphs which are $C(m_1, n_1)$ and $C(m_2, n_2)$ are discussed.

New results in all of the above areas are discussed and the three main methods of proof analysed.

1. C(m, 0)

"Independent discoveries" have often occured in mathematics. This present problem provides a prime example of two groups of workers on separate continents who have produced, in many cases, identical results almost simultaneously.

In 1979 M.D. Plummer visited the University of Melbourne and together he and D.A. Holton decided to work on the $C(m, n)$ property. A graph G is said to be $C(m, n)$ if given any pair of disjoint sets $M, N \subseteq VG$ with $|M| = m$ and $|N| = n$ there is a cycle C in G such that $M \subseteq VC$ and $N \cap VC = \phi$. So for instance, any hamiltonian graph of order p is $C(p, 0)$ and a hypohamiltonian graph of order p is $C(p-1, 1)$. This means that the Petersen graph P is $C(9, 1)$. As we shall see later this last result is of great importance.

This concept of cycles through prescribed and proscribed vertices evolved from papers of Dirac [2], Mesner and Watkins [15], Plummer [18], [19], Plummer and Wilson [20], Wilson, Hemminger and Plummer [22] and others. In these papers cycles through specified vertices were considered as well as the path analogue of C(m, n).

Holton and Plummer considered two types of problems both of which have their origins in the work mentioned in the last paragraph. The first problem was to try to decide for any 3-connected cubic graph G, the largest possible value of m such that G ε C(m, 0). (If we drop the "cubic" restriction, there are many graphs that are not C(4, 0).). By Dirac it can be seen m is at least three. The Petersen graph shows that m is at least three. The Petersen graph shows that m cannot exceed nine. Where between 3 and 9 does m lie?

Using fairly primitive techniques (see section 3) they were able to push m up to 6 and possible even 7, but at this stage it was clear that although their method would probably work for larger values of m, it would not be feasible to attempt this proof by hand. Hence recourse to a computer was made and B.D. McKay organised this. With a lemma or two for assistance the three were then able to prove the following result.

Thereom 1.1. *(The Nine Vertex Theorem) Let* A *be any set of* 9 *or fewer vertices in a* 3*-connected cubic graph* G. *Then* G *contains a cycle* C *such that* A $\subseteq$ VC.

As mentioned already the proof of this was somewhat messy. It was C. Thomassen who saw how to produce a nicer proof and the Nine Vertex Theorem app eared in [9] with this proof.

Meanwhile, somewhere in Russia, A.K. Kelmans and M.V. Lomonosov were also working on the same problem. Using different techniques, they also proved the Nine Point Theorem. Their announcement appeared in [11].

A natural extension of this result is to try to decide whether or not the Petersen graph is essentially the only graph stopping any 10 vertices to lie on a cycle in a 3-connected cubic graph, (see [9]). M.N. Ellingham, D.A. Holton and C.H.C. Little were able to settle this question [4] in the latter half of 1981. In early 1982, Kelmans and Lomonosov announced this result in [13].

The result obtained by both groups is Theorem 1.2.

Theorem 1.2. *(The Ten Vertex Theorem). Let* A *be any set of* 10 *vertices in a* 3-*connected cubic graph* G. *Then either there is a cycle* C *in* G *such that* $A \subseteq VC$ *or there is a contraction* $\alpha : G \to P$ *with* $\alpha(A) = VP$.

In other words, the theorem says that any 10 vertices lie on a cycle unless they are arranged in a Petersen-like configuration.

If we restrict our attention to 3-connected cubic planar graphs then the number of vertices which can lie on a common cycle increases tremendously.

Thereom 1.3. *If* G *is a* 3-*connected cubic planar graph then any* 16 *vertices lie on a common cycle.*

This result is however, not thought to be best poosible. We do know that 24 cannot lie on a common cycle in a 3-connected cubic graph. This comes via a graph of Grunbaum and Walther [6]. At the moment, therefore, the magic number lies between 16 and 23.

Theorem 1.3 can be proved using the same techniques as in [4].

During their work on the Ten Vertex Theorem, Ellingham et. al. found it necessary to consider cycles through specified vertices which avoid a given edge. If we define an unavoidable edge given A to be an edge that any cycle through the vertices A must contain, then they proved [3] the following theorems. The graph Q is shown in Figure 1.1.

Theorem 1.4. *(The Five Vertex Excluding One Edge Theorem). Let* G *be a* 3-*connected cubic graph, let* $A \subseteq VG$ *with* $|A| \leq 5$ *and let* $e \in EG$. *Then there is a cycle* C *in* G *with* $A \subseteq VC$ *and* $e \notin EC$.

Thereom 1.5. *(The Six Vertex excluding One Edge Theorem).*

Let G *be a* 3-*connected cubic graph and let* $A \subseteq VG$ *with* $|A| = 6$. *Further let* X *be the set of unavoidable edges given* A.

Then (i) $|X| = 0$, 1 or 3;

(ii) $|X| = 1$ if and only if there is a contraction $\alpha : G \to Q$ with $\alpha(A) = \{a_1, a_2, b_5, b_6, b_7, b_8\}$ and $\alpha(X) = \{a_1a_2\}$;

(iii) $|X| = 3$ if and only if there is a contraction $\beta \to G \to P$ with $\beta(A) = \{a_1, a_2, b_5, b_6, b_7, b_8\}$ and $\beta(X) = \{a_1a_2, b_5b_7, b_6b_8\}$.

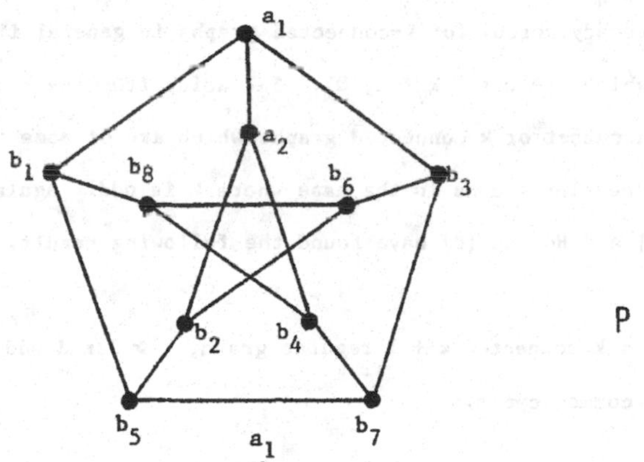

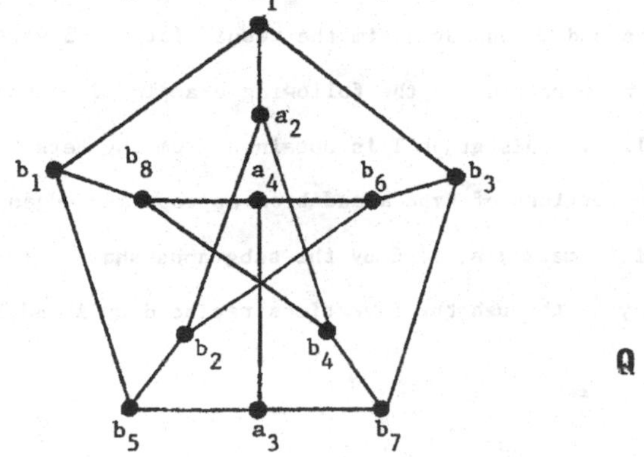

Figure 1.1.

If interesting results can be produced for 3-connected cubic graphs, why not consider k-connected k-regular graphs? So define f(k) so that if G is any k-connected k-regular graph ($k > 3$), f(k) is the largest value of m such that G is C(m, 0). As a result of the Nine Vertex Theorem we know that $f(3) = 9$. In [14], Meredith shows examples of non-hamiltonian k-connected k-regular graphs. Using these examples it is possible to show that $g(k) \leq 10k - 11$. In separate papers Kelmans and Lomonosov [11] and Holton [8] made a small step up from Dirac's result.

<u>Theorem 1.6.</u> *In a k-connected k-regular graph any* $k + 4$ *vertices lie in a common cycle.*

Hence we know that $k + 4 \leq f(k) \leq 10k - 11$. *The Nine Vertex Theorem suggests that if there is a simple formula for* f(k), *then it is* $f(k) = 3k$.

As we have already noted, for k-connected graphs in general it is easy to find many graphs which are not C(k + 1, 0). But aside from the k-regular graphs, there is a subset of k-connected graphs which are of some interest. These are the k+1-regular graphs in the case where k is odd. Again, Kelmans and Lomonosov [12] and Holton [8] have found the following result.

<u>Thereom 1.7.</u> In a k-connected k + 1 regular graph, $k > 3$ and odd, any $k + 2$ vertices lie on a common cycle.

In fact Kelmans and Lomonosov claim the result for k + 3 vertices but this does not appear to be true as the following example of Holton and Plummer shows (see Figure 1.2). This graph T is obtained from the Herschel graph by first inserting the vertices of type a and b on appropriate edges and then replacing the vertices marked A, B, C by the subgraphs shown. In Herschel's graph there is no cycle through the 6 vertices replaced by A and C in T. This is because these 6 vertices are one part of a bipartition of a graph of order 11.

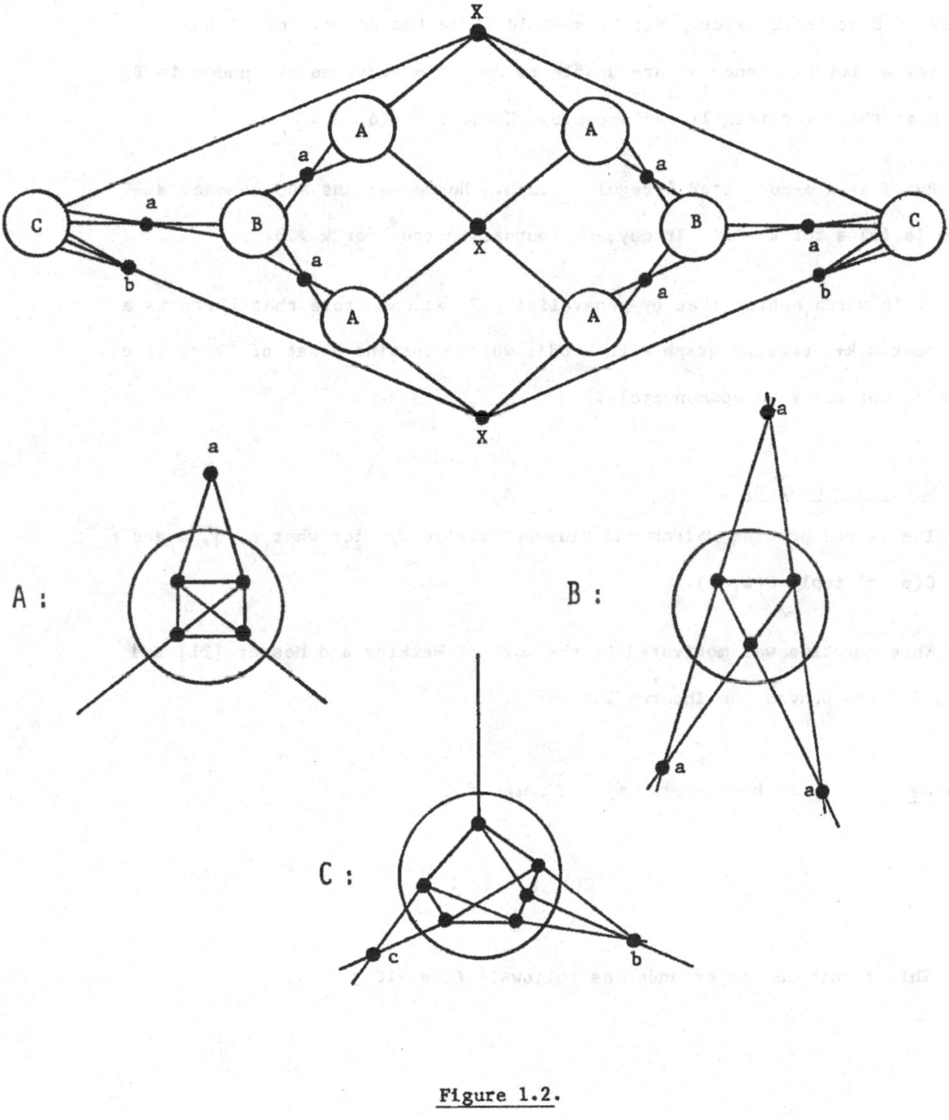

Figure 1.2.

Suppose we chose 6 vertices from VT - one from each copy of A and one from each copy of C. Let D be a cycle containing all 6 vertices. Since Herschel's graph is bipartite, D must enter and leave one of the vertices X or one of the graphs B, at least twice. Hence D must enter and leave one of the copies of B at least twice. But this would force the use of one of the vertices a, twice. Hence we are unable to choose 6 vertices at random in T such that these vertices lie on a cycle. Hence $T \notin C(6, 0)$.

But T is a 3-connected 4-regular graph. Hence Kelmans and Lomosov's claim is false for $k = 3$. It may, of course, be true for $k > 3$.

It is worth noting that by generalising T, we can prove that there is a k-connected k+1-regular graph with k odd, which contains a set of 2k vertices which do not lie on a common cycle.

2. <u>$C(p, q) \to C(m, n)$</u>

The second problem Holton and Plummer tackled was for what p, q, m and n does C(p, q) imply C(m, n).

This question was motivated by the work of Watkins and Mesner [21] and Halin [7] who proved the Theorem 2.1.

<u>Theorem 2.1.</u> G *is* k-*connected if and only if*

$$G \in C(2, k-2) .$$

This result can be extended as follows. (see [10]).

Theorem 2.2. *For each* r *such that* $2 \leq r \leq k$,

$$C(2, k - 2) \textit{ implies } C(r, k - r) .$$

Of course there are some obvious implications. Certainly $C(p, q) \rightarrow C(p, q - 1)$ and $C(p, q) \rightarrow C(p - 1, q)$. These results suggest the construction of the table of implications of Figure 2.1. Every arrow in this table represents an implication. Clearly not every known implication is inlcuded. The original problem now turns into filling in the table of implications.

From Theorem 2.1 we know that $C(2, 1) \rightarrow C(3, 0)$. Further the Petersen graph is a graph which is $C(9, 1)$ but not $C(10, 0)$. So the first part of the

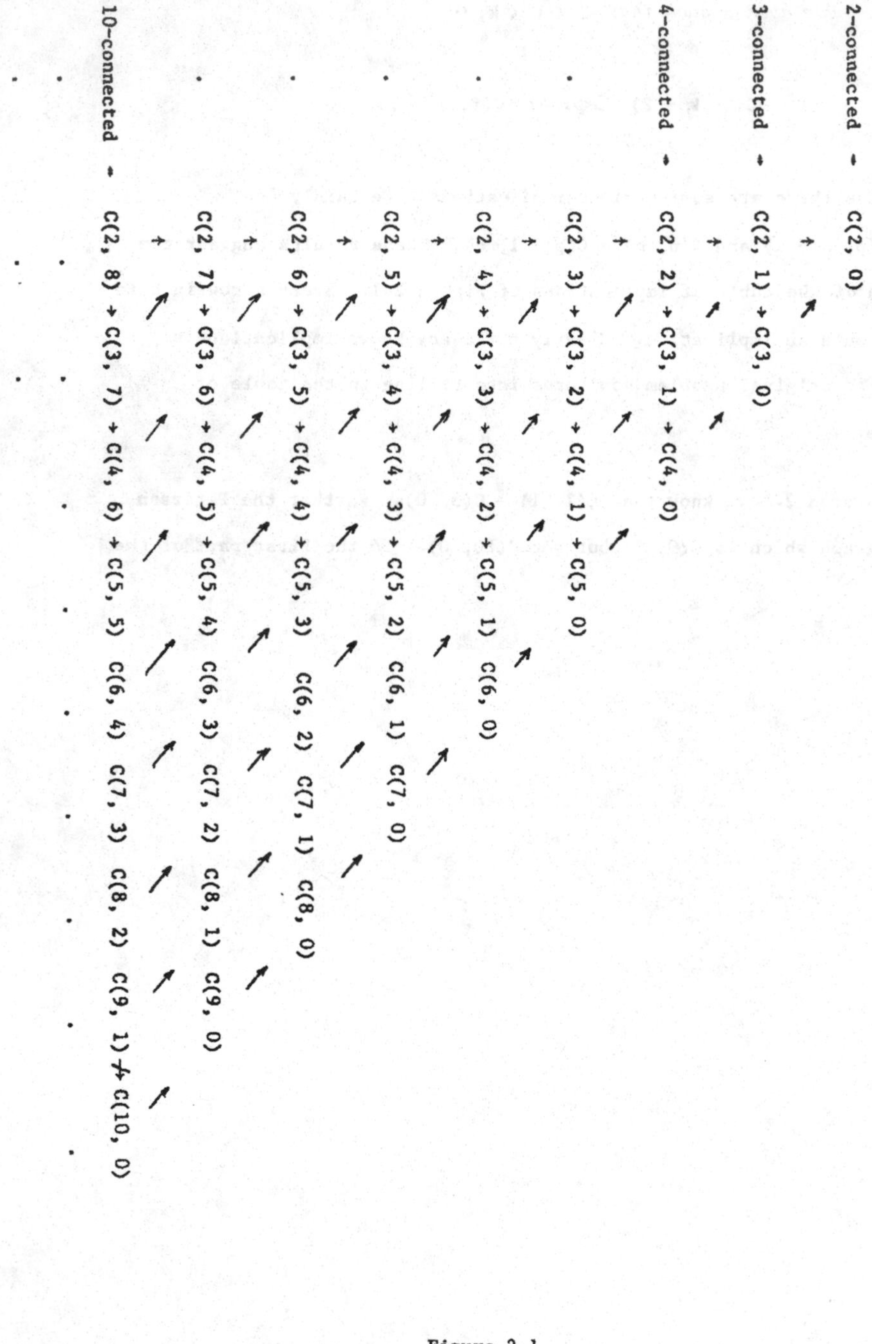
2-connected
C(2, 0)
3-connected
C(2, 1)
C(3, 0)
4-connected
C(2, 2)
C(3, 1)
C(4, 0)
C(2, 3)
C(3, 2)
C(4, 1)
C(5, 0)
C(2, 4)
C(3, 3)
C(4, 2)
C(5, 1)
C(6, 0)
C(2, 5)
C(3, 4)
C(4, 3)
C(5, 2)
C(6, 1)
C(7, 0)
C(2, 6)
C(3, 5)
C(4, 4)
C(5, 3)
C(6, 2)
C(7, 1)
C(8, 0)
C(2, 7)
C(3, 6)
C(4, 5)
C(5, 4)
C(6, 3)
C(7, 2)
C(8, 1)
C(9, 0)
10-connected
C(2, 8)
C(3, 7)
C(4, 6)
C(5, 5)
C(6, 4)
C(7, 3)
C(8, 2)
C(9, 1)
C(10, 0)

Figure 2.1

problem that was tackled was to try to determine for what p, $C(p, 1) \to C(p + 1, 0)$.

In [10] the next results were established.

Theorem 2.3 *(i)* $C(3,1) \to C(4, 0)$.

(ii) $C(4,1) \to C(5, 0)$.

Building on this work Gardiner and Holton [5] were able to extend Theorem 2.2 as follows.

Theorem 2.4. *For each* r *such that* $2 \leq r \leq k$,

(i) $C(3, k - 3) \to C(r, k - r)$,

and *(ii)* $C(4, k - 4) \to C(r, k - r)$.

Once again Kelmans and Lomonosov had also been working on this problem. Thereom 2.4 can be deduced directly from Theorem 3 of their announcement [12].

Theorem 2.4 shows that the properties $C(3, k - 3)$ and $C(4, k - 4)$ are stronger than the properties which follow them in their respective rows in Figure 2.1. This suggests that for each value of k, k-connected, $C(3, k - 3)$, $C(4, k - 4)$ constitutes a genuine hierarchy, with $C(3, k - 3)$ stronger than $(k - 1)$ - connected and $C(4, k - 4)$ stronger-than $(k - 2)$ - connected.

An attempt to extend Theorem 2.4 to $C(5, k - 5)$ by Aldred has proven partially successful. His result can be found in [1].

One more result from [5] is worth noting here.

Theorem 2.5. *Let* p, m > 2, q, n > 0 *and* r > 0. *Then*

$$C(p, q) \rightarrow C(m, n) \text{ if and only if } C(p, q + r) \rightarrow C(m, n + r) .$$

This theorem shows that any proven implication between any pair of properties in Figure 2.1 forces all the corresponding implications both above and below the proven one. A similar statement applies for non-implications. Hence we now know that $C(3, 2) \rightarrow C(4, 1)$ and that $C(9, 2) \rightarrow C(10, 1)$.

3. Techniques

Three main techniques have been used in the production of the results of the first two sections. We outline these techniques here and give an indication of the way in which they have bee used.

3.1 Perfect's Theorem

This theorem was the basis of the Holton, McKay and Plummer proof of Theorem 1.1 and it was also used extensively in [5] and [10]. We note that two paths P and Q are openly disjoint if they are disjoint except for end vertices.

Theorem 3.1 *(Perfect's Theorem) Let* G *be a* k*-connected graph and let* $b_1, b_1, \ldots b_{k-1}$ *be distinct vertices in* $B \subseteq VG$, *where* $|B| > k$.
Let $P_1, P_2, \ldots, P_{k-1}$ *be* k − 1 *openly disjoint paths from a vertex* $a \in B$ *to vertices* $b_1, b_2, \ldots, b_{k-1}$, *respectively.*

Then there exist k openly disjoint paths $Q_1, Q_2, \ldots, Q_k$ from a to B, k − 1 of whose end vertices are $b_1, b_2, \ldots, b_{k-1}$.

<u>Proof:</u> The original result is in [17]. A rather nicer proof can be found in [16].

We now show how to use Perfect's Theorem to prove that $C(3, 1) \to C(4, 0)$.

<u>Theorem 2.3.</u> *(i)* $C(3, 1) \to C(4, 0)$.

<u>Proof:</u> Let $u_1, u_2, u_3, u_4 \in VG$. Since $G \in C(3, 1)$ then $G \notin C(3, 0)$ and so there is a cycle C in G such that $u_1, u_2, u_3 \in VC$. Suppose $u_4 \in VC$.

Now since $G \in C(3, 1)$ then $G \in C(2, 1)$ and so G is 3-connected. Hence there exist openly disjoint paths P_1, P_2, P_3 from u_4 to C. These paths together with C contain a cycle through u_1, u_2, u_3, u_4 unless we have the subgraph H of G shown in Figure 3.1. This graph contains many vertices of degree 2 which we have not included in the figure.

H :

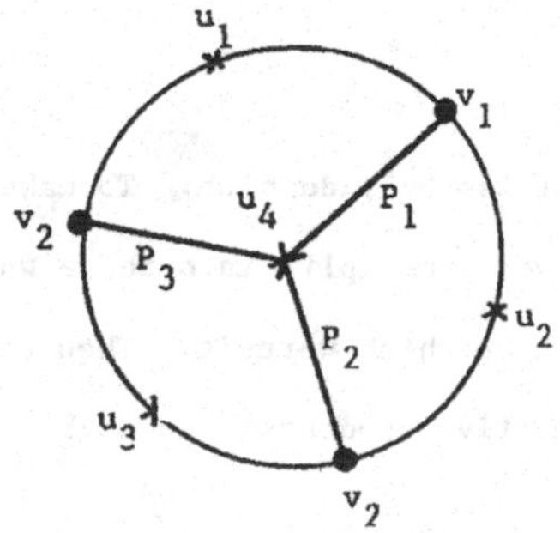

<u>Figure 3.1</u>

It is at this stage that we invoke Perfect's Theorem. Let B = VH $\{u_1\}$. Now G is 3-connected and currently we have two openly disjoint paths from a = u, to v_1, v_3 in B. By the theorem there exist three openly disjoint paths from u_1 to B, two of which meet B at v_1 and v_3 while the other Q meets B at w, say. Although the two new paths from u_1 to v_1 and v_3 may not be those shown in H, we may assume without loss of generality here that they are. Hence we may think of adding the third path Q from u_1 to w only to the subgraph H. Unless w = v_2 we obtain the required cycle through $\{u_1, u_2, u_3, u_4\}$.

If w = v_2, then we consider a = u_2 and B' = VH ∪ VQ ∖ $\{u_2\}$ and apply Perfect's Theorem again. This time we obtain the desired cycle unless the third path Q' from u_2 to B' meets B' at v_3. In this case we apply Perfect's Theorem to a = u_3 and B" = VH ∪ VQ ∪ VQ' ∖ $\{u_3\}$. The theorem is now complete unless the third path from u_3 meets B" at v_1.

The subgraph of G now produced is isomorphic to $K_{3,3}$ which is not C(3, 1). Hence addition or paths are forced which show that u_1, u_2, u_3, u_4 must lie on a common cycle.

3.2 Reductions

In [3], [4] and [9] the proof was by induction. To make the induction work the 3-connected graphs involved were split into those which were cyclically 4-edge connected and those which weren't. Then certain reductions were applied in order for the inductive hypothesis to work.

We note that a graph G is cyclically 4-edge connected if it takes the removed of at least four edges betore G breaks up into two or more components, at least two of which contain a cycle.

Let G be a 3-connected cubic graph which is not cyclically 4-edge connected. Then G has a coboundary containing 3 edges u_1v_1, u_2v_2, u_3v_3 (see Figure 3.2(a)). The graphs H and J of Figure 3.2(b) are called the <u>3-cut reductions</u> of G corresponding to this coboundary.

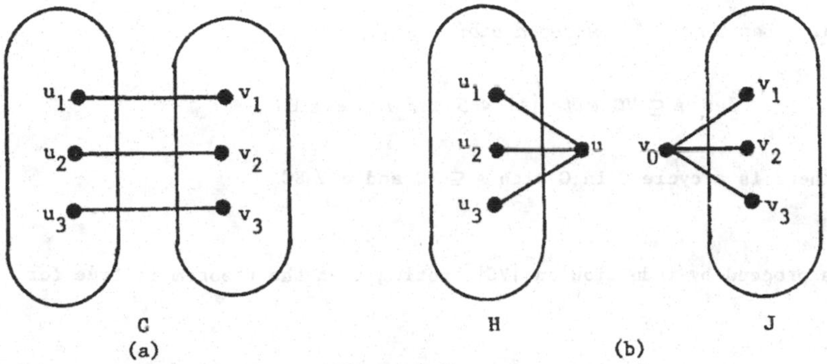

Figure 3.2

Another useful reduction is the e-reduction of G. This is illustrated in Figure 3.3. Essentially an

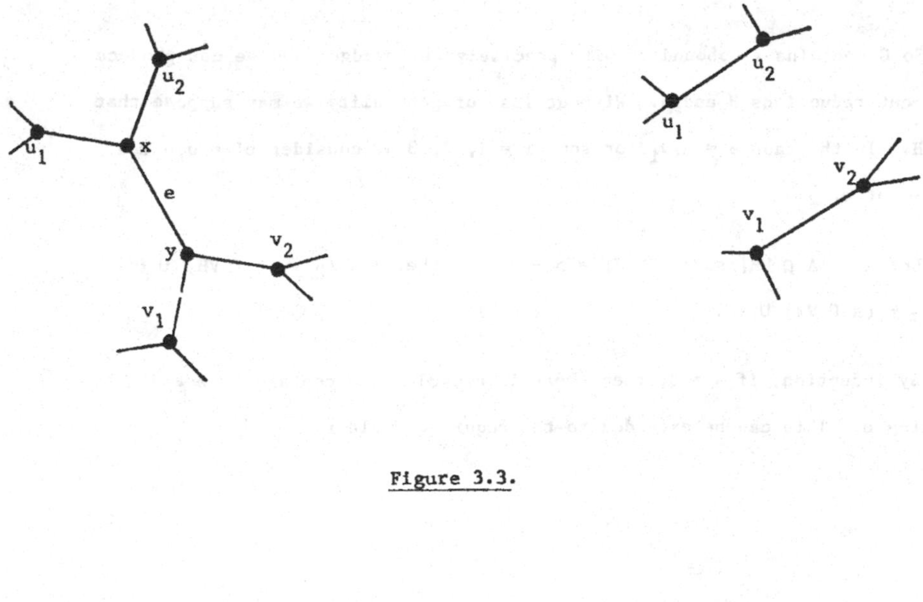

Figure 3.3.

e-reduction of G is the removal of the edge e from G. Its endvertices x, y which are now of degree 2 are removed while their respective neighbours are joined by an edge. We will now give an outline of the proof of the Five Vertex Excluding One Edge Theorem.

Theorem 1.4. *Let G be a 3-connected cubic graph,*

let $A \subseteq VG$ *with* $|A| \leq 5$ *and let* $e \in EG$.

Then there is a cycle C in G with $A \subseteq VC$ and $e \notin EC$.

Proof: We proceed by induction on $|VG|$, noting that the theorem is true for $|VG| = 4$.

Now we may suppose that $|VG| > 6$ that $|A| = 5$ (if $|A| < 5$ then simply add more vertices) and that the theorem holds for any 3-connected cubic graph with fewer vertices than G.

Case 1. G is not cyclically 4-edge connected.

So G contains a coboundary with precisely three edges and we can produce the 3-cut reductions H and J. Without loss of generality we may suppose that $e \in EH$. In the case $e = u_i v_i$ for some $i = 1, 2, 3$ we consider $e' = u_i u \in H$ instead of e.

Let $a = |A \cap VH|$ so $|A \cap VJ| = 5 - a$. Further let $A_H = (A \cap VH) \cup \{u\}$ and $A_J = (A \cap VJ) \cup \{v\}$.

By induction, if $a = 5$, then there is a cycle in H containing A and avoiding e. This can be extended to the required cycle in G.

If $1 \leq a \leq 4$, then $|A_H| \leq 5$. Hence there is a cycle C in H which containes A_H and avoids e. Without loss of generality we may assume that $u_1u, u_2u \in EC$. Now in J, $|A_J| \leq 5$ and so there is a cycle D in J which avoids v v_3 and which passes through the vertices of A_J. Combining C and D gives the required cycle in G.

If a = 0 then we use induction in J to give the desired cycle in G.

Case 2. G is cyclically 4-edge connected.

Suppose $f \in EG$ and f is not incident with a vertex of A. It is not difficult to show that the f-reduction F of G is 3-connected.

If f = e then by induction F has a cycle C with $A \subseteq VC$.

If $f \neq e$ but f has a vertex u in common with e, then let e = uv, f = uw and let the third edge incident to u be ux. The f contains a cylce C with $A \subseteq VC$ which avoids the new edge vx.

If f and e are independent edges, then $e \in EF$ and by induction F contains a cycle C with $A \subseteq VC$ and $e \notin EC$.

In each case C can be extended to the required cycle in G.

So suppose every edge of G is incident with a vertex of A. A simple count shows that $|EG| \leq 15$, so $|VG| \leq 10$. Consideration of all 3-connected cubic cyclically 4-edge connected graphs on 10 or fewer vertices reveals that they satisfy the theorem.

The proof techniques employed above are typical of those used in [3], [4] and [9]. The main results are always proved by using induction and dividing the graphs into two classes, those that are cyclically 4-edge connected and those that are not. In the latter case 3-cut reductions are used to produce the required result. In the former case the problem can be reduced to that of considering a finite set of graphs.

3.3 T-separators.

Kelmans and Lomonosov's work revolves around the notion of T-separators. Their methods are on extension of the work of Watkins and Mesner [20].

The only mention of T-separators currently in the literature is to be found in [11]. This seems to be a somewhat difficult paper to read. In what follows I am indebted to discussions with R.E.L. Aldred and correspondance from A. Gardiner.

The definition of a T-separator comes a little out of the blue. We will give the definition and then provide some motivation. First note that $\langle Y_i \rangle$ in G is the subgraph of G induced by the vertex set of Y_i and ∂Y_i is the set of vertices in Y_i adjacent to vertices in $VG \setminus X$ for $X \subseteq VG$.

Let $T \subseteq VG$. Then a disjoint collection $\{X; Y_1, Y_2, \ldots, Y_p\}$ of subsets of VG T not all of which are empty is called a <u>T-separator</u> of G if

(i) $G \setminus (X \cup \bigcup_{i=1}^{p} E\langle Y_i \rangle)$ has at least $1 + |X| + \sum_{i=1}^{p} \lfloor \frac{1}{2} |\partial Y_i| \rfloor$

components S_j with $S_j \cap T \neq \emptyset$, where none of the graphs $\langle Y_i \rangle$ are isolated vertices

and (ii) each Y_i is a component of $G \setminus (X \cup \cup ES_j)$.

The aim of this definition is to produce the following theorem.

<u>Theorem 3.2.</u> *Let* G *be a* k*-connected graph for* $k > 2$, *let* $m < k + 2$ *and let* $T \subseteq VG$ *with* $|T| = m$.

Then there is a cycle in G containing all of the vertices of T if and only if G has no T'-separator for any $T' \subseteq T$.

The idea behind the notion of T-separator is that a bipartite graph B with one part smaller than m, cannot have a cycle through m vertices in the other part. Hence it cannot be C(m, 0). Built into the definition then is this sort of biparticity.

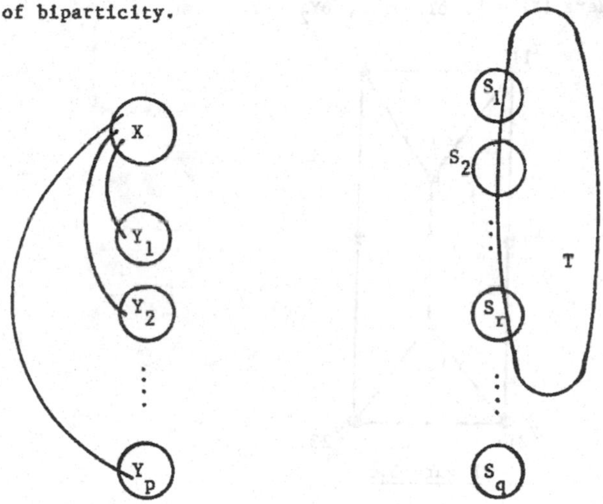

<u>Figure 3.4.</u>

Figure 3.4 shows what is going on. The X and Y_i's are on one side of the graph and the S_j components on the other. The set T intersects $S_1, S_2, \ldots, S_r$ where $r > |+|X| + \sum_{t=1}^{p} \lfloor \frac{1}{2} |\partial Y_i| \rfloor$. Between the X and Y_i's on one side and the S_j's on the other, are an unspecified collection of edges.

Because there are no edges linking the S_j's, a cycle through the vertices of T must go across to the X and Y_i's side and back. Hence we can loosely think of the vertices of X and the vertices of the Y_i adjacent to the S_j as forming one part of a bipartite graph and the S_j as forming the other part. If there are too many S_j's which intersect T then we cannot produce a cycle through the vertices of T for the same kind of reason that the bipartite graph B was not C(m, 0). The factor of $\frac{1}{2}$ arises sinces a potential cycle may enter and leave each Y_i.

Consider the graph G of Figure 3.5. Here $T = \{t_1, t_2, t_3\}$, $X = \emptyset$, $Y_1 = \{y_{11}, y_{12}, y_{13}\}$ and $Y_2 = \{y_{21}, y_{22}, y_{23}\}$. Here $|X| = 0$, $\partial Y_1 = Y_1$, $\partial Y_2 = Y_2$

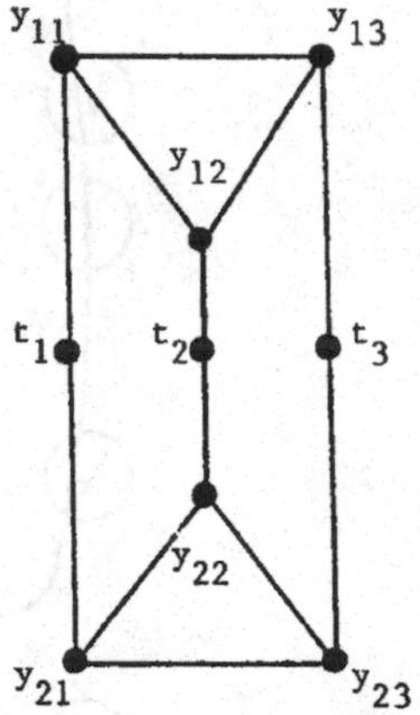

Figure 3.5

and so $1 + |X| + \sum_{t=1}^{2} \lfloor \frac{1}{2} |\partial Y_i| \rfloor = 3$. The components S_j are the isolated vertices t_1, t_2, t_3 so there are (at least) three of these components. Finally each Y_i is a component of $G \setminus (X \cup \bigcup_{j=1}^{3} ES_j)$. $\{\emptyset; \{y_{11}, y_{12}, y_{13}\}, \{y_{21}, y_{22}, y_{23}\}\}$ is a T-separator in G. It is clear that there is no cycle through the vertices of T.

Kelmans and Lomonosov give no proofs in their work but the T-separators seem to be used in the way described below.

Theorem 2.4 *(ii)* $C(4, k - 4) \to C(k, 0)$.

Proof: Let $G \varepsilon C(4, k - 4)$ $C(k, 0)$ and let T be a set of k vertices in G. Assume that $G \varepsilon C(k - 1, 0)$.

By Theorem 3.2 there exists a T-separator in G. Since there is no smaller T-separator ($G \varepsilon C(k - 1, 0)$) then $1 + |X| + \sum_{t=1}^{p} \lfloor \frac{1}{2} |\partial Y_i| \rfloor = n$. But now $G \varepsilon C(4, k - 4)$. So take a subset of size 4 in T and of size $k - 4$ in X. Then we are able to show that $1 + |X| + \sum_{t=1}^{p} \lfloor \frac{1}{2} |\partial Y_i| \rfloor > n + 1$. Hence we have a contradiction.

Note that we called this last theorem Theorem 2.4'(ii) since Theorem 2.4(ii) follows from it as an immediate corollary.

4. Open Questions

We present here some of the more interesting open problems in this area of graph theory.

4.1. f(k).

Let G be a k-connected k-regular graph. In section 1 we defined f(k) to be the largest value of m for which $G \varepsilon C(m, 0)$. Thus $f(3) = 9$.

(a) Determine f(k).

This is currently too difficult a question.

(b) Determine respectable bounds on f(k).

In section 1 we saw that $k + 4 \leq f(k) \leq 10k - 11$. Both these bounds are most likely far from the truth, if the evidence of $k = 3$ can be relied upon. Is $f(k) \geq 2k$ for all k?

Failing this how are $f(k_1)$ and $f(k_2)$ related?

(c) Show that $f(k_1) > f(k_2)$ for $k_1 > k_2$.

At the moment we cannot even show that $f(4) > f(3)$.

4.2. g(k).

Let G be a k-connected (k + 1)-regular graph for k odd. Define g(k) to be the largest value of m for which $G \in C(m, 0)$.

(a) Determine g(k).

If this is too hard, try (b).

(b) Determine respectable bounds for g(k).

Currently we are in better shape than we are with f(k). There seems good evidence to suggest that $k + 2 \leq 2k - 1$. Can we do better?

But failing this we again ask (c).

(c) Is g(k) a monotonically increasing function? Now $g(3) = 5$ and $g(5) \geq 7$. Hence we at least know that $g(5) > g(3)$. It is possible, though we believe unlikely, that $g(7) \geq g(9)$.

4.3. $C(m, n) \to C(p, q)$.

If we consider $C(m, 1) \to C(m + 1, 0)$, then the smallest value of m for which this implication is known to fail is $m = 9$.

(a) Does $C(m, 1) \rightarrow C(m + 1, 0)$ for $2 \leq m \leq 8$?

The Petersen graph shows that $C(9, 1) \rightarrow C(10, 0)$ and this is because P is hypohamiltonian. In other words, P is not hamiltonian but every vertex deleted subgraph of P is.

(b) Does $C(m - 1) \rightarrow C(m + 1, 0)$ if and only if there exists a hypohamiltonian graph of order m?

Is there a more than superficial relation between $C(m, 1) \rightarrow C(m + 1, 0)$ and hypohamiltonian graphs?

4.4. Critical graphs

The Petersen graph is in some sense critical. It is essentially the graph which stops all 3-connected cubic graphs from being $C(10, 0)$. Let $R \in C(m + 1, n)$ and let G be in a class of graphs which is $C(m, n)$. Define R(X) to be an (m, n) - critical graph if R is in the given class and G is $C(m + 1, n)$ unless there is a contraction of G to R in the appropriate way.

We know that $K_{3,3} \in C(3, 1)$.

(a) Is $K_{3,3}$ (3, 1)-critical for the class of 3-connected cubic graphs?

(b) Are there (m, n)-critical graphs for k-connected k-regular graphs?

(c) Are there (m, n)-critical graphs for k-connected (k + 1)-regular graphs?

References

[1] R.E.L. Aldred, C(m, n) properties in graphs, this volume.

[2] G.A. Dirac, In abstrakten Graphen vorhandene vollstandige 4-Graphen und ihre Unterteilungen, *Math. Nachr.*, 22, 1960, 61-85.

[3] M.N. Ellingham, D.A. Holton and C.H.C. Little, Cycles through six vertices excluding one edge in 3-connected cubic graphs, University of Melbourne, Mathematics Research Report No. 11, 1982.

[4] M.N. Ellingahm, D.A. Holton and C.H.C. Little, Cycles through ten vertices in 3-connected cubic graphs, University of Melbourne, Mathematics Research Report, No. 17. 1982.

[5] A. Gardiner and D.A. Holton, Cycles with prescribed and proscribed vertices, University of Melbourne, Mathematics Research Report No. 6, 1981.

[6] B. Grunbaum and H. Walther, Shortness exponent of graphs, *J. Comb. Th.*, 14A, 1973, 364-385.

[7] R. Halin, Zur Theorie der n-fach zusammenhangenden Graphen, *Abh. Math. Sem Hamburg*, 33, 1969, 133-164.

[8] D.A. Holton, Cycles through specified vertices in k-connected regular graphs, *Ars Comb.*, to appear.

[9] D.A. Holton, B.D. McKay, M.D. Plummer and C. Thomassen, A nine point theorem for 3-connected graphs, *Combinatorica*, 2, 1982, 53-62.

[10] D.A. Holton and M.D. Plummer, Cycles through proscribed and forbidden sets, *Annals of Discrete Math.*, 16, 1982, 129-147.

[11] A.K. Kelmans and M.V. Lomonosov, When m vertices in a k-connected graph cannot be walked round along a simple cycle, *Discrete Math.*, 38, 1982, 317-322.

[12] A.K. Kelmans and M.V. Lomonosov, On cycles through given vertices of a graph, *Amer. Math. Soc Abstracts*, No. 82T-05-245, 3, 1982, 255.

[13] A.K. Kelmans and M.V. Lomonosov, A cubic 3-connected graph having no cycles through given 10 vertices has the "Petersen form", *Amer. Math. Soc. Abstracts*, No. 82T-05-260, 3, 1982, 283.

[14] G.H.J. Meredith, Regular n-valent n-connected non-hamiltonian non-n-edge colorable graphs, *J. Comb. Th.*, 14B, 1973, 55-60.

[15] D.M. Mesner and M.E. Watkins, Some theorems about n-vertex connected graphs, *J. Math. Mech.*, 16, 1966, 321-326.

[16] U.S.R. Murty, A simple proof of Perfect's Theorem, *Ars Comb.*, to appear.

[17] H. Perfect, Applications of Menger's Theorem, *J. Math. Anal. Appl.*, 22, 1968, 96-111.

[18] M.D. Plummer, On path properties versus connectivity I, Proc. 2nd S.E. Conf. on Combinatorics, Graph Theory and Computing, L.S.U., Baton Range, 1971, 458-472.

[19] M.D. Plummer, On the (m^+, n^-) connectivity of 3-polytopes, Proc. 3rd. S.E. Conf. on Combinatorics, Graph Theory and Computing, F.A.U., Boca Raton, 1972, 393-408.

[20] M.D. Plummer and E.L. Wilson, On cycles and connectivity in planar graphs, *Canad. Math. Bull.*, 16, 1973, 283-288.

[21] M.E. Watkins and D.M. Mesner, Cycles and connectivity in graphs, *Canad. J. Math.*, 19, 1967, 1319-1328.

[22] E.L. Wilson, R.L. Hemminger and M.D. Plummer, A family of path properties for graphs, *Math. Ann.*, 197, 1972, 107-122.

SEQUENCEABLE GROUPS, GENERALIZED COMPLETE MAPPINGS, NEOFIELDS AND BLOCK DESIGNS

A.D. KEEDWELL

Recently, a number of new connections between complete mappings, sequencings of groups, and the construction of neofields and block designs have come to light. Also, some progress has been made in determining classes of groups which are sequenceable or R-sequenceable. We survey these results, point out their inter-connections and indicate some unsolved problems.

1. COMPLETE MAPPINGS AND SEQUENCEABILITY

I should like to discuss several properties of finite groups which are closely connected with latin squares and which turn out to be of value in the construction of neofields and block designs of Mendelsohn type.

We shall be interested in row complete latin squares and in pairs of orthogonal latin squares. Both these types of square are of interest to statisticians in connection with the design of experiments in which the effects of influences extraneous to those to be compared are to be eliminated.

We shall start with some definitions.

Definition 1. A latin square of order n is an n × n matrix involving n distinct symbols with the property that each of the n symbols occurs exactly once in each row and exactly once in each column of the matrix.

For example, the Cayley multiplication table of any finite group of order n is a latin square of order n. However, not all latin squares can be so obtained.

Definition 2. A transversal of a latin square of order n is a set of n cells taken one from each row and one from each column of the square whose entries are all different.

Definition 3. Two latin squares L_1 and L_2 of the same order are said to be orthogonal if, when they are placed in juxtaposition (so that each cell then contains one symbol from each of the two squares), each of the n^2 ordered pairs obtainable from n distinct symbols occurs in just one of the n^2 cells. The square L_2 is said to be an orthogonal mate for the square L_1, and vice versa.

A latin square L_1 has an orthogonal mate if and only if its cells can be separated into n non-overlapping transversals. The symbols of the square L_2 can be regarded as labelling these n transversals. An example of smallest possible size is given in FIG.1.

$$L_1 = \begin{array}{|ccc} 1 & 2 & 3 \\ 2 & 3 & 1 \\ 3 & 1 & 2 \end{array} \qquad L_2 = \begin{array}{|ccc} 1 & 2 & 3 \\ 3 & 1 & 2 \\ 2 & 3 & 1 \end{array} \qquad \begin{array}{|ccc} 1_1 & 2_2 & 3_3 \\ 2_3 & 3_1 & 1_2 \\ 3_2 & 1_3 & 2_1 \end{array}$$

FIG.1

Definition 4. A permutation $g \to \theta(g)$ of the elements of a finite group G of order n is said to be a complete mapping of G if the mapping $g \to \phi(g)$, where $\phi(g) = g\theta(g)$ is again a permutation of G. The complete mapping is in canonical form if $\theta(e) = e$, where e is the identity element of G.

If θ' is a complete mapping of G which is not in canonical form, then the mapping $\theta : g \to \theta'(g).[\theta'(e)]^{-1}$ is in canonical form.

If θ is a complete mapping of G which is in canonical form then, when the permutation ϕ is written as a product of cycles $\phi = (e)(g_{11}\ g_{12} \cdots g_{1k_1})(g_{21}\ g_{22} \cdots g_{2k_2}) \cdots (g_{s1}\ g_{s2} \cdots g_{sk_s})$, we have that $\theta(g_{ij}) = g_{ij}^{-1}\, g_{i,j+1}$, where the second suffix is taken modulo k_h in the (h+1)th cycle, $h = 1,2,\ldots,s$. In the special case when G possesses a complete mapping for which $s = 1$, we shall say that G is R-sequenceable. (See also definition 5 below.)

If and only if the group G possesses a complete mapping, the latin square L_G formed by its Cayley multiplication table, has an orthogonal mate. For, suppose that (a,b) denotes the cell which lies in the ath row and bth column of G. Then the cells $[g_i, \theta(g_i)]$, $i = 1,2,\ldots,n$, form a transversal of G (because $g_i\theta(g_i) = \phi(g_i)$ and so the entries in these cells are all different) and so also the cells $[g_i, \theta(g_i)g_h]$, where g_h is any fixed element of G, form a transversal of G which is disjoint from the first. In this way, we separate L_G into n disjoint transversals.

<u>Example 1</u> The mapping $\phi = (e)(\alpha_{\alpha}\,\alpha^2{}_{\beta}\,\beta\alpha^2{}_{\alpha^3}\,\beta\alpha_{\alpha^2}\,\beta\alpha^3{}_{\beta\alpha^2}\,\alpha^3{}_{\beta\alpha^3}\,\beta_{\beta\alpha})$ of the dihedral group $D_4 = gp\{\alpha, \beta : \alpha^4 = \beta^2 = e,\ \alpha\beta = \beta\alpha^{-1}\}$ shows that it is R-sequenceable. The corresponding complete mapping θ can be expressed as

$$\theta = \begin{pmatrix} e & \alpha & \alpha^2 & \beta\alpha^2 & \beta\alpha & \beta\alpha^3 & \alpha^3 & \beta \\ e & \alpha & \beta & \alpha^3 & \alpha^2 & \beta\alpha^2 & \beta\alpha^3 & \beta\alpha \end{pmatrix} \text{ or, in cycle form as } \theta = (e)(\alpha)(\alpha^2\ \beta\ \beta\alpha)(\alpha^3\ \beta\alpha^3\ \beta\alpha^2).$$

The transversal of the Cayley table which is defined by the set of cells $[g_i, \theta(g_i)]$ is shown underlined in FIG.2.

	e	α	α^2	α^3	β	$\beta\alpha$	$\beta\alpha^2$	$\beta\alpha^3$
e	$\underline{e}$	α	α^2	α^3	β	$\beta\alpha$	$\beta\alpha^2$	$\beta\alpha^3$
α	α	$\underline{\alpha^2}$	α^3	e	$\beta\alpha^3$	β	$\beta\alpha$	$\beta\alpha^2$
α^2	α^2	α^3	e	α	$\underline{\beta\alpha^2}$	$\beta\alpha^3$	β	$\beta\alpha$
α^3	α^3	e	α	α^2	$\beta\alpha$	$\beta\alpha^2$	$\beta\alpha^3$	$\underline{\beta}$
β	β	$\beta\alpha$	$\beta\alpha^2$	$\beta\alpha^3$	e	$\underline{\alpha}$	α^2	α^3
$\beta\alpha$	$\beta\alpha$	$\beta\alpha^2$	$\underline{\beta\alpha^3}$	β	α^3	e	α	α^2
$\beta\alpha^2$	$\beta\alpha^2$	$\beta\alpha^3$	β	$\underline{\beta\alpha}$	α^2	α^3	e	α
$\beta\alpha^3$	$\beta\alpha^3$	β	$\beta\alpha$	$\beta\alpha^2$	α	α^2	$\underline{\alpha^3}$	e

<u>FIG.2.</u>

When a group G is R-sequenceable its elements $a_0 = e, a_1, a_2, \ldots, a_{n-1}$ can be ordered in such a way that the partial products $b_0 = a_0$, $b_1 = a_0a_1, \ldots, b_{n-2} = a_0a_1 \ldots a_{n-2}$ are all different and so that $b_{n-1} = b_0 = e$. In fact, in the notation of definition 4 we have that

$$\phi = (e)(e_{a_1}\,b_1{}_{a_2}\,b_2{}_{a_3}\,b_3 \ldots b_{n-3}{}_{a_{n-2}}\,b_{n-2}{}_{a_{n-1}}) \text{ where } b_{i+1} = \phi(b_i) = b_i\theta(b_i)$$

and $\theta(b_i) = a_{i+1}$ is the corresponding complete mapping. A closely related idea is that of <u>sequenceability</u>.

<u>Definition 5.</u> A group (G,.) is said to be <u>R-sequenceable</u> if its elements $a_0 = e, a_1, a_2, \ldots, a_{n-1}$ can be orderd in such a way that the

partial products $b_0 = a_o$, $b_1 = a_0a_1$, $b_2 = a_0a_1a_2$, ..., $b_{n-2} = a_0a_1...a_{n-2}$ are all different and so that $b_{n-1} = b_0 = e$. It is said to be sequenceable if its elements $a_0 = e$, a_1, a_2, ..., a_{n-1} can be ordered in such a way that all of the partial products $b_0 = a_0$, $b_1 = a_0a_1$, $b_2 = a_0a_1a_2$, ..., $b_{n-1} = a_0a_1...a_{n-1}$ are different.

B. Gordon [8] showed in 1961 that the Cayley multiplication of a finite group (G,.) could be written in the form of a row-complete latin square if and only if the group is sequenceable.

Definition 6. A latin square $L = (\ell_{ij})$ is row complete if the n(n-1) ordered pairs $(\ell_{ij},\ell_{i,j+1})$ are all distinct. It is column complete if the n(n-1) ordered pairs $(\ell_{ij},\ell_{i+1,j})$ are all distinct.

With the notation of definition 5, it is easy to see that if G is a sequenceable group then the latin square L whose (i,j)th cell contains $b_i^{-1}b_j$, i,j = 0,1,...,n-1, is both row complete and column complete.

Example 2. The sequencing 0,1,2,3 of the cyclic group C_3 written additively gives $b_0 = 0$, $b_1 = 1$, $b_2 = 3$, $b_3 = 2$. Then L has $-b_i + b_j$ as entry in its (i,j)th cell and is as follows : L =

0	1	3	2
3	0	2	1
2	3	1	0
1	2	0	3

Just as an R-sequenceable group may be thought of as one possessing a special kind of complete mapping, so a sequenceable group may be regarded as having a special kind of near complete mapping. The latter concept was introduced by D.F. Hsu and A.D. Keedwell. (See [12].) By using the two concepts of complete mapping and near complete mapping together, these authors have been able to charaterize left neofields completely.

Definition 7. A finite group (G,.) is said to have a near complete mapping θ if its elements can be arranged in such a way as to form a single non-cyclic sequence of length h and s cyclic sequences of lengths $k_1,k_2,...,k_s$, say $[g_1' \; g_2' \dots g_h'](g_{11} \; g_{12} \dots g_{1k_1})(g_{21} \; g_{22} \dots g_{2k_2}) \dots (g_{s1} \; g_{s2} \dots g_{sk_s})$ in such a way that the elements $\theta(g_i') = g_i'^{-1}g_{i+1}'$ and $\theta(g_{ij}) = g_{ij}^{-1}g_{i,j+1}$ together with the elements $\theta(g_{ik_i}) = g_{ik_i}^{-1}g_{i1}$ comprise the non-identity elements of G. The mapping θ maps $G\sim\{g_h'\}$ one-to-one onto $G\sim\{e\}$ and the mapping ϕ defined by $\phi(g) = g\theta(g)$ for all $g \in G$ maps $G\sim\{g_h'\}$ one-to-one onto $G\sim\{g_1'\}$. If $g_1' = e$, the near complete mapping is in canonical form.

Note that a complete mapping in canonical form may be regarded as a special case of a near complete mapping in canonical form : namely one for which the non-cyclic sequence has length one and comprises the identity element alone. On the other hand, a near complete mapping in canonical form for which H = ord G exists if and only if G is sequenceable. In the notation of definition 5, the non-cyclic sequence is the sequence $[b_0 \; b_1 \; b_2 \dots b_{n-1}]$.

Before discussing the application of these concepts to neofields and to the construction of designs, let us summarize the progress which has been made in deciding which finite groups are sequenceable or R-sequenceable. (In [4], C.V. Eynden has proved that all countably infinite groups are sequenceable.)

For abelian groups, the question of which ones are sequenceable is completely answered. A finite abelian group is sequenceable if and only if it has a unique element of order 2. The necessity that there is a unique element of order 2 follows from the fact that the product (in any order) of all the elements of a finite abelian group is equal to the identity element except when the group has a unique element t of order 2. In the latter case, the product is equal to t. This was proved by L.J. Paige [23] in 1947. So, a finite abelian group has $b_{n-1} = a_0a_1..a_{n-1} = t$ or the identity for any ordering of its elements. It can only be sequenceable if $b_{n-1} = t$ and can only be R-sequenceable if $b_{n-1} = e$ (the identity). B. Gordon [8] showed in 1961 that, if $b_{n-1} = t$, the group is sequenceable. However, it is not known whether all finite abelian groups for which

$b_{n-1} = e$ are R-sequenceable. Some progress in solving this question has recently been made by R. Friedlander, B. Gordon and M.D. Miller [6]. These authors have shown that the answer is in the affirmative for a number of types of abelian group.

For non-abelian groups, not too much is known. A finite dihedral group is R-sequenceable if and only if its order is a multiple of 4, see [17]. The question as to which dihedral groups are sequenceable seems to be more difficult. The groups D_3 and D_4 of orders 6 and 8 respectively are not sequenceable. The groups D_n, $3 < n < 37$, n odd, are sequenceable [10]. The groups D_p, p prime and $p \equiv 1 \bmod 4$ are sequenceable [5]. Also, the groups D_p, p prime, $p \equiv 7 \bmod 8$ and for which 2 belongs to the exponent $\frac{1}{2}(p-1)$ are sequenceable [10]. D_6 and D_8 are sequenceable. It seems very likely that all dihedral groups of singly even order except D_3 are sequenceable. The same conjecture may be true for dihedral groups of doubly even order other than D_4 but evidence to date is scanty. Non abelian groups of order pq, p and q primes greater than 2, $p < q$ and such that 2 belongs to the exponent p-1, are both sequenceable and R-sequenceable. (See [18] and [17].) Indeed such groups have a stronger property. In the language of [19], they are super P-groups. So far as the author is aware, no other classes of non-abelian group have been successfully investigated as yet.

2. COMPLETE MAPPINGS AND NEOFIELDS

Definition 8. A set N on which two binary operations (+) and (.) are defined is called a left neofield if

(i) (N,+) is a loop, with identity element 0 say;

(ii) $(N\setminus\{0\},.)$ is a group;

and (iii) $a(b+c) = ab + ac$ for all $a,b,c \in N$.

If also the right distributive law $(b+c)a = ba + ca$ holds for all $a,b,c \in N$, then (N,+,.) is a neofield.

Neofields were first introduced by L.J. Paige [24] who hoped to use them to construct projective planes. If a neofield had additional properties sufficient to enable the points of a projective plane to be co-ordinatized by homogeneous co-ordinates taken from it, the neofield was called planar. However, it turns out that no proper finite planar neofields exist [20].

Since in a left neofield (or in a two-sided neofield), $x + y = x(1+x^{-1}y)$ for all $x \neq 0$, it is evident that a left-neofield with given multiplicative group $(G,.)$ is completely determined by its presentation function ψ given by $\psi(w) = 1 + w$, a fact which was first pointed out and used in [16].

Defintion 9. Let θ be a near complete mapping in canonical form of a group $(G,.)$. The element g_h' of G which has no image under θ will be called its exdomain element and from now on we shall denote it by η.

In [12], the following theorems have been proved:

Theorem 1. *Let $(N,+,.)$ be a finite left neofield with multiplicative group $(G,.)$, where $G = N - \{0\}$. Then, if $1 + 1 = 0$ in N, N defines a complete mapping (in canonical form) of G. If $1 + 1 \neq 0$ but $1 + \eta = 0$, N defines a near complete mapping of G with η as ex-domain element.*

Conversely, let $(G,.)$ be a finite group with identity element 1 which possesses a complete mapping θ in canonical form. Let 0 be a symbol not in the set G and define $N = G \cup \{0\}$. Then $(N,+,.)$ is a left neofield, where we define $\psi(w) = 1 + w = w\theta(w)$ for all $w \neq 0,1$ and $\psi(1) = 0$. Also $x + y = x(1+x^{-1}y)$ for $x \neq 0$, $0 + y = y$ and $0.x = 0 = x.0$ for all $x \in N$.

Alternatively, let $(G,.)$ possess a near complete mapping θ in canonical form. Then, with N defined as before, $(N,+,.)$ is a left neofield where we define $\psi(w) = 1 + w = w\theta(w)$ for all $w \neq 0,\eta$, where η is the ex-domain element of θ, and $\psi(0) = 1$, $\psi(\eta) = 0$. Also $x + y = x(1+x^{-1}y)$ for $x \neq 0$ as before, $0 + y = y$ and $0.x = 0 = x.0$ for all $x \in N$.

Theorem 2. *A finite left neofield constructed as above from a group (G,.) is a neofield if and only if the mapping θ maps conjugacy classes of G to conjugacy classes and, in the case when θ is a near complete mapping, if and only if we have additionally that the exdomain element η is in the centre of G.*

We note that every finite field and every finite nearfield [27] is a finite left neofield and that, in these cases, the exdomain element η is the element which is usually denoted by -1 and is an element of multiplicative order 2.

More generally, if η is the exdomain element of a near complete mapping of any finite abelian group, it has multiplicative order 2. This ceases to be true for non-abelian groups. One consequence is that there exist (infinitely many) finite left neofields for which $(-1)^2 \neq 1$. However, if the additive loop of a left neofield for which $1 + 1 \neq 0$ has the left or right inverse property or is commutative or is associative then $(-1)^2 = 1$. Detailed proofs of these resulcs are in [20].

Definition 10. A left neofield (N,+,.) for which $1 + 1 = 0$ and for which the presentation function ψ defines a permutation of $N \setminus \{0,1\}$ which consists entirely of cycles of length k is said to be a left neofield of pseudo-characteristic k.

A left neofield (N,+,.) for which $1 + 1 \neq 0$ and for which the presentation function ψ defines a permutation of N which consists entirely of cycles of length k is said to be a left neofield of characteristic k.

The concept of characteristic(*) of a neofield was first introduced in [16] and that of pseudo-characteristic in [12]. A field of characteristic p, $p \neq 2$, is an example of a neofield of characteristic p but there exist

(*) The term "characteristic" has been used in a much weaker sense by D.R. Hughes [14]. For example, Hughes regards every neofield for which $1 + 1 = 0$ as having characteristic 2.

examples of neofields and left neofields of characteristic p which are not fields. Likewise, a field of characteristic 2 is an example of a neofield of characteristic and pseudo-characteristic 2.

A left neofield (N,+,.) for which $1 + 1 = 0$ has pseudo-characteristic k if and only if it gives rise to a k-regular complete mapping of the multiplicative group $(N \setminus \{0\}, .)$ as defined in [7].

Example 3. The identity mapping $\theta(g) = g$ of the cyclic group $C_7 = gp\{a : a^7 = e\}$ is a 3-regular complete mapping in canonical form because the mapping $\phi(g) = g.\theta(g) = g^2$ has cycle decomposition $(e)(a\ a^2\ a^4)(a^3\ a^6\ a^5)$. The presentation function of the corresponding neofield of order 8 is $\psi = (0\ 1)(a\ a^2\ a^4)(a^3\ a^6\ a^5)$.

Friedlander, Gordon and Tannenbaum's paper [7] may be regarded as an investigation into which pseudo characteristics are possible for a finite left neofield whose multiplication group is abelian, though this is not the way in which these authors themselves thought of their investigation.

We observe that R-sequenceable and sequenceable groups allow the construction of left neofields of maximal pseudo-characteristic and maximal characteristic respectively.

3. GENERALIZED COMPLETE MAPPINGS AND BLOCK DESIGNS.

The concepts of complete mapping and near complete mapping can be generalized as follows:

Definition 11. A (K,λ) complete mapping, where $K = \{k_1, k_2, \ldots, k_s\}$ and the k_i are integers such that $\sum_{i=1}^{s} k_i = \lambda(|G|-1)$, is an arrangement of the non-identity elements of G (each used λ times) into s cyclic sequences of lengths $k_1, k_2, \ldots, k_s$, say

$$(g_{11}\ g_{12} \ldots g_{1k_1})(g_{21}\ g_{22} \ldots g_{2k_2}) \ldots \ldots \ldots (g_{s1}\ g_{s2} \ldots g_{sk_s}),$$

such that the elements $g_{ij}^{-1} g_{i,j+1}$ (where $i = 1,2,\ldots,s$; and the second suffix j is added modulo k_i) comprise the non-identity elements of G each counted λ times.

A <u>(K,λ) near complete mapping</u>, where $K = \{h_1,h_2,\ldots,h_r;k_1k_2\ldots,k_s\}$ and the h_i and k_j are integers such that $\sum_{i=1}^{r} h_i + \sum_{j=1}^{s} k_j = \lambda|G|$, is an arrangement of the elements of G (each used λ times) into r sequences with lengths $h_1,h_2,\ldots,h_r$ and s cyclic sequences with lengths $k_1,k_2,\ldots,k_s$, say

$$[g'_{11}\ g'_{12}\ \ldots\ g'_{1h_1}]\ \ldots\ [g'_{r1}\ g'_{r2}\ \ldots\ g'_{rh_r}](g_{11}\ g_{12}\ \ldots\ g_{1k_1})\ \ldots\ (g_{s1}\ g_{s2}\ \ldots\ g_{sk_s})$$

such that the elements $(g'_{ij})^{-1}g'_{i,j+1}$ and $g_{ij}^{-1}g_{i,j+1}$ together with the elements $g_{ik_i}^{-1}g_{i1}$ comprise the non-identity elements of G each counted λ times. (We have $\Sigma(h_i-1)+\Sigma k_j = \lambda(|G|-1)$ so it is immediate from the definition itself that $r = \lambda$.)

<u>Example 4.</u> $(a\ a^3)(a^2\ a^6)(a^4\ a^5)(a\ a^2\ a^4)(a^3\ a^6\ a^5)$ is a $(K,2)$ complete mapping of the cyclic group $C_7 = gp\{a : a^7=e\}$, where $K = \{2,2,2,3,3\}$.

<u>Example 5.</u> $[e\ ba][e\ ba^2](a^2\ b\ ba^2\ a)(a\ b\ ba\ a^2)$ is a $(K,2)$ near complete mapping of the dihedral group $D_3 = gp\{a,b : a^3=b^2=e,\ ab=ba^{-1}\}$, where $K = \{2,2;4,4\}$.

<u>Definition 12.</u> A <u>(k,λ) complete mapping</u> is a (K,λ) complete mapping such that $K = \{k,k,\ldots,k\}$. For such a generalized complete mapping, $s = \lambda(|G|-1)/k$.

A <u>(k,λ) near complete mapping</u> is a (K,λ) near complete mapping such that $K = \{h,h,\ldots,h;k,k,\ldots,k\}$ and $k - h = 1$.

<u>Example 6.</u> $[e\ a^4][e\ a^4](a\ a^2\ a^7)(a^3\ a^6\ a^5)(a\ a^7\ a^6)(a^2\ a^3\ a^5)$ is a $(3,2)$ near complete mapping of the cyclic group $C_8 = gp\{a : a^8=e\}$.

In recent years, although interest in the well-known concept of balanced incomplete block designs has been maintained, a considerable interest in a related type of design in which cyclic order within blocks is significant has grown up. Such designs, originally introduced by N.S. Mendelsohn in [21], are connected in several ways with the subject matter of the present paper. We shall show a direct connection with the generalized complete mappings which we have just introduced and also a connection with R-sequenceability and orthogonal latin squares.

Definition 13. A block design of Mendelsohn type comprises a set G of v elements and a collection of b cyclically ordered subsets of G called blocks of cardinalities $k_1, k_2, \ldots, k_b$ respectively with the property that every ordered pair of elements of G are consecutive in exactly λ of the blocks. We call such a design a (v,K,λ) Mendelsohn design, where K is the set formed by the distinct integers among $k_1, k_2, \ldots, k_b$. More briefly, we shall write it as a (v,K,λ)-MD and we shall denote the set of blocks by B.

Example 7. Let $G = C_{14} \cup \{\infty\}$ and $B = \bigcup_{i=0}^{13} \{(i\ 7{+}1\ \infty),\ (1{+}i\ 2{+}i\ 6{+}i\ 11{+}i\ 3{+}i),$ $(4{+}i\ 12{+}i\ 9{+}i\ 5{+}i),\ (8{+}i\ 10{+}i\ 13{+}i)\}$, where all addition is modulo 14. Then (G,B) is a $(15,K,1)$-MD with $K = \{5,4,3\}$.

Definition 14. A (v,k,λ)-MD is a (v,K,λ)-MD such that $k_1 = k_2 = \ldots = k_b = k$.

Example 8. Let $G = C_{13}$ and $B = \bigcup_{i=0}^{12} \{(1{+}i\ 4{+}i\ 3{+}i\ 12{+}i\ 9{+}i\ 10{+}i),$ $(2{+}i\ 6{+}i\ 4{+}i\ 11{+}i\ 7{+}i\ 9{+}i),\ (5{+}i\ 2{+}i\ 10{+}i\ 8{+}i\ 11{+}i\ 3{+}i),\ (6{+}i\ 1{+}i\ 7{+}i\ 12{+}i\ 5{+}i)\}$, where addition is modulo 13. Then (G,B) is a $(13,6,2)$-MD.

In [13], D.F. Hsu and A.D. Keedwell have shown that every generalized complete mapping gives rise to a block design of the above type. We now summarize their main results but first we require a further definition.

Definition 16. Let $D = (G,B)$ be a (v,K,λ)-MD and let L be a group of v permutations $\alpha_1, \alpha_2, \ldots, \alpha_v$ of G such that $G = \bigcup_{i=1}^{v} \alpha_i(g)$, where g is any fixed element of G. Suppose further that there exists a subset $B^* = \{B_1, B_2, \ldots, B_f\}$ of blocks of B such that $B = \bigcup_{i=1}^{v} \{\alpha_i(B_1), \alpha_i(B_2), \ldots, \alpha_i(B_f)\}$. Then it follows that each permutation of L permutes the blocks of B among themselves and is an automorphism of D. We shall say that D admits L as a regular group of automorphisms with the blocks of B^* as basis blocks.

With the aid of this definition, we may state the following:

Theorem 3. *If* $(g_{11}\ g_{12} \ldots g_{1k_1}), (g_{21}\ g_{22} \ldots g_{2k_2}), \ldots, \ldots, (g_{s1}\ g_{s2} \ldots g_{sk_s})$ *is a* (K,λ) *complete mapping of a group* $(G,.)$ *of order* v *with* $K = \{k_1,k_2,\ldots,k_s\}$ *then the blocks of the set*

$$B = \bigcup_{g\in G}\{(gg_{11}\ gg_{12} \ldots gg_{1k_1}),(gg_{21}\ gg_{22} \ldots gg_{2k_2}),\ldots,\ldots,(gg_{s1}\ gg_{s2} \ldots gg_{sk_s})\}$$

form a $(v,K.\lambda)$*-MD which admits the left regular representation* L_G *of* $(G,.)$ *as a regular group of automorphsims such that the cyclic sequences which define the* (K,λ) *complete mapping of* G *are its base blocks.*

Conversely, let L_G *be the left regular representation of a group* $(G,.)$ *of order* v *and suppose that there exists a* (v,K,λ)*-MD defined on the set* G *which admits* L_G *as a regular group of automorphisms with the blocks* $(g_{11}\ g_{12} \ldots g_{1k_1}), (g_{21}\ g_{22} \ldots g_{2k_2}), \ldots, \ldots, (g_{s1}\ g_{s2} \ldots g_{sk_s})$ *as basis blocks, where* $K = \{k_1,k_2,\ldots,k_s\}$ *and where one element* x *of* G *does not occur at all among the elements of the basis blocks but every other element of* G *occurs exactly* λ *times, then the cyclic sequences* $(x^{-1}g_{11}\ x^{-1}g_{12} \ldots x^{-1}g_{1k_1}), (x^{-1}g_{21}\ x^{-1}g_{22} \ldots x^{-1}g_{2k_2}), \ldots, \ldots, (x^{-1}g_{s1}\ x^{-1}g_{s2} \ldots x^{-1}g_{sk_s})$ *form a* (K,λ) *complete mapping of the group* $(G,.)$.

Corollary. *If the group* $(G,.)$ *of order* v *is R-sequenceable then there exists a* $(v,v-1,1)$*-MD which admits the group* L_G *of permutations of the left regular representation of* $(G,.)$ *as a regular group of automorphisms.*

Theorem 4. *If* $[g'_{11}\ g'_{12} \ldots g'_{1h_1}] \ldots [g'_{\lambda 1}\ g'_{\lambda 2} \ldots g'_{\lambda h_\lambda}](g_{11}\ g_{12} \ldots g_{1k_1}) \ldots (g_{s1}\ g_{s2} \ldots g_{sk_s})$ *is a* (K,λ) *near complete mapping of a group* $(G,.)$ *of order* $v-1$ *with* $K = \{h_1,h_2,\ldots h_\lambda;\ k_1,k_2,\ldots,k_s\}$ *then the blocks of the set*

$$B = \bigcup_{g\in G}\{(gg'_{11}\ gg'_{12} \ldots gg'_{1h_1}\ \infty), (gg'_{21}\ gg'_{22} \ldots gg'_{2h_2}\ \infty), \ldots, (gg'_{\lambda 1}\ gg'_{\lambda 2} \ldots gg'_{\lambda h_\lambda}\ \infty), (gg_{11}\ gg_{12} \ldots gg_{1k_1}), (gg_{21}\ gg_{22} \ldots gg_{2k_2}), \ldots, (gg_{s1}\ gg_{s2} \ldots gg_{sk_s})\}$$

form a (v,K^*,λ)*-MD on the set* $G^* = G\cup\{\infty\}$, *where* $K^* = \{h_1+1,h_2+1,\ldots,h_\lambda+1,$ $k_1,k_2,\ldots k_s\}$. *Moreover the group of permutations*

$$\bigcup_{g\in G}\begin{pmatrix} g_1 & g_2 & \cdots & g_{v-1} & \infty \\ gg_1 & gg_2 & \cdots & gg_{v-1} & \infty \end{pmatrix} \quad \text{where } G = \{g_1,g_2,\ldots,g_{v-1}\}, \text{ acts}$$

as a regular group of automorphisms on this design with base blocks $(g'_{11},g'_{12},\ldots,g'_{1h_1}\ \infty)$, $\ldots$, $(g'_{\lambda 1}\ g'_{\lambda 2}\ \cdots\ g'_{\lambda h_\lambda}\ \infty)$, $(g_{11}\ g_{12}\ \cdots\ g_{1k_1})$, $\ldots$, $(g_{s1}\ g_{s2}\ \cdots\ g_{sk_s})$.

<u>Corollary.</u> *If the group* $(G,.)$ *of order* $v-1$ *is sequenceable then there exists a* $(v,v,1)$*-MD which admits the group of permutations*

$$\bigcup_{g\in G}\begin{pmatrix} g_1 & g_2 & \cdots & g_{v-1} & \infty \\ gg_1 & gg_2 & \cdots & gg_{v-1} & \infty \end{pmatrix} \text{ as a regular group of automorphisms.}$$

The proofs of theorems 3 and 4 are in [13] and the following illustration of theorem 4 is from the same source.

<u>Example 9.</u> The $(K,2)$ near complete mapping of the dihedral group D_3 given in example 5, where $K = \{2,2;4,4\}$, defines a $(7,K^*,2)$-MD with $K^* = \{3,4\}$ whose blocks are

$(e\ ba\ \infty)$, $(e\ ba^2\ \infty)$, $(a^2\ b\ ba^2\ a)$, $(a\ b\ ba\ a^2)$

$(a\ b\ \infty)$, $(a\ ba\ \infty)$, $(e\ ba^2\ ba\ a^2)$, $(a^2\ ba^2\ b\ e)$

$(a^2\ ba^2\ \infty)$, $(a^2\ b\ \infty)$, $(a\ ba\ b\ e)$, $(e\ ba\ ba^2\ a)$

$(b\ a\ \infty)$, $(b\ a^2\ \infty)$, $(ba^2\ e\ a^2\ ba)$, $(ba\ e\ a\ ba^2)$

$(ba\ e\ \infty)$, $(ba\ a\ \infty)$, $(b\ a^2\ a\ ba^2)$, $(ba^2\ a^2\ e\ b)$

$(ba^2\ a\ \infty)$, $(ba^2\ e\ \infty)$, $(ba\ a\ e\ b)$, $(b\ a\ a^2\ ba)$

The design admits a regular group of automorphisms isomorphic to D_3.

In [1], the concept of <u>resolvability</u> of a certain type of Mendelsohn design was introduced according to the following definition.

Definition 16. If the blocks of a (v,k,1)-MD for which $v \equiv 1 \bmod k$ can be partitioned into v sets each containing $(v-1)/k$ blocks which are pairwise disjoint (as sets), we say that the (v,k,1)-MD is resolvable and any such partition is called a resolution of the design.

Each set of $(v-1)/k$ pairwise disjoint blocks together with the singleton which is the only element not in any of its blocks is called a parallel class of the resolution. Any resolution of this kind has v parallel classes.

Hsu and Keedwell [13] have extended the concept of resolvability to cover also (v,k,1)-MD's for which $v \equiv 0 \bmod k$.

Definition 17. If the blocks of a (v,k,1)-MD for which $v \equiv 0 \bmod k$ can be partitioned into v-1 sets each containing v/k blocks which are pairwise disjoint (as sets). we shall again say that the (v,k,1)-MD is resolvable.

Each set of v/k pairwise disjoint blocks will be called a parallel class.

It is immediate to see that every (v,k,1)-MD obtained from a (k,1) complete mapping in the manner of theorem 3 is resolvable in the sense of definition 16 and that every (v,k,1)-MD obtained from a (k,1) near complete mapping in the manner of theorem 4 is resolvable in the sense of definition 17.

Another relevant concept is that of a perfect cyclic design, first introduced in [22].

Definition 18. Let S be a given set and let $B_i = (a_{i1}\ a_{i2}\ \dots\ a_{ik})$ be a cyclically ordered subset of k elements of S. Then the elements a_{ir} and $a_{i,r+t}$, where addition of the second suffix is modulo k, are said to be t-apart in the cyclic k-tuple B_i.

A (v,k,λ) Mendelsohn design (G,B) is said to be <u>ℓ-fold perfect</u> if each ordered pair (x,y) of elements of G appears t-apart in exactly λ of the blocks of B for all $t = 1,2,\ldots,\ell$. If $\ell = k-1$, the design (G,B) is said to be <u>perfect</u>. We call such a design a (v,k,λ)-PMD.

In [13], the concept of ℓ-perfect design has been linked with generalized complete mappings and orthogonal latin squares in the following way.

<u>Definition 19.</u> Let $(g_{11}\ g_{12}\ \cdots\ g_{1k})(g_{21}\ g_{22}\ \cdots\ g_{2k})\ \cdots\ (g_{s1}\ g_{s2}\ \cdots\ g_{sk})$ be a (k,λ) complete mapping of a group $(G,.)$ of order v such that for each value of t, $t = 1,2,\ldots,\ell$, the elements $g_{ij}^{-1}\ g_{i,j+t}$ (where $i = 1,2,\ldots,s$; and the second suffix j is added modulo k) comprise the non-identity elements of G each counted λ times. Then the complete mapping is said to be an <u>ℓ-fold perfect (k,λ) complete mapping.</u> If $\ell = k-1$, the mapping is said to be a <u>perfect (k,λ) complete mapping</u>.

<u>Definition 20.</u> Let $[g'_{11}\ g'_{12}\ \cdots\ g'_{1h}]\ \cdots\ [g'_{\lambda 1}\ g'_{\lambda 2}\ \cdots\ g'_{\lambda h}](g_{11}\ g_{12}\ \cdots\ g_{1k})$ $\ldots\ (g_{s1}\ g_{s2}\ \cdots\ g_{sk})$, where $k-h = 1$, be a (k,λ) near complete mapping of a group $(G,.)$ of order v and let B^* denote the set of $\lambda+s$ k-tuples $(g'_{11}\ g'_{12}\ \cdots\ g'_{1k})(g'_{21}\ g'_{22}\ \cdots\ g'_{2k})\ \cdots\ (g'_{\lambda 1}\ g'_{\lambda 2}\ \cdots\ g'_{\lambda k})(g_{11}\ g_{12}\ \cdots\ g_{1k})\ \cdots$ $(g_{s1}\ g_{s2}\ \cdots\ g_{sk})$, where $g'_{ik} = \infty$ for $i = 1,2,\ldots,\lambda$. If, for each value of t, $t = 1,2,\ldots,\ell$, the elements of the set

$$\left(\bigcup_{i=1}^{\lambda}\{g'^{-1}_{ij}\ g'_{i,j+t} : j = 1,2,\ldots,k-1\}\right)\cup\left(\bigcup_{i=1}^{s}\{g_{ij}^{-1}\ g_{i,j+t} : j = 1,2,\ldots,k\}\right)$$

comprise all the non-identity elements of the set $G\cup\{\infty\}$ each counted λ times where we define $g^{-1}\infty = \infty$, then the near complete mapping is called <u>ℓ-fold perfect</u>. If $\ell = k-1$, the mapping is called <u>perfect</u>.

<u>Theorem 5.</u> *If there exists an ℓ-fold perfect (k,λ) complete mapping of a group $(G,.)$ of order v then, by the construction of theorem 3, there exists a (v,k,λ)-MD which is ℓ-fold perfect and on which the left regular representation L_G of G acts as a regular group of automorphisms. Likewise, if there exists an ℓ-fold perfect (k,λ) near complete mapping of the group*

then, by the construction of theorem 4, there exists a $(v+1,k,\lambda)$-MD with these same properties.

An ℓ-fold perfect $(v-1,1)$ complete mapping of a group $(G,.)$ of order v is an R_ℓ-sequencing of the group, as defined in [17]. It is there shown that, when a group is R_ℓ-sequenceable, it is possible to construct at least $\ell+1$ mutually orthogonal latin squares based on the Cayley table of G. Hence, for example, we get the following theorem.

Theorem 6. *The elementary abelian group of order p^n has a $(p^n-1,1)$ perfect complete mapping.*

In [13], a number of constructions for generalized complete mappings are given. We mention some of these below.

4. CONSTRUCTIONS FOR GENERALIZED COMPLETE MAPPINGS.

(i) If we repeat the cycles and sequences of a (K,λ) generalized complete mapping h times, we get a $(K,h\lambda)$ generalized mapping. Such a construction is called trivial and the generalized complete mapping so obtained is called decomposable.

(ii) If we reverse the order of the elements in all cycles and sequences of a generalized complete mapping, we get another such mapping. In the case of a complete or near complete mapping with $\lambda = 1$, the neofields constructed from the original mapping and from the reversed mapping may have different algebraic properties. See [13] for examples.

(iii) If we adjoin a (K,λ) generalized complete mapping to its reverse, we get a $(K,2\lambda)$ generalized mapping. A mapping which can be so constructed is called patterned.

(iv) If there exists a (k_1,λ_1) complete mapping of the group G_1 and a (k_2,λ_2) complete mapping of the group G_2, then there exists a (K,λ) complete mapping of the group $G_1 \times G_2$, where $\lambda = \lambda_1\lambda_2$ and $K = \{k,k,\ldots,k,k_1,k_1,\ldots,k_1, k_2,k_2,\ldots,k_2\}$, k being the least common multiple of k_1 and k_2.

<u>Definition 21.</u> A (K,λ) generalized complete mapping is called tight (or <u>pure</u>) if the subsets of elements formed by the members of each of its cycles are distinct.

For example, the adjunction of a (k_1,λ_1) mapping to a (k_2,λ_2) mapping gives a tight $(K,\lambda_1+\lambda_2)$ mapping, $K = \{k_1,k_2\}$.

(v) If a group of order n is R_ℓ-sequenceable, it possesses a tight (K,λ) complete mapping where λ is the number of distinct integers in the set $\{(n-1)/i : i = 1,2,\ldots,\ell\}$.

The proofs of (iv) and (v) are in [13]. Some further methods of construction will be found both there and in [11].

5. UNSOLVED PROBLEMS

We end this survey by listing some unsolved problems. In most cases we make some relevant comments and sometimes we offer our own conjectures regarding the solutions.

(i) <u>Which finite non-abelian groups possess complete mappings?</u>

It was proved by Paige [23] and [25] and later also by Carlitz [3] that the only finite abelian groups which have no complete mapping are those which have a unique element of order 2. It is also known that any finite group which has a cyclic Sylow 2-subgroup does not possess a complete mapping. On the other hand, a finite soluble group whose Sylow 2-subgroups are not cyclic does possess a complete mapping. The proofs of these results are in [9]. It is a widely held conjecture that every finite group whose Sylow 2-subgroups are not cyclic has a complete mapping.

(ii) <u>Does every non-abelian group have a near complete mapping?</u>

It is easy to see that an abelian group G cannot have a near complete mapping unless it has a unique element of order 2. For suppose that the near complete mapping is in canonical form $\phi = [e\ g_1'\ g_2'\ \ldots\ g_{h-2}'\ \eta](g_{11}\ g_{12}\ \ldots\ g_{1k_1})\ \ldots\ (g_{s1}\ g_{s2}\ \ldots\ g_{sk_s})$. Then the elements g_1', $g_1'^{-1}g_2'$, $g_2'^{-1}g_3'$, $\ldots$, $g_{h-2}'^{-1}\eta$, $g_{11}^{-1}g_{12}$, $g_{12}^{-1}g_{13}$, $\ldots$, $g_{1k_1}^{-1}g_{11}$, $\ldots$, $g_{s1}^{-1}g_{s2}$, $g_{s2}^{-1}g_{s3}$, $\ldots$, $g_{sk_s}^{-1}g_{s1}$ are the whole set of elements of G. Their product is $\eta \neq e$. Consequently η

must be the unique element of order 2 in G : for, if G did not have a unique element of order 2, the product of all its elements would be the identity element e. On the other hand, if an abelian group has a unique element of order 2, it is sequenceable and so it certainly has a near complete mapping.

In the case of non-abelian groups, the author does not know of any one which lacks a near complete mapping and so he conjectures that the answer to the question is "Yes".

(iii) Does every finite group possess either a complete mapping or else a near complete mapping?

In virtue of the foregoing remarks, the author conjectures that the answer is "Yes".(*)

(iv) Is it true that, with the exception of D_3, all dihedral groups of singly even order are sequenceable?

For orders up to at least 70 the answer is "Yes" as already remarked in section 1.

(v) Are all dihedral groups except those of orders 4,6 and 8 sequenceable?

The author conjectures that the answer is "Yes". Question (v) seems likely to be more difficult to solve than question (iv).

(vi) Are all non-abelian groups of order pq, where p and q are odd primes, sequenceable?

The answer is "Yes" if p<q and p is a prime such that GF[p] has 2 as a primitive root, as already remarked in section 1.

(vii) Which other classes of finite non-abelian groups are sequenceable?

(viii) Are all finite non-abelian groups of order greater than or equal to nine sequenceable?

The author thinks it quite likely that the answer is "Yes". This question is linked to question (iii) since a sequenceable group has a near complete mapping.

(*) This is equivalent to the conjecture that the Cayley table of every finite group of order n has a transversal of length at least n-1.

(ix) Is it true that all finite abelian groups whose Sylow 2-subgroups are not cyclic are R-sequenceable?

R. Friedlander, B. Gordon and M.D. Miller [6] conjecture that the answer is "Yes", see section 1.

(x) Are all non-abelian groups of order pq, where p and q are odd primes, R-sequenceable?

The answer is "Yes" if $p<q$ and p is a prime such that GF[p] has 2 as a primitive root, see section 1.

(xi) Which further classes of finite non-abelian groups are R-sequenceable?

Since an R-sequencing is a complete mapping, a finite group whose Sylow 2-subgroups are cyclic cannot be R-sequenceable.

(xii) For which finite orders do left neofields whose multiplicative group is non-abelian exist?

(xiii) For which finite orders do two-sided neofields whose multiplicative group is non-abelian exist?

The answers to these questions depend on the answers to questions (iii) to (xi). Since every finite cyclic group of even order is sequenceable and every one of odd order is R-sequenceable (see [6] and [17]), there exist two-sided neofields with abelian multiplication group of all finite orders. The properties of such (cyclic) neofields are treated in detail in [11].

(xiv) For which values of v, k and λ do (v,k,λ) Mendelsohn designs exist? Which of these designs are resolvable and which perfect?

The first part of this question is equivalent to the question "For which values of v and k can the complete directed multigraph λK_v^* on v vertices be separated into edge-disjoint k cycles?" Here, λK_v^* denotes the complete graph in which each two vertices are joined by 2λ edges, λ of which are directed in one sense and an equal number in the other. Clearly, a necessary condition is that $\lambda v(v-1) \equiv 0 \bmod k$. This condition is sufficient in many cases but not in all. In particular, no design exists when $\lambda = 1$ and $v = k = 4$ or $v = k = 6$ or $v = 6$ and $k = 3$. The problem is surveyed in [2]. An additional more recent result is in [26].

A sufficient condition for the existence of a resolvable (v,k,1)-MD for v ≡ 0 mod k is that there exist a group of order v which has a k-regular complete mapping. For v ≡ 1 mod k, a sufficient condition is that there exist a group of order v-1 which has a (k,1) near complete mapping. However, these conditions may not be necessary. Certainly, there exist resolvable designs obtained in other ways. Indeed, in [1], F.E. Bennett, E. Mendelsohn and N.S. Mendelsohn have shown that when v is a prime power and v ≡ 1 mod k there exists a resolvable perfect (v,k,1)-MD.

The question of existence of perfect (v,k,1) Mendelsohn designs (which are not necessarily resolvable) has been treated in [22]. The case $\lambda > 1$ does not seem to have been much considered.

<u>ADDENDUM:</u>

Since this paper was prepared for submission, T. Evans has sent the author a copy of a recent Ph.D. thesis [15] of one of his students, C.P. Johnson. In this, the concept of an "almost transversl" is introduced which is effectively the dual concept to that of a near complete mapping. It is used to construct right neofields. Among other things, the author gives a constructive proof that every finite dihedral group of doubly even order has a complete mapping and that every finite dihedral group of singly even order has a near complete mapping. This is relevant to problem (ii) above. Unfortunately, the near complete mappings constructed by Johnson do not correspond to sequencings of the group and so they do not contribute to the solution to problem (iv).

In Johnson's thesis, the concept of characteristic is used in the same weak sense as in [14], see the footnote on page 8.

REFERENCES

[1] F.E. Bennett, E. Mendelsohn and N.S. Mendelsohn, Resolvable perfect cyclic designs, *J. Combinatorial Theory*, (A) 29 (1980), 142-150.

[2] J.C. Bermond and D. Sotteau, Graph decompositions and G-designs. *Proc. Fifth British Combinatorial Conf., Aberdeen* 1975. (Congressus Numerantium XV, Utilitas Math., 1976.) pp. 53-72.

[3] L. Carlitz, A note on abelian groups, *Proc. Amer. Math. Soc.* 4 (1953), 937-939.

[4] C.V. Eynden, Countable sequenceable groups, *Discrete Math.*, 23 (1978), 317-318.

[5] R.J. Friedlander, Sequences in non-abelian groups with distinct partial products, *Aequationes Math.*, 14 (1976), 59-66.

[6] R.J. Friedlander, B. Gordon and M.D. Miller, On a group sequencing problem of Ringel. *Proc. Ninth S.E. Conf. on Combinatorics, Graph Theory and Computing. Florida Atlantic Univ. Boca Raton, 1978.* (Congressus Numerantium XXI, Utilitas Math., 1978), pp. 307-321.

[7] R.J. Friedlander, B. Gordon and P. Tannenbaum, Partitions of groups and complete mappings. *Pacific J. Math.*, 92 (1981), 283-293.

[8] B. Gordon, Sequences in groups with distinct partial products, *Pacific J. Math.*, 11 (1961), 1309-1313.

[9] M. Hall and L.J. Paige, Complete mappings of finite groups, *Pacific J. Math.*, 5 (1955), 541-549.

[10] G.B. Hoghton and A.D. Keedwell, On the sequenceability of dihedral groups, *Annals of Discrete Math.*, 15 (1982), 253-258.

[11] D.F. Hsu, "*Cyclic Neofields and Combinatorial Dêsigns*". (Springer-Verlag, 1980, Lecture Notes in Mathematics, No.824.)

[12] D.F. Hsu and A.D. Keedwell, Generalized complete mappings, neofields, sequenceable groups and block designs, I.

[13] D.F. Hsu and A.D. Keedwell, Generalized complete mappings, neofields, sequenceable groups and block designs, II.

[14] D.R. Hughes, Planar division neorings, *Trans, Amer. Math. Soc.* 80 (1955), 502-527.

[15] C.P. Johnson, Constructions of neofields and right neofields. Ph.D. Thesis, Emory University, 1981.

[16] A.D. Keedwell, On property D neofields, *Rend. Mat. e Appl.*, (5) 26 (1967), 383-402.

[17] A.D. Keedwell, On R-sequenceability and R_h-sequenceability of groups. Atti del Convegno Internazionale Geometrie Combinatorie e Loro Applicazioni, Roma, 7-12 Giugno, 1981.

[18] A.D. Keedwell, On the sequenceability of non-abelian groups of order pq. *Discrete Math.*, 37 (1981), 203-216.

[19] A.D. Keedwell, On the existence of super P-groups, *J. Combinatorial Theory*, Ser. A, to appear.

[20] A.D. Keedwell, On left neofields and the non-existence of proper finite planar neofields.

[21] N.S. Mendelsohn, A natural generalization of Steiner triple systems, "Computers in Number Theory", *Proc. Sci. Res. Council Atlas Sympos. No. 2 Oxford 1969*, (Academic Press, 1971.) pp. 323-338.

[22] N.S. Mendelsohn, Perfect cyclic designs, *Discrete Math.*, 20 (1977), 63-68.

[23] L.J. Paige, A note on finite abelian groups, *Bull Amer. Math. Soc.*, 53 (1947), 590-593.

[24] L.J. Paige, Neofields, *Duke Math. J.*, 16 (1949), 39-60.

[25] L.J. Paige, Complete mappings of finite groups, *Pacific J. Math.*, 1 (1951), 111-116.

[26] T.W. Tillson, A Hamiltonian decomposition of K^*_{2m}, $2m \geq 8$. *J. Combinatorial Theory*, (B) 29 (1980), 68-74.

[27] H. Zassenhaus, Über endliche Fastkörper, *Abhandl. Math. Sem. Univ. Hamburg*, 11 (1935), 187-220.

CLIQUE COVERINGS OF GRAPHS - A SURVEY

NORMAN J. PULLMAN

The problem of covering and partitioning the edge set of a simple graph with a minimum number of complete subgraphs has been studied by several writers over the years. This paper surveys some of the progress made so far and presents a number of open problems.

1. INTRODUCTION

For our purposes, graphs are finite, loopless and have no multiple edges. If G is a graph we call its complete subgraphs *cliques*. A *clique covering* of G is a family C of cliques whose edge sets "cover" the edge set of G. That is, each edge of G belongs to at least one member of C. If C partitions the edge set of G (that is, each edge belongs to exactly one member of C) then C is a *clique partition* of G. If G has no edges then we take $\emptyset$ to be its clique partition and clique covering.

A *minimum* clique covering (clique partition) of G is one whose cardinality is least among all clique coverings (clique partitions) of G. The common cardinality of its minimum clique coverings is called the *clique covering number* of G, denoted *cc(G)*. The common cardinality of its minimum clique partitions is called the *clique partition number* of G, denoted *cp(G)*.

These concepts emerged in the literature in a paper of P. Erdös, A.W. Goodman and L. Pósa [5] in 1966 and L. Lovász [10] in 1968. They have antecedents in studies of the intersection patterns of finite sets (M. Hall, Jr. [9] 1941), and E. Spilrajn-Marczewski [18] 1945) and parallels in (0,1)-matrix theory (H. Ryser [17] 1973). Recently F. Roberts [16] has compiled a survey paper listing many applications of these and related ideas.

The earliest work concerned global bounds.

Theorem 1.1 *(P. Erdös, A.W. Goodman and L. Pósa [5]). If* G *has* n *vertices, then*

$$cp(G) \leq \lfloor n^2/4 \rfloor, \tag{1.1}$$

with equality holding $G = K_{\lfloor n/2 \rfloor, \lceil n/2 \rceil}$. *Moreover,* G *has a clique partition each of whose members has fewer than four vertices.*

Notice that $\lfloor n^2/4 \rfloor$ is also a global upper bound on the clique covering number because $cc(G) \leq cp(G)$. Erdös, Goodman and Pósa posed the problem of improving the bound when the number of edges is also given. In 1968 the following solution was given.

Theorem 1.2 *(L. Lovász [10]).*

Suppose G *has* n *vertices and* m *edges.*

If $k = \binom{n}{2} - m$ *and* $t = \max\{s \in Z : s^2 - s \leq k\}$, *then*

(1.2) $$cc(G) \leq k + t.$$

There is as yet, no analogue of Theorem 1.2 for the clique partition number.

Later work deals with the algorithmic complexity of the problem (see [15] or it emphasized the computation or estimation of $cc(G)$ and $cp(G)$ for specific classes of graphs. For example, J. Orlin ([11]) in 1977 determined $cc(G^*)$ and $cp(G^*)$ for the line graph G^* of an arbitrary graph.

Theorem 1.3. *Suppose* G *is any connected graph on* n *vertices, other than* K_3. *Let* p *be the number of vertices of* G *of degree* 2 *and* q *be the number of those* 3*-cycles in* G *each having exactly two vertices of degree* 2. *Then*

(1.3) $$cc(G^*) = n - p - q \quad \textit{and}$$

(1.4) $$cp(G^*) = n - p.$$

In this paper we survey results on evaluating and estimating $cp(G)$ and/or $cc(G)$ for complements of cliques (in Section 2) and regular graphs (in Section 3). Although we omit most methods of proof, we discuss some techniques in Section 4 where we also sketch the outlines of some linear-time algorithms for computing $cp(G)$ and $cc(G)$ when the maximum of the degrees of the vertices of G is less than 5. A number of open problems are presented. Many of the results announced in the paper were not in print when it was written.

2. COMPLEMENTS OF CLIQUES

If H is a subgraph of G, let $G \setminus H$ denote the *complement of* H *in* G, the graph obtained by deleting the edges, but not the vertices, of H from G. In particular, $K_n \setminus K_m$ denotes the complement of an *m-clique* (clique on m vertices) in an n-clique.

Evidently, $cc(K_n \setminus K_m) = m$ for all $1 \leq m < n$ and $cc(K_n \setminus K_n) = 0$. The corresponding problem of evaluating $cp(K_n \setminus K_m)$ is considerably harder. It is a corollary of the next theorem that, if we could determine the exact value of $cp(K_{111} \setminus K_{11})$, or even show that $cp(K_{111} \setminus K_{11}) \geq 111$, then we could determine whether or not there is a projective plane of order 10.

Theorem 2.1 *(N.G. de Bruijn and P. Erdös [3], 1948).*

If C *is a clique partition of* K_n, *then* $|C| > n$, $C = \{K_n\}$, C *consists of one* $(n-1)$*-clique and* $n - 2$ *2-cliques, or* $n = m^2 + m + 1$ *and* C *consists of* n *copies of* K_{m+1}.

Corollary 2.1.1 *(J. Orlin [11], 1977).*

(a) *If* $n > 2$, *then* $cp(K_n \setminus K_2) = n - 1$.

(b) *If* $n \geq m > 1$, *then* $cp(K_n \setminus K_{m+1}) = m^2 + m$ *if and only if a projective plane of order* m *exists.*

Corollary 2.1.1 can be reformulated as

Corollary 2.1.2 *If* $2 < r < m^2 + m + 1$, *then*

$$cp(K_{m^2+m+1} \setminus K_r) \geq m^2 + m \tag{2.1}$$

with equality holding in (2.1) if and only if a projective plane of order m *exists and* $r = m + 1$.

In a 1981 paper P. Erdös, R.C. Mullin, V.T. Sós and D.R. Stinson [6] extended Theorem 2.1 considerably. Translating some of their results into clique partition language we obtain the following theorem.

Theorem 2.2. *Suppose* $5 \leq m^2 - m + 2 \leq n \leq m^2 + m + 1$.

(a) $cp(K_n \setminus K_{m+1}) \geq m^2 + m - \varepsilon$, *where* $\varepsilon = 2$ *if* $n = m^2 - m + 2$, $\varepsilon = 1$ *if* $m^2 - m + 2 < n \leq m^2 + 1$ *and* $\varepsilon = 0$ *otherwise.*

(b) *Equality holds in (a) if* m *is the order of some projective plane.*

(c) *If equality holds in (a), then* m *is the order of a projective plane for* $n = m^2 + m - 2$, *for* $m^2 - 3 \leq n \leq m^2 + m + 1$, *and for* $n = m^2 - \alpha$ $(\alpha \geq 0)$ *provided* $\alpha^2 + \alpha(2m-3) - (2m^2-2m) < 0$ *when* m *is even and* $\alpha^2 + \alpha(2m+1) - (2m^2+2m) < 0$ *when* m *is odd.*

(d) $cp(K_n \setminus K_m) = cp(K_n \setminus K_{m+1})$ *if* $n \leq m^2 + m$ *and* m *is the order of a projective plane.*

(e) *If equality does not hold in (a) and* $2 < r < n$, *then*

$$cp(K_n \setminus K_r) \geq n - 1 + ((3-\sqrt{5})/2)m$$
$$\geq n + (.36)\sqrt{n} \quad \textit{when} \quad n \geq 44.$$

D.R. Stinson [personal communication] has pointed out the following consequences of the work done in [6] and a result of Erdös, Sós and Wilson [7].

Theorem 2.3. *Suppose* $5 \leq m^2 - m + 2 \leq n \leq m^2 + m + 1$ *and* $3 \leq k \leq n - 1$.

(a) If $m^2 + 2 \leq n \leq m^2 + m + 1$, *then* $cp(K_n \setminus K_k) \geq m^2 + m$ *with equality iff there exists a projective plane of order* m *and* $n - m^2 \leq k \leq m + 1$. *If* $k < n - m^2$, *then* $cp(K_n \setminus K_k) > m^2 + m$.

(b) If $n = m^2 + m + 1$ *and* $k \neq m + 1$, *then* $cp(K_n \setminus K_k) \geq m^2 + 2m$ *with equality iff* $k = m$ *and* m *is the order of a projective plane.*

(This is from [7].)

(c) If $n = m^2 + 1$, *then* $cp(K_n \setminus K_k) \geq m^2 + m - 1$ *with equality iff* $m \leq k \leq m + 1$ *and* m *is the order of a projective plane.*

(d) If $n = m^2$, *then* $cp(K_n \setminus K_k) \geq m^2 + m - 1$ *with equality iff* $m - 1 \leq k \leq m + 1$ *and* m *is the order of a projective plane.*

(e) $m^2 - m + 3 \leq n \leq m^2 + 1$ *and* $n - (m^2-m+1) \leq k \leq m + 1$, *then* $cp(K_n \setminus K_k) \geq m^2 + m - 1$ *with equality if there exists a plane of order* m. *If* $k < n - (m^2-m+1)$, $\alpha = m^2 - n \geq 0$ *and* $\alpha^2 + \alpha(2m-3) - (2m^2-2m) < 0$ *when* m *is even and* $\alpha^2 + \alpha(2m+1) - (2m^2+2m) < 0$ *when* m *is odd, then* $cp(K_n \setminus K_k) \geq m^2 + m$ *with equality if there is a projective plane of order* m. *Otherwise, if* $k < n - (m^2+m+1)$, *then* $cp(K_n \setminus K_k) > m^2 + m - 1$.

(f) If $n = m^2 - m + 2$ *and* $m - 1 \leq k \leq m + 1$, *then* $cp(K_n \setminus K_k) \geq m^2 + m - 2$ *with equality iff there exists a projective plane of order* m. *If* $3 \leq k < m - 1$, *then* $cp(K_n \setminus K_k) \geq m^2 + m - 1$ *with equality if there exists a projective plane of order* m.

The next result completes our list of values of m and n for which the exact value of $cp(K_n \setminus K_m)$ is known. It is a corollary of a theorem of A. Donald and myself ([15], Theorem 3), see Theorem 4.3 of Section 4 of this survey.

Theorem 2.3. *If* $\lfloor (n+1)/2 \rfloor \leq m \leq n$, *then*

(2.2) $$cp(K_n \setminus K_m) = (n-m)(3m-n+1)/2.$$

Table 2.1 presents the exact values of $cp(K_n \setminus K_m)$ known for $2 \leq m < n \leq 21$. It was constructed using Theorems 2.1, 2.2, 2.3 and 2.4.

n \ m	2	3	4	5	6	7	8	9	10	11	12	13	14	15	16	17	18	19	20
3	2																		
4	3	3																	
5	4	5	4																
6	5	6	7	5															
7	6	6	9	9	6														
8	7	10	10	12	11	7													
9	8	11	11	14	15	13	8												
10	9	11	11	15	18	18	15	9											
11	10	12	12		20	22	21	17	10										
12	11	12	12		21	25	26	14	19	11									
13	12	12	12			27	30	30	27	21	12								
14	13	18	18	18		28	33	35	34	30	23	13							
15	14	19	19	19			35	39	40	38	33	25	14						
16	15	19	19	19			36	42	45	45	42	36	27	15					
17	16	19	19	19				44	49	51	50	46	39	29	16				
18	17	20	20	20				45	52	56	57	55	50	42	31	17			
19	18	20	20	20					54	60	63	63	60	54	45	33	18		
20	19		20	20					55	63	68	70	69	65	58	48	35	19	
21	20		24	20						65	72	76	77	75	70	62	51	37	20

TABLE 2.1

We can estimate $cp(K_n \setminus K_m)$ by using the following lemma whose straight-forward proof is omitted. Theorem 2.2(e) can also be used for lower bounds.

<u>Lemma 2.1</u>. *For all* $2 \leq m < n - 1$,

(a) $cp(K_{n+1} \setminus K_m) \geq cp(K_n \setminus K_m)$

and *(b)* $cp(K_{n+1} \setminus K_{m+1}) \geq cp(K_n \setminus K_m)$.

For example Lemma 2.1 implies that $15 \leq cp(K_{11} \setminus K_5) \leq 21$.

In 1981, W.D. Wallis [19] proved the following asymptotic result.

<u>Theorem 2.4</u>. *If* H *has* $O(\sqrt{n})$ *vertices, then*

$$\lim_{n\to\infty} cp(K_n \setminus H)/n = 1.$$

<u>Corollary 2.4.1</u>. *For each fixed* $m \geq 2$,

$$\lim_{n\to\infty} cp(K_n \setminus K_m)/n = 1.$$

3. REGULAR GRAPHS

When a k-regular graph G has n vertices it has $kn/2$ edges. Therefore

$$cc(G) \leq cp(G) \leq kn/2 \tag{3.1}$$

with equality if and only if G contains no triangles (3-cliques).

D. de Caen and I determined when equality holds in (3.1) in the following

Theorem 3.1 *([14], Theorem 3.3). There exists a k-regular connected graph on* n *vertices containing no triangles if and only if*

(a) $2k \leq n$ *when* n *is even*

and (b) $5k/2 \leq n$ *when* n *is odd.*

It should be noted here that the proof of this theorem given in [14] has a gap: the sufficiency of (b) when $n \equiv 2,4 \pmod 5$ was omitted. However N. Wormald [private communication] has a construction which leads to a proof that covers all cases and is simpler than the one in [14].

Let $G(k,n)$ denote the set of all k-regular, connected graphs G on n vertices. This set is empty when $1 \leq n \leq k$ and when k and n are both odd. Also $G(k,k+1) = \{K_{k+1}\}$ and $G(2,n)$ consists of one graph, the simple cycle on n vertices, when $n \geq 3$.

Let $MP(k,n) = \max\{cp(G) : G \in G(k,n)\}$ and $MC(k,n) = \max\{cc(G) : G \in G(k,n)\}$ for $k > n$. We assume n is even when k is odd. Theorem 3.1 implies that $MC(k,n) = MP(k,n) = kn/2$ when $n \geq (9-(-1)^n)k/4$.

What can be said about these maxima when $k + 1 < n < (9-(-1)^n)k/4$? The first k for which a problem arises is $k = 4$. Table 3.1 gives the solution for $k = 4$ (see [14], Theorem 5.3).

n	6	7	9
MC(4,n)	4	7	14
MP(4,n)	4	10	14

TABLE 3.1

Problem 3.1. Determine $MP(k,n)$ and $MC(k,n)$ for $k > 4$ when $k+1 < n < (9-(-1)^n)k/4$.

Next, we consider $SP(k,n)$, the smallest value of $cp(G)$ taken over all G in $G(k,n)$.

Let V denote the vertex set of $G \in \mathcal{G}(k,n)$ and C a clique partition of G. Suppose the number of j-cliques in C incident with the vertex v is $a_{j-1}(v)$ and c_j is the total number of j-cliques in C, for $2 \leq j \leq k$. Then $jc_{j+1} = \Sigma\{a_j(v): v \in V\}$ for all $1 \leq j \leq k-1$ and $\Sigma\{ja_j(v): 1 \leq j \leq k-1\} = k$ for all $v \in V$. Therefore $|C| = \Sigma\{a_j(v)/j: 1 \leq j \leq k-1, v \in V\}$.

Let $\mu(k) = \text{glb}\left\{\sum_{j=1}^{k-1} \chi_j/j: \chi_j \geq 0 \text{ and } \sum_{j=1}^{k-1} j\chi_j = k\right\}$. It was shown in [14], Lemma 2.2.1, that $\mu(k) = 4/(k+2)$ when k is even and $\mu(k) = 4(k+2)/(k+1)(k+3)$ when k is odd. Therefore

<u>Theorem 3.2</u> *([14], Theorem 2.2). For all* $3 \leq k+1 < n$

(3.1) $$SP(k,n) \geq \lceil n\mu(k) \rceil.$$

How good is this bound? It was shown in [14] Theorem 3.4 that for every $k \geq 3$, equality holds in (3.1) for some n. But P.J. Robinson has shown [private communication] that equality holds in (3.1) for some n whenever $n\mu(k)$ is an integer. Moreover when k is even,

(3.2) $$0 \leq SP(k,n) - \lceil n\mu(k) \rceil \leq 2$$

for all sufficiently large n.

Here are some more specific results. The findings for $k = 3,4$ were obtained in [14]. Those for $k = 5,6,7$ were obtained by P.J. Robinson [private communication], using a computer assisted refinement of the technique of [14].

<u>$k = 3$</u> For all even $n \geq 6$, $SP(3,n) = \lceil n\mu(3) \rceil + \varepsilon$ where $\varepsilon = 1$ if $n/2 \equiv 1 \pmod 3$ and $\varepsilon = 0$ otherwise.

<u>$k = 4$</u>

n	6	7	8	9	10	11	12	13
SP(4,n)	4	8	6	6	9	9	8	11

TABLE 3.2

Table 3.2 gives the values of $SP(4,n)$ for $6 \leq n \leq 13$.

For all $n > 13$, $SP(4,n) = \lceil n\mu(4) \rceil + \varepsilon$ where $\varepsilon = 1$ if $n \not\equiv 0 \pmod 3$ and $\varepsilon = 0$ otherwise.

<u>$k = 5$</u> For all even $n \geq B$, $SP(5,n) = \lceil n\mu(5) \rceil + \varepsilon$ if $n \equiv i \pmod{12}$ and i, B, ε are as in Table 3.3

i	0	2	4	6	8	10
B	12	14	16	18	20	22
ε		1	2	2	2	1

TABLE 3.3

<u>$k = 6$</u> For all $n \geq B$, $SP(6,n) = \lceil n\mu(6) \rceil + \varepsilon$ where:

$B = 12$ and $\varepsilon = 0$ if n is even and

$B = 21$ and $\varepsilon = 1$ if n is odd.

<u>$k = 7$</u> For all even $n \geq 20$, $SP(7,n) = \lceil n\mu(7) \rceil + \varepsilon$ where:

$$\varepsilon = \begin{cases} 0 & \text{if } n \equiv 0 \pmod{20}. \\ 1 & \text{if } n \equiv 2 \pmod{20} \text{ or } n = 24, 30 \text{ or } 36. \\ 3 & \text{if } n \equiv 8 \pmod{20} \text{ and } n \neq 48. \\ 2 & \text{otherwise.} \end{cases}$$

The bound given in 3.1 is quite sharp, at least for even k (see (3.2)). But we are not yet sure about the case of odd k (when $n\mu(k)$ is not an integer).

<u>Problem 3.2</u>. Do there exist constants a and b such that

$$0 \leq SP(k,n) - \lceil n\mu(k) \rceil \leq ak + b$$

for all sufficiently large even n, when k is odd?

Next, we will consider the possible values of $cp(G)$ when G is a connected, k-regular graph on n vertices. Let $S_k(n) = \{cp(G) : G \in G(k,n)\}$. Thus $S_k(n) = \emptyset$ for $1 \leq n \leq k$ and $S_k(k+1) = \{1\}$. Also $S_2(n) = \{n\}$ for all $n \geq 4$ and $S_k(n) = \emptyset$ when both k and n are odd.

The sets $S_3(n)$ and $S_4(n)$ were determined by P. Eades, P.J. Robinson and myself in [4]. We found that for all $m \geq 3$, $S_3(2m) = \{\chi \equiv m \pmod 2 : 5m/3 \leq \chi \leq 3m\}$. Let $J_n = \{\chi \in Z : \lceil 2n/3 \rceil + 1 \leq \chi \leq 2n-4\}$. We also showed that when $n > 13$,

$$S_4(n) = \begin{cases} J_n \cup \{2n-2, 2n\}, & \text{if } n \not\equiv 0 \pmod 3) \\ J_n \cup \{2n/3, 2n-2, 2n\}, & \text{if } n \equiv 0 \pmod 3). \end{cases}$$

The sets $S_4(n)$ for $5 \leq n \leq 13$ were also determined.

<u>Problem 3.3</u>. Determine $S_k(n)$ for all $n \geq k + 2 \geq 7$.

Only even n need be considered when k is odd.

<u>Problem 3.4</u>. For $k > 4$ (and n even when k is odd) does there exist a parameter C depending only on k, such that for all n, there is an interval of consecutive integers J_n such that

$$|S_k(n) \setminus J_n| \leq C \qquad ?$$

Problem 3.5. We know that $\lim_{m\to\infty} |S_3(2m)|/2m = 2/3$

and $\lim_{n\to\infty} |S_4(n)|/n = 4/3$.

Are there analogues for $k > 4$?

Now we turn our attention to clique covering numbers of regular graphs. D. de Caen and I found the following lower bound.

Theorem 3.3 *([13], Theorem 2.3). If* $n > k + 1$, *then*

$$(3.7) \qquad SC(k,n) \geq \begin{cases} 3n/(k+1) & \text{if } 2 \leq k \leq 4 \\ kn/(k-1)(k-2) & \text{if } k > 4. \end{cases}$$

We also showed in [13], Theorem 2.4, that equality holds in (3.7) when $2 \leq k \leq 4$ and $n \equiv 0 \bmod (k+1)$. (In the statement of Theorem 2.3 in [13] the constraint should have read "$n > k + 1 \geq 3$").

We also showed that

$$(3.8) \qquad SC(3,n) = \lfloor (3n+2)/4 \rfloor \text{ for all even } n > 8 \text{ and}$$

$$(3.9) \qquad SC(4,n) = \begin{cases} \lceil 3n/5 \rceil, & \text{if } 7 < n \not\equiv 3 \pmod 5 \text{ or } n \in \{13,18\} \\ 1 + \lceil 3n/5 \rceil & \text{for all other } n > 7. \end{cases}$$

Problem 3.6. Find $SC(k,n)$ for $n > k + 2 > 6$ (n is even when k is odd). Note, $SC(n-2,n) = \min\left\{q\colon n \leq 2\binom{q-1}{\lceil q/2 \rceil}\right\}$ for even n. See (3.12).

What are the possible values of $cc(G)$ when G is a connected, k-regular graph on n vertices?

Let $T_k(n) = \{cc(G)\colon G \in G(k,n)\}$. Then $T_k(n) = \emptyset$ when $1 \leq n \leq k$ and also when both k and n are odd. As with S_k, $T_k(k+1) = \{1\}$ and $T_2(n) = \{n\}$ for all $n \geq 4$.

In [1] and [2] L. Caccetta and I determined $T_3(n)$ and $T_4(n)$ for all n. We found that

(3.10) for all even $n > 8$,

$$T_3(n) = \{\chi \in Z\colon \lfloor (3n+2)/4 \rfloor \leq \chi \leq (3n-4)/2\} \cup \{3n/2\}, \text{ and}$$

(3.11) for all $n > 11$,

$$T_4(n) = \{\chi \in Z\colon \lceil 3n/5 \rceil + \varepsilon \leq \chi \leq 2n-2\} \cup \{2n\}$$

where $\varepsilon = 0$ if $n \not\equiv 3 \pmod 5$ or $n \in \{13,18\}$, and $\varepsilon = 1$ otherwise.

Problem 3.7. Determine $T_k(n)$ for all $n > k + 2 \geq 7$. Only even n need be considered when k is odd.

Problem 3.8. For $k > 4$ (and n even when k is odd), is $T_k(n) \cup \{(kn/2)-1\}$ an interval of integers for all sufficiently large n? It is when $k = 3,4$.

Problem 3.9. We have $\lim_{m\to\infty} |T_3(2m)|/2m = 3/4$

and $\lim_{n\to\infty} |T_4(n)|/n = 7/5$.

Are there analogues for $k > 4$?

Finally we examine the special case of a k-regular graph on $k + 2$ vertices. For such a graph to exist, k must be even and the graph must be the complement of a perfect matching in K_{k+2}. Call that graph T_{2n}.

In his 1977 paper [11], J. Orlin asked for an asymptotic estimate of $cc(T_{2n})$, explaining that it arises in an optimization problem in the theory of Boolean functions.

Using the set-intersection formulation of the clique covering problem (reversing the procedure of [5]), D. Gregory and I [8] were able to compute $cc(T_{2n})$ exactly. We found that

$$(3.12) \qquad cc(T_{2n}) = \min\left\{k: n \leqslant \binom{k-1}{\lceil k/2 \rceil}\right\} .$$

From which it follows that

$$(3.13) \qquad \lim_{n\to\infty} cc(T_{2n})/\log_2(n) = 1.$$

J. Orlin also conjectured in [11] that $\lim_{n\to\infty} cp(T_{2n})/n = 1$. As far as I know, this is still an open problem. He showed ([11], Prop. 3.4) that

$$(3.14) \qquad n + 1 \leqslant cp(T_{2n}) \quad \text{for all} \quad n \geqslant 2.$$

It remains to find a good upper bound on $cp(T_{2n})$.

Problem 3.10. What can be said about $cc(G)$ and $cp(G)$ when G is the complement of a k-factor in K_n?

4. METHODS OF COMPUTATION AND ESTIMATION.

Here we'll mention some devices which may be useful in computing or estimating $cc(G)$ and $cp(G)$. The first provides a lower bound on $cp(G)$ which may be reasonable when $cp(G)$ is close to n. We could have used it to establish (for example) that $cp(K_n \setminus K_2) = n - 1$ or that $cp(K_n \setminus K_3) = 6$ rather than appealing to the de Bruijn/Erdös theorem (Theorem 2.1). In fact in [14], Lemma 3.3(a), we found it useful in a situation where Theorem 2.1 did not seem applicable.

Theorem 4.1 *([14], Theorem 2.3). Suppose* G *is a graph on* n *vertices* $v_1, v_2, \ldots, v_n$. *For* $1 \leq i \leq n$, *let* a_i *be the cardinality of a largest independent set of vertices adjacent to* v_i.

Let $\mathcal{B}$ *denote the set of all diagonal* $n \times n$ *matrices* B *such that* $b_{ii} \geq a_i$ *for* $1 \leq i \leq n$. *If* A *is the vertex-to-vertex adjacency matrix of* G, *then*

(4.1) $$\mathrm{cp}(G) \geq \min\{\mathrm{rank}(A+B) : B \in \mathcal{B}\}.$$

The Cartesian product is useful in constructing graphs with special properties. Fortunately $\mathrm{cp}(G_1 \times G_2)$ and $\mathrm{cc}(G_2 \times G_2)$ are readily expressable in terms of $\mathrm{cp}(G_i)$ and $\mathrm{cc}(G_i)$. Recall that the vertex set of $G_1 \times G_2$ is $V_1 \times V_2$ where V_i is the vertex set of G_i and $(u,v)(w,x)$ is an edge of $G_1 \times G_2$ if $u = w$ and vx is an edge of G_2 or $v = x$ and uw is an edge of G_1. We have

Theorem 4.2 *([14], Lemma 2.4.1).*

(a) $\mathrm{cp}(G_1 \times G_2) = n_1\mathrm{cp}(G_2) + n_2\mathrm{cp}(G_1)$

(b) $\mathrm{cc}(G_1 \times G_2) = n_1\mathrm{cc}(G_2) + n_2\mathrm{cc}(G_1)$.

Graphs with large edge-free subgraphs also have turned out to be very tractable. If G_1 and G_2 are graphs on disjoint vertex sets V_1, V_2 then $G_1 \vee G_2$ is the graph G with vertex set $V_1 \cup V_2$ and xy is an edge of G if xy is an edge of G_1 or G_2 or $x \in V_1$ and $y \in V_2$.

A. Donald and I found the next theorem useful in dealing with complements.

Theorem 4.3 *([15], Theorem 3). Suppose* H *is a graph on* p *vertices and* m *edges and* $\bar{K}_q$ *is the edge-free graph on* q *vertices. If* $q \geq \chi'(H)$ *(the edge-chromatic number of* H*), then*

(a) $\mathrm{cp}(H \vee \bar{K}_q) = pq - m$ *and*

(b) any minimum clique partition of $H \vee \bar{K}_q$ *consists entirely of* r*-cliques with* $r \leq 3$.

For example, Theorem 2.4 is a corollary of this theorem. It also shows that if $m = \lfloor (n+1)/2 \rfloor$, then the graph $K_n \setminus K_m$ which has a large number of edges $\left(\text{about } .75\binom{n}{2}\right)$, still requires its minimum clique partitions to contain no r-clique with $r > 3$.

Problem 4.1. Is there any graph on n vertices with more than

$\binom{n}{2} - \binom{\lfloor (n+1)/2 \rfloor}{2}$ edges such that a (every?) minimum clique partition contains no r-clique with $r > 3$?

Problem 4.2. What is the largest number of edges m_0 a graph on n vertices can have and still possess a minimum clique partition containing no r-clique with $r > r_0$?

If $r_0 = 2$, the answer is $m_0 = \lfloor n^2/4 \rfloor$.

If $r_0 = 3$, then $m_0 \geq \binom{n}{2} - \binom{\lfloor (n+1)/2 \rfloor}{2}$.

Frequently we found it useful to break up a graph G into a family of mutually edge-disjoint subgraphs G_i on which the clique covering number and clique partition number functions are additive ($cc(G) = \sum_i cc(G_i)$, $cp(G) = \sum_i cp(G_i)$).

If G has a subgraph H such that for every clique K, every edge of K lies in H or no edge of K lies in H, then we say that H *separates the cliques* of G. In that case, $cp(G) = cp(H) + cp(G \setminus H)$ and $cc(G) = cc(H) + cc(G \setminus H)$. For example, to prove Theorem 4.2, observe that $G_1 \times G_2$ is the edge-disjoint union of n_1 vertex-disjoint isomorphs $\{v\} \times G_2$ of G_2 (one for each v in G_1) with n_2 vertex-disjoint isomorphs of G_1. Also, the graphs $\{v\} \times G_2$ and $G_1 \times \{w\}$ separate the cliques of $G_1 \times G_2$. This is easy to verify using the fact that H separates the cliques of G if for every 3-clique K, every edge of K lies in H or none of its edges lies in H.

A subgraph H of G is *proper* if $H \neq G$. It is *empty* if H has no edges. (We assume that all graphs have vertices.) If G has a proper, nonempty subgraph separating the cliques of G, then G is said to be *clique-separable*. Otherwise G is *clique-inseparable*.

If a subgraph B separates the cliques of G, but no proper, nonempty subgraph of B does so, then B is a *clique-block* of G. The clique-blocks of G can also be thought of as the maximal clique-inseparable subgraphs of G (see [12]). If $\mathcal{B}$ is the set of all clique blocks in G, then $cp(G) = \sum_{B \in \mathcal{B}} cp(B)$ and $cc(G) = \sum_{B \in \mathcal{B}} cc(B)$.

A systematic way then of computing cc(G) or cp(G) is to (1) locate a clique block B, (2) compute cc(B), cp(B) and add them to a running subtotal, then (3) delete the edges of B from G before beginning the cycle again. The process stops when the current graph is empty. The subtotals are then the clique covering and partition numbers of G. Step (1) can be done in several ways. One way is to find the *triangle graph* T(G). That is, the graph whose vertices are the 3-

cliques of G with 3-cliques K,L deemed adjacent in T(G) if $K \cap L$ is a 2-clique. It turns out (see [12]) that the clique blocks of G are either 2-cliques contained in no 3-clique of G (*whiskers*) or they are whisker-free subgraphs B of G such that T(B) is a connected component of the triangle graph. Thus the task of finding clique blocks in G is transformed into the task of locating the connected components of the triangle graph of G.

When the maximum degree of the vertices of G, $\Delta(G)$ is less than 5, then cp(G) and cc(G) can be computed in time proportional to the number of vertices in G. This was done in several ways in [12]. The algorithms were based on the curious fact that for all $n \geq 1$, there exist at most 6 clique-inseparable, nonempty graphs G with $\Delta(G) \leq 4$ having n vertices. In fact, for each $n > 7$, there are only 3 such graphs (their triangle graphs are simple paths or cycles). Theorem 1 in [12] catalogues all the clique inseparable graphs G with $\Delta(G) \leq 4$.

Problem 4.3. Is it possible to devise a linear time algorithm for graphs G with $\Delta(G) = 5$?

ACKNOWLEDGEMENTS

This work was supported in part by the Natural Sciences and Engineering Research Council of Canada under grants A4041 and T1821. Thanks are also due for the hospitality of the Departments of Mathematics of the University of Newcastle (N.S.W.) and the University of Queensland.

REFERENCES

[1] Cacetta, L. and Pullman, N.J. Clique Covering Numbers of Cubic Graphs. *Proceedings Combinatorial Mathematics X, (1982).*

[2] Cacetta, L. and Pullman, N.J. Clique Covering Numbers of Regular Graphs. (To appear.)

[3] de Bruijn, N.G. and Erdös, P. On a combinatorial problem. *Indag. Math.*, 10 (1948), 421-423.

[4] Eades, P., Pullman, N.J. and Robinson, P.J. On Clique Partition Numbers of Regular Graphs. *Proceedings of the Waterloo Conference on Combinatorial Mathematics*, University of Waterloo, (1982).

[5] Erdös, P., Goodman, A.W. and Pósa, L. The representation of a graph by set intersections. *Can. J. Math.*, 18 (1966), 106-112.

[6] Erdös, P., Mullin, R.C., Sós, V.T. and Stinson, D.R. Finite linear spaces and projective planes. *Discrete Mathematics*. (To appear.)

[7] Erdös, P., Sós, V.T. and Wilson, R. On t-designs. (To appear.)

[8] Gregory, D.A. and Pullman, N.J. On a clique covering problem of Orlin. *Discrete Mathematics*. (To appear.)

[9] Hall, M. Jr. A problem in partitions. *Bull. Amer. Math. Soc.*, 47 (1941), 804-807.

[10] Lovász, L. On covering of graphs. *Theory of Graphs*, 231-236. (Proc. Colloq. Tihany (1966). Academic Press, New York and London (1968).)

[11] Orlin, J. Contentment in graph theory. *Indag. Math.*, 39 (1977), 406-424.

[12] Pullman, N.J. Clique coverings of graphs IV: algorithms. (To appear.)

[13] Pullman, N.J. and de Caen, D. Clique coverings of graphs III: clique coverings of regular graphs. *Cong. Numer.*, 29 (1980), 795-808.

[14] Pullman, N.J. and de Caen, D. Clique coverings of graphs I: clique partitions of regular graphs. *Utilitas Math.*, 19 (1981), 177-205.

[15] Pullman, N.J. and Donald, A. Clique coverings of graphs II: complements of cliques. *Utilitas Math.*, 19 (1981), 207-213.

[16] Roberts, F.S. Applications of edge coverings by cliques. (To appear.)

[17] Ryser, H.J. Intersection properties of finite sets. *J. Comb. Th. (A)*, 14 (1973), 79-92.

[18] Spilrajn-Marczewski, E. Sur deux proprietes des classes d'ensembles. *Fund. Math.*, 33 (1945), 303-307.

[19] Wallis, W.D. Asymptotic values of clique partition numbers. *Combinatorica*. (To appear.)

ROOM SQUARES AND SUBSQUARES

D. R. Stinson

If a Room square of side s *contains a Room subsquare of side* t*, then* $s \geq 3t + 2$. *For* $t = 3$ *or* 5, *there is no Room square of side* t*, yet one can construct (incomplete) Room squares of side* s *"missing" subsquares of side* 3 *or* 5 *(the same bound* $s \geq 3t + 2$ *holds).*

It has been conjectured that if s *and* t *are odd,* $s \geq 3t + 2$ *and* $(s,t) \neq (5,1)$, *then there exists a Room square of side* s *containing (or missing, if* $t = 3$ *or* 5) *a subsquare of side* t. *Substantial progress has been made toward proving this conjecture. In this paper we show that there exists a Room square of odd side* s *containing or missing a subsquare of odd side* t *provided* $s \geq 6t + 41$. *For odd* $t \geq 127$ *and odd* $s \geq 4t + 29$, *there exists a Room square of side* s *containing a subsquare of side* t.

1. INTRODUCTION

A *Room square* of side s is a square array R of side s, which satisfies the following properties:

(1) each cell of R either is empty or contains an unordered pair of elements (called *symbols*) chosen from a set S of size $s + 1$,

(2) each symbol occurs in precisely one cell of each row and column of R,

(3) every unordered pair of symbols occurs in a unique cell of R.

Suppose R is a Room square of side s (on symbol set S). A square t by t subarray T of R is said to be a *Room subsquare* of side t provided it is itself a Room square of side t on a subset of S of size $t + 1$. We will refer to a Room subsquare simply as a *subsquare*.

In this paper we study the embedding question: for which ordered pairs (s,t) does there exist a Room square of side s containing a subsquare of side t?

Mullin and Wallis proved that a Room square of side t exists precisely for odd positive t other than 3 or 5 (see [5]). However the embedding question can be altered slightly to include the cases $t = 3$ and 5. An *incomplete* Room square of side s *missing* a subsquare of side t is a square array R of side s, which satisfies:

(1) each cell of R either is empty or contains an unordered pair of symbols (from a symbol set S of size $s + 1$),

(2) there is a t by t subarray T of empty cells contained in S(we call T the *missing subsquare*),

(3) every symbol occurs in precisely one cell of each row and column not meeting T,

(4) there is a set $S' \subseteq S$ of size $t + 1$, such that every row and column meeting T contains precisely the symbols of $S \setminus S'$ once each,

(5) the pairs occurring in R are precisely those $\{x,y\}$ with $(x,y) \in (S \times S) \setminus (S' \times S')$.

An incomplete Room square of side s missing a subsquare of side t can exist only for s and t odd. If $t \neq 3$ or 5, then the incomplete Room square is equivalent to a Room square of side s containing a subsquare of side t, since the subsquare may be inserted or removed at will. However, for $t = 3$ or 5, the incomplete Room square can exist, whereas there does not exist any Room square containing a subsquare of side 3 or 5. Thus, in this paper we study the existence of incomplete Room squares.

We will refer to an incomplete Room square of side s missing a subsquare of side t as an <u>(s,t)-incomplete Room square</u>.

<u>Lemma 1.1</u> ([2]) *If an (s,t)-incomplete Room square exists, then* $s \geq 3t + 2$.

<u>Proof</u>. Pick a symbol x not in the symbol set of the missing subsquare. The symbol x occurs t times in rows, and t more times in columns meeting the missing subsquare. Also, x must occur with each of the $t + 1$ symbols of the missing subsquare. We have counted $3t + 1$ occurrences of x, so $s \geq 3t + 1$. However t and s are odd, so $s \geq 3t + 2$. □

It is known that for any odd t, there is a constant $S(t)$ such that, for any odd $s > S(t)$, there exists an (s,t)-incomplete Room square. For $t \neq 3$ or 5, $S(t)$ was first shown to be finite by Wallis [9], although no bound for $S(t)$ was explicitly determined. Also, note that $S(1) = 5$, since any Room square has a subsquare of side 1. The following was proved in [7].

<u>Theorem 1.2</u> *Let* $t \geq 1$ *be an odd integer. Then for all odd* $s \geq \max\{t+644, 6t+9\}$, *there exists an* (s,t)*-incomplete Room square. Hence* $S(t) < \max\{t+644, 6t+9\}$.

In this paper we shall improve this result. For all odd t, we can show that $S(t) \leq 6t + 39$; while for odd $t \geq 127$, we prove that $S(t) \leq 4t + 27$. For small t, we have $S(1) = 5$, $S(3) \leq 39$, $S(5) \leq 67$, and $S(7) \leq 53$.

It is conjectured [6] that $S(1) = 5$, and $S(t) = 3t$ for odd $t \geq 3$. Evidence that this conjecture is correct is the proof of the existence of $(3t + 2, t)$-incomplete Room squares for all odd $t \geq 3$ (see [10]).

2. CONSTRUCTIONS

In this section we present several recursive constructions for incomplete Room squares. A certain generalization of a Room square, known as a frame, is of fundamental importance.

Let $\{S_1,\ldots,S_n\}$ be a partition of a symbol set S. An $\{S_1,\ldots,S_n\}$*-frame* is an $|S|$ by $|S|$ array F, indexed by S, which satisfies the properties:

(1) every cell either is empty or contains an unordered pair of symbols of S,

(2) the subarrays $S_i \times S_i$, $1 \leq i \leq n$, are empty (these are referred to as *holes*),

(3) each symbol of $S \setminus S_i$ occurs precisely once in row (or column) $s \in S_i$,

(4) the pairs occurring in F are those $\{s,s'\}$ where $(s,s') \in (S \times S) \setminus \bigcup_{i=1}^{n} (S_i \times S_i)$.

The *type* of an $\{S_1,\ldots,S_n\}$-frame is the multiset $\{|S_1|,\ldots,|S_n|\}$. We will use the notation $t_1^{u_1}\ldots t_k^{u_k}$ to describe a multiset which contains u_i elements equal to t_i, $1 \leq i \leq k$.

A frame can be thought of as a Room square from which a special set of subsquares have been removed. Suppose R is a Room square of side s, on symbol set S. Pick any $\infty \in S$, and index R by $S\setminus\{\infty\}$ so that $R(s,s) = \{\infty,s\}$ for any $s \in S\setminus\{\infty\}$. If the contents of cells $R(s,s)$ are then deleted, one obtains a frame of type 1^s. This procedure is reversible, so one can construct a Room square of side s from a frame of type 1^s. More generally, a frame of type $1^{s-t}t^1$ is equivalent to an (s,t)-incomplete Room square.

The following construction indicates that frames can often be completed to Room squares, by filling in the holes and possibly adding a few new rows and columns.

Construction 2.1 ([6]) (1) Suppose there exists a frame of type $t_1^{u_1} \dots t_k^{u_k}$. Let $w \geq 0$, and suppose there exists a Room square of side $t_i + w$ containing a subsquare of side w, for $1 \leq i \leq k$. Then there exists a Room square of side $s + w$ ($s = \sum_{i=1}^{k} t_i u_i$), which contains subsquares of sides $t_i + w$ ($1 \leq i \leq k$), and w.

(2) Suppose there exists a frame of type $t_1^{u_1} \dots t_k^{u_k}$, where $u_k = 1$. Let $w \geq 0$, and suppose there exists a $(t_i + w, w)$-incomplete Room square for $1 \leq i \leq k - 1$. Then there exists an $(s + w, t_k + w)$-incomplete Room square, where $s = \sum_{i=1}^{k} t_i u_i$.

The above construction is particularly useful in the cases $w = 0$ and $w = 1$. When $w = 0$, we adopt the convention that any Room square has a subsquare of side 0. Also, note that any filled cell of a Room square is a subsquare of side 1.

The following result records the existence of frames which have holes all of size 2.

Lemma 2.2 ([3]) *There exists a frame of type* 2^u *if and only if* $u \geq 5$.

These frames are of considerable use in a recursive construction, which we now briefly describe. A *group-divisible design* (or GDD) is a triple (X, G, A), where X is a finite set (of *points*), G is a partition of X into subsets (called *groups*), and A is a set of subsets (called *blocks*) of X, such that (1) every unordered pair of points, not contained in a group, is contained in a unique block, and (2) a group and a block contain at most one common point. A *weighting* of a GDD is a map $w = X \to \mathbb{Z}^+ \cup \{0\}$. For a weighting w, and a subset $Y \subseteq X$, let $w(Y)$ denote the multiset $\{w(x) : x \in Y\}$. The following construction is proved in [6].

Construction 2.3 Suppose (X, G, A) is a GDD and w is a weighting. Suppose that, for every block $A \in A$, there is a frame of type $w(A)$. Then there is a frame of type $\{\sum_{x \in G} w(x) : G \in G\}$.

The above construction is most easily applied to transversal designs. A *transversal design* TD(k,n) is a GDD which has nk points, k groups each of size n, and n^2 blocks each of size k. It is well-known that a TD(k,n) is equivalent to $k-2$ mutually orthogonal Latin squares (MOLS) of order n.

In section 3, we will apply Constructions 2.1 and 2.3 (making use of the frames of Lemma 2.2) to obtain our general results concerning the existence of incomplete Room squares.

3. INCOMPLETE ROOM SQUARES

The following is the main construction for attacking the general problem of constructing incomplete Room squares.

Lemma 3.1 *Suppose* (X,G,A) *is a* GDD *such that every block has size at least* 5. *Denote* $G = \{G_1,\ldots,G_k\}$, *and suppose* $|G_i| \geq 3$ *for* $1 \leq i \leq k-1$. *Then*

(1) *there is a* $(2|X|+1,\ 2|G_k|+1)$*-incomplete Room square*

(2) *if* $|G_k| \geq 3$, *there is a* $(2|X|+1,\ 2|G_i|+1)$*-incomplete Room square for any* i, $1 \leq i \leq k$.

Proof. Define a weighting w by $w(x) = 2$ for all $x \in X$. Apply Construction 2.3, noting that a frame of type 2^t exists for all $t \geq 5$ (Lemma 2.2). A frame is constructed, of type $\{2|G_1|,\ldots,2|G_k|\}$. Now apply Construction 2.1 with $w = 1$. The results follow. □

Using transversal designs, we obtain

Lemma 3.2 *Suppose there is a* TD(k,n). *For* $6 \leq i \leq k-1$, *let* $d_i = 0$ *or* $3 \leq d_i \leq n$. *Also, let* $0 \leq d_k \leq n$. *Then there exists a* $(2u+1,\ 2d_k+1)$*-incomplete Room square, where* $u = 5n + \sum_{i=6}^{n} d_i$. *If* $d_k \neq 1$ *or* 2, *then for* $6 \leq i \leq k$, *there is a* $(2u+1,\ 2d_i+1)$*-incomplete Room square, and also there exists a* $(2u+1,\ 2n+1)$*-incomplete Room square.*

Proof. Let (X,G_1,A_1) be a TD(k,n), where $G_1 = \{G_1,\ldots,G_k\}$. For $6 \leq i \leq k$, delete $n-d_i$ points from G_i. Thus obtain a GDD (X,G,A) which satisfies the hypotheses of Lemma 3.1. The results follow. □

Corollary 3.3 *Let* $v \geq 1$ *be an integer, and suppose there is a* $TD(k,n)$, *where* $n \geq v$ *and* $k \geq 7$. *Then for all* u *satisfying* $5n + v + 3 \leq u \leq (k-1)n+v$, *there is a* $(2u+1, 2v+1)$*-incomplete Room square.*

Proof. Apply Lemma 3.2 with $d_k = v$. The remaining d_i's can be suitably chosen for all values of u in the stated interval. □

The following result is proved similarly.

Corollary 3.4 *Suppose there exists a* $TD(k,n)$, *where* $k \geq 7$. *Then, for all* u *satisfying* $5n + 3 \leq u \leq kn$, *there is a* $(2u+1, 2n+1)$*-incomplete Room square.*

Let T_7 be the set of positive integers defined by

$$n \in T_7 \text{ if } \begin{cases} n \geq 63 \\ \text{or } n \geq 7 \text{ is a prime power} \\ \text{or } n = 12, 50, 55, 56, 57, \text{ or } 58. \end{cases}$$

Brouwer [1] has shown that there is a $TD(7,n)$ for all $n \in T_7$.

The following facts concerning T_7 are also pertinent.

Lemma 3.5 *Let* $T_7 = \{n_1, n_2, \ldots\}$ *where* $n_1 < n_2 < \ldots$. *Then*

(1) $\max\{n_{i+1} - n_i\} = 5$

(2) For all $i \geq 1$, *there is a* $TD(k_i, n_i)$ *where* $k_i \geq 1 + \frac{5n_{i+1}+2}{n_i}$.

Proof. (1) is easily verified. (2) is clearly true for $n_i \geq 62$, since $k_i \geq 7$. For $n_i < 62$ this condition can be checked from Brouwer's list of lower bounds on the number of MOLS ([1]). □

Theorem 3.6 *Let* $t = 2v + 1$ *be a positive odd integer, and suppose* $n \in T_7$, *where* $v \leq n$. *Then, for all odd* $s \geq 6t + 1 + 12(n-v)$, *there is an* (s,t)*-incomplete Room square. Further, if* $v \in T_7$, *then for all odd* $s \geq 5v + 2$, *there is an* (s,t)*-incomplete Room square.*

Proof. Let s be odd, $s \geq 6t + 1 + 10(n-v)$. Choose n_i to be the largest element of T_7 such that $s \geq 6t + 1 + 10(n_i - v)$, (note $n_i \geq n \geq v$). We apply Corollary 3.3, checking that $2(5n_i + v + 3) + 1 \leq s \leq 2((k_i - 1)n_i + v) + 1$.

First, $s \geq 6t + 1 + 10(n_i - v) = 10n_i + 2v + 7 = 2(5n_i + v + 3) + 1$. Also

$s \leq 6t + 10(n_{i+1} - v) - 1$ (since s is odd)

$\leq 6t + 10(\frac{(k_i - 1)n_i - 2}{2} - v) - 1$ (Lemma 3.5(2))

$= 2((k_i - 1)n + v) + 1,$

as desired.

Now, suppose $v \in T_7$. We need only show that there is an (s,t)-incomplete Room square for $5t + 2 \leq s \leq 6t - 1$, s odd. This follows immediately from Corollary 3.4. □

Corollary 3.7

(1) $s(3) \leq 87$, $s(5) \leq 89$.

(2) *For odd* $t \geq 7$ *and odd* $s \geq 6t + 41$, *there is an* (s,t)*-incomplete Room square. Hence* $S(t) \leq 6t + 39$ *for odd* $t \geq 7$.

(3) *For odd* $t \geq 127$ *and odd* $s \geq 5t + 2$, *there is an* (s,t)*-incomplete Room square. Hence* $S(t) \leq 5t$ *for odd* $t \geq 127$.

Proof. From Lemma 3.5(1), noting that $n \in T_7$ if $n \geq 63$. □

In [4], several (s,t)-incomplete Room squares are constructed for $t = 3$, 5, and 7. The result for $t = 3$ is of importance to the proof of our bound $S(t) \leq 4t + 27$ for odd $t \geq 127$.

Lemma 3.8 $S(3) \leq 39$, $S(5) \leq 67$, *and* $S(7) \leq 53$.

We will use another construction which employs transversal designs and frames. This was first described in [8].

Lemma 3.9 *Suppose there is a* TD(5,n), *and* $0 \leq v \leq 3n$. *Then there is a frame of type* $4n^4\, 2v^1$.

Proof. Let (X,G,A) be the TD(5,n), where $G = \{G_1,G_2,G_3,G_4,G_5\}$. Define a weighting. $w : X \to \{0,2,4,6\}$ by $w(x) = 4$ if $x \in X \setminus G_5$ and so that $\sum_{x \in G_5} w(x) = 2v$. Apply Construction 2.3 (the required input frames, of types 4^4, $4^4 2^1$, 4^5, and $4^4 6^1$ are shown to exist in [6]). □

Lemma 3.10 *Suppose* $n \neq 2,3,6,10$, *or* 14, *and* $0 \leq v \leq 3n$. *Then*

(1) *There is a* $(16n+2v+1, 2v+1)$*-incomplete Room square. If* $v \neq 1$ *or* 2, *there is a* $(16n+2v+1, 4n+1)$*-incomplete Room square.*

(2) *If there is a* $(4n+3, 3)$*-incomplete Room square then there is a* $(16n+2v+3, 2v+3)$*-incomplete Room square. If, further, there is a* $(2v+3, 3)$*-incomplete Room square, then there is a* $(16n+2v+3, 4n+3)$*-incomplete Room square.*

Proof. For the $n \neq 2,3,6,10$ or 14, a TD(5,n) is known to exist [1]. Apply Lemma 3.9, constructing a frame. Then apply Construction 2.1. For (1), use $w = 1$; and for (2), use $w = 3$. The result follows. □

Lemma 3.11 *Suppose* $t \equiv 1 \bmod 4$, $t \neq 5,9,25,41$, *or* 57. *Then there exists an* (s,t)*-incomplete Room square for all odd* s *satisfying* $4t + 3 \leq s \leq \frac{11t-9}{2}$.

Proof. Apply Lemma 3.10(1) with $n = \frac{t-1}{4}$, $3 \leq v \leq 3n$. Since $n \neq 1$, we construct $(4t-3+2v, t)$-incomplete Room squares for $3 \leq v \leq 3n$. □

In a similar fashion, we prove

Lemma 3.12 *Suppose* $t \equiv 3 \bmod 4$, $t \geq 47$, $t \neq 59$. *Then there is an* (s,t)*-incomplete Room square for odd* s, $4t + 29 \leq s \leq \frac{11t-27}{2}$.

Proof. Here we apply Lemma 3.10(2), with $n = \frac{t-3}{4}$, $19 \leq v \leq 3n$. There exist $(2v+3,3)$- and $(4n+3,3)$-incomplete Room squares by Lemma 3.8. Thus we construct $(4t-9+2v, t)$-incomplete Room squares for $19 \leq v \leq 3n$. □

Theorem 3.13 *For odd* $t \geq 127$, *and odd* $s \geq 4t + 29$, *there is an* (s,t)*-incomplete Room square.*

Proof. Corollary 3.7(3), and Lemmata 3.11 and 3.12. □

We now present a table of upper bounds for $S(t)$, $t \leq 125$. All these values are immediate consequences of Theorem 3.5 and Lemmata 3.8, 3.11 and 3.12.

TABLE 1 Upper bounds for S(t), $t \le 125$

t	S(t)	t	S(t)	t	S(t)
1	5	43	277	85	519
3	39	45	279	87	375
5	67	47	215	89	563
7	53	49	303	91	565
9	83	51	231	93	567
11	85	53	327	95	407
13	87	55	247	97	591
15	75	57	351	99	423
17	69	59	295	101	407
19	95	61	375	103	637
21	135	63	279	105	679
23	115	65	261	107	455
25	125	67	441	109	663
27	135	69	443	111	471
29	193	71	445	113	453
31	195	73	447	115	487
33	133	75	327	117	469
35	175	77	491	119	503
37	231	79	493	121	735
39	195	81	495	123	519
41	275	83	359	125	759

REFERENCES

[1] A. E. Brouwer, The number of mutually orthogonal Latin squares - a table up to order 10000, *Research report ZW 123/79*, Mathematisch Centrum, Amsterdam, 1979.

[2] R. J. Collens and R. C. Mullin, Some properties of Room squares - a computer search, *Proc. First Louisiana Conference on Combinatorics, Graph Theory and Computing*, Baton Rouge, 1970, 87-111.

[3] J. H. Dinitz and D. R. Stinson, Further results on frames, *Ars Combinatoria* 11 (1981), 275-288.

[4] J. H. Dinitz, D. R. Stinson and W. D. Wallis, Room squares with small holes, preprint.

[5] R. C. Mullin and W. D. Wallis, The existence of Room squares, *Aequationes Math.* 13 (1975), 1-7.

[6] D. R. Stinson, Some constructions for frames, Room squares, and subsquares, *Ars Combinatoria* 12 (1981), 229-267.

[7] D. R. Stinson, Some results concerning frames, Room squares, and subsquares, *J. Austral. Math. Soc.* A 31 (1981), 376-384.

[8] D. R. Stinson, The spectrum of skew Room squares, *J. Austral. Math. Soc.* A 31 (1981), 475-480.

[9] W. D. Wallis, Supersquares, *Combinatorial Mathematics, Proc. of the Second Australian Conf,* 143-148.

[10] W. D. Wallis, All Room squares have minimal supersquares, *Congressus Numerantium,* to appear.

GEOMETRIES IN FINITE PROJECTIVE SPACES : RECENT RESULTS

J.A. THAS

We survey recent results on the embedding of generalised quadrangles, partial geometries, semi partial geometries, and (0,α)-geometries in the finite projective space PG(n,q).

1. INTRODUCTION

Let S=(P,B,I) be a finite incidence structure. We say that S is embedded in the projective space PG(n,q) if B is a line set of PG(n,q), if P is the set of all points of PG(n,q) on these lines, and if I is the natural incidence. The geometry S is fully embedded in PG(n,q) if no proper subspace PG(n',q) of PG(n,q), n'<n, contains P. Here we shall be concerned with the problem : determine all geometries of a certain type which are fully embedded in PG(n,q). Successively we shall consider generalized quadrangles, partial geometries, semi partial geometries and (0,α)-geometries.

Notice that any geometry S embedded in PG(n,q) satisfies the axiom of Pasch : if L_1, L_2, M_1, M_2 are distinct lines such that $L_1 I x I L_2$, $x \not I M_i$ and L_i is concurrent with M_j, i,j=1,2, then necessarily M_1 is concurrent with M_2.

2. GENERALIZED QUADRANGLES, PARTIAL AND SEMI PARTIAL GEOMETRIES, (0,α)-GEOMETRIES

2.1. Generalized quadrangles

A (finite) generalized quadrangle (GQ) is an incidence structure S=(P,B,I) satisfying the following axioms :

(i) each point is incident with 1+t (t≥1) lines and two distinct points are incident with at most one line;

(ii) each line is incident with 1+s (s≥1) points and two distinct lines are incident with at most one point;

(iii) if x is a point and L is a line not incident with x, then there is a unique point x' and a unique line L' for which xIL'Ix'IL.

We have |P|=v=(1+s)(1+st) and |B|=b=(1+t)(1+st). D.G. Higman proved that for s>1 and t>1, there holds $t \leq s^2$ and dually $s \leq t^2$ [17]. Moreover s+t|st(1+s)(1+t) [17].

The classical generalized quadrangles. (a) Consider a non-singular quadric Q of projective index 1 [18] of the projective space PG(d,q), with d=3,4 or 5. Then the points of Q together with the lines of Q (which are the subspaces of maximal dimension on Q) form a GQ Q(d,q) with parameters

$s=q,\ t=1,\ v=(q+1)^2,\ b=2(q+1)$, when $d=3$;

$s=t=q,\ v=b=(q+1)(q^2+1)$, when $d=4$;

$s=q,\ t=q^2, v=(q+1)(q^3+1),\ b=(q^2+1)(q^3+1)$, when $d=5$.

(b) Let H be a non-singular hermitian variety of the projective space $PG(d,q^2)$, d=3 or 4. Then the points of H together with the lines on H form a GQ $H(d,q^2)$ with parameters

$s=q^2,\ t=q,\ v=(q^2+1)(q^3+1),\ b=(q+1)(q^3+1)$, when $d=3$;

$s=q^2,\ t=q^3,\ v=(q^2+1)(q^5+1),\ b=(q^3+1)(q^5+1)$, when $d=4$.

(c) The points of PG(3,q), together with the totally isotropic lines with respect to a symplectic polarity, form a GQ W(3,q) with parameters

$s=t=q,\ v=b=(q+1)(q^2+1)$.

All these GQ (all of which are associated with classical simple groups) are due to J. Tits. Clearly they are embedded in a projective space.

Literature. GQ were introduced in 1959 by J. Tits [36]. For a survey on GQ we refer to S.E. Payne [23], J.A. Thas and S.E. Payne [35], and J.A. Thas [29,31].

2.2. Partial geometries

A (finite) partial geometry (PG) is an incidence structure $S=(P,B,I)$ satisfying (i) and (ii) of 2.1., and also

(iii)' if x is a point and L is a line not incident with x, then there are exactly α ($\alpha \geq 1$) points $x_1,\ldots,x_\alpha$ and α lines $L_1,\ldots,L_\alpha$ such that $x I L_i I x_i I L$, $i=1,2,\ldots,\alpha$.

We have $|P|=v=(1+s)(st+\alpha)/\alpha$ and $|B|=b=(1+t)(st+\alpha)/\alpha$. There holds $\alpha(s+t+1-\alpha) \mid st(s+1)(t+1)$ [17], $(t+1-2\alpha)s \leq (t+1-\alpha)^2(t-1)$ [5] and dually $(s+1-2\alpha)t \leq (s+1-\alpha)^2(s-1)$. The PG with $\alpha=1$ clearly are the GQ.

The four classes of partial geometries. (a) The PG with $\alpha=s+1$, or, dually, $\alpha=t+1$. The PG with $\alpha=s+1$ clearly are the 2-(v,s+1,1) designs.

(b) The PG with $\alpha=s$, or, dually, $\alpha=t$. The PG with $\alpha=t$ clearly are the nets of order $s+1$ and degree $t+1$ (or deficiency $s-t+1$).
(c) The GQ (here $\alpha=1$).
(d) The proper PG, i.e. the PG with $1<\alpha<\min(s,t)$.

The partial geometry H_q^n . Here P is the set of all points of $PG(n,q)$ which are not contained in a given subspace $PG(n-2,q)$ $(n\geq 2)$; B is the set of all lines of $PG(n,q)$ which do not have a point in common with $PG(n-2,q)$; I is the natural incidence. Then (P,B,I) is a PG with parameters $s=q$, $t=q^{n-1}-1$, $\alpha=q$, which is fully embedded in $PG(n,q)$. This well-known dual net is denoted H_q^n . It is the only known PG with $\alpha\notin\{1,s+1,t+1\}$ and satisfying the axiom of Pasch. In [33] J.A. Thas and F. De Clerck prove that H_s^n is the only dual net of degree $s+1$ (≥ 3) and deficiency $t-s+1$ (>0) which satisfies the axiom of Pasch (independently C.C. Sims proved the same result, but with the restriction $t+1>s^2$).

Literature. PG were introduced by R.C. Bose [2]. For a survey on PG we refer to F. De Clerck [9] and J.A. Thas [29].

2.3. Semi partial geometries and $(0,\alpha)$-geometries

A (finite) semi partial geometry (SPG) is an incidence structure $S=(P,B,I)$ satisfying (i) and (ii) of 2.1., and also

(iii)" if a point x and a line L are not incident, then there are 0 or α $(\alpha\geq 1)$ points which are collinear with x and incident with L (i.e. there are 0 or α points x_i and respectively 0 or α lines L_i such that $x\,I\,L_i\,I\,x_i\,I\,L$), and

(iv) if two points are not collinear, then there are μ $(\mu>0)$ points collinear with both.

We have $|P|=v=1+(t+1)s(1+t(s-\alpha+1)/\mu)$, and $v(t+1)=b(s+1)$ with $|B|=b$. There holds $\alpha^2\leq\mu\leq(t+1)\alpha$, $\mu\mid st(t+1)(s-\alpha+1)$, $s+1\mid t(t+1)(\alpha t+\alpha-\mu)$, $\alpha\mid st(s+1)$, $\alpha\mid st(t+1)$, $\alpha\mid\mu$, $\alpha^2\mid\mu st$, $\alpha^2\mid t((t+1)\alpha-\mu)$, and $b\geq v$ if $\mu\neq(t+1)\alpha$ [7]. Moreover $D=(t(\alpha-1)+s-1-\mu)^2+4((t+1)s-\mu)$ is a square, except in the case $\mu=s=t=\alpha=1$ where $D=5$ (and then S is the pentagon), and $((t+1)s+(v-1)(t(\alpha-1)+s-1-\mu+\sqrt{D})/2)/\sqrt{D}$ is an integer [7].

A SPG with $\alpha=1$ is called a partial quadrangle (PQ). PQ were introduced and studied by P.J. Cameron [4]. A SPG is a PG iff $\mu=(t+1)\alpha$. A proper SPG is a SPG with $\mu<(t+1)\alpha$. The dual of a SPG S is again a SPG iff $s=t$ or S is a PG [6].

If we write " → " for "generalizes to", then we have the scheme

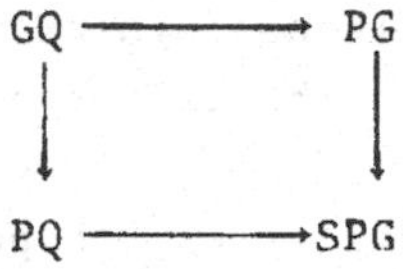

A $(0,\alpha)$-geometry ($\alpha>1$) is a connected incidence structure $S=(P,B,I)$ satisfying (iii)" of 2.3., and also

(i)' two distinct points are incident with at most one line, and

(v) each line is incident with at least two points, and each point is incident with at least two lines.

It is easy to prove that a $(0,\alpha)$-geometry S satisfies (i) and (ii) of 2.1. Hence any $(0,\alpha)$-geometry satisfying (iv) of 2.3. is a SPG, and clearly any SPG with $\alpha>1$ is a $(0,\alpha)$-geometry. If a $(0,\alpha)$-geometry does not satisfy (iv) then it is called proper. Notice also that the point graph of a $(0,\alpha)$-geometry is strongly regular iff it is a SPG.

Examples of SPG and $(0,\alpha)$-geometries embedded in PG(n,q)

(a) Let U be a set with m elements, $U_2=\{T\subset U \,\|\, |T|=2\}$, $U_3=\{R\subset U \,\|\, |R|=3\}$, and I the inclusion. Then $U_{2,3}(m)=(U_2,U_3,I)$ is a SPG with $s=\alpha=2$, $t=m-3$ and $\mu=4$. For certain values of m, there are geometries isomorphic to $U_{2,3}(m)$ which are embedded in PG(n,q) (see section 5).

(b) In PG(n,q) we consider a symplectic polarity W of rank 2k, i.e. rad $W=PG(n-2k,q)$. Let P be the set of points of PG(n,q)-rad W, let B be the set of all lines of PG(n,q) which are not totally isotropic in the polarity W and let I be the natural incidence. Then $\overline{W(n,2k,q)}=(P,B,I)$ is a $(0,\alpha)$-geometry with $s=\alpha=q$, $t=q^{n-1}-1$. If $k=1$, then $\overline{W(n,2,q)}$ is the PG H_q^n. If $2k=n+1$, so n is odd, then the symplectic polarity is non-singular. In this case the geometry $\overline{W(n,n+1,q)}$ is a SPG with $\mu=q^{n-1}(q-1)$ and will shortly be denoted $\overline{W(n,q)}$ [7]. In all other cases $\overline{W(n,2k,q)}$ is a proper $(0,\alpha)$-geometry.

(c) Consider a (possibly singular) quadric Q in PG(n,2), $n\geq 3$. Let B be the set of non-intersecting lines of Q, let P be the set of all points incident in PG(n,2) with the elements of B, and let I be the incidence of PG(n,2). Then (P,B,I) is a (0,2)-

geometry, unless Q consists of one or two hyperplanes, or Q is the (non-singular) hyperbolic quadric in PG(3,2), or $n\geq 4$ and Q is the cone with vertex PG(n-4,2) which projects a (non-singular) hyperbolic quadric of a threespace skew to PG(n-4,2) (in the last two cases the geometry is not connected).

If n=2d-1 and Q is a non-singular elliptic (resp. hyperbolic) quadric Q^- (resp. Q^+), then (P,B,I) is a SPG, denoted by $NQ^-(2d-1,2)$ (resp. $NQ^+(2d-1,2)$), with parameters $s=\alpha=2$, $t+1=2^{2d-3}+2^{d-2}$ (resp. $t+1=2^{2d-3}-2^{d-2}$), $\mu=2^{2d-3}+2^{d-1}$ (resp. $\mu=2^{2d-3}-2^{d-1}$)[37].

If n=2d and Q is non-singular, then (P,B,I) is a SPG, denoted NQ(2d,2), which is isomorphic to $\overline{W(2d-1,2)}$.

In all other cases the geometry is a proper (0,2)-geometry.

(d) Consider a (non-singular) hyperbolic quadric Q^+ of $PG(3,2^h)$ ($h\geq 2$). Then the construction of (c) gives us a $(0,2^{h-1})$-geometry $NQ^+(3,2^h)$ with $s=2^h$, $t+1=2^{h-1}(2^h-1)$. This geometry is never a SPG.

Literature. SPG were introduced by I. Debroey and J.A. Thas [7], and $(0,\alpha)$-geometries by F. De Clerck and J.A. Thas [11]. Further we refer to I. Debroey [6], which is a survey on SPG.

3. GENERALIZED QUADRANGLES AND PARTIAL GEOMETRIES IN FINITE PROJECTIVE SPACES

3.1. Generalized quadrangles embedded in PG(n,q)

The following fundamental and beautiful result of F. Buekenhout and C. Lefèvre [3] is well-known.

Theorem 1. *If the GQ S is embedded in a projective space, then S is of classical type (i.e. is one of* Q(d,q), d=3,4 *or* 5, $H(d,q^2)$, d=3 *or* 4, *or* W(3,q)).

Remark. Partial solutions to the problem are in [21, 22, 34].

3.2. Partial geometries embedded in PG(n,q)

Theorem 2. (F. De Clerck and J.A. Thas [10]). *If the PG* S=(P,B,I) *with parameters* $s=q, t, \alpha$ *is fully embedded in* PG(n,q), *then the following cases may occur :*

(a) $\alpha=s+1$ *and* S *is the design of points and lines of* PG(n,q);
(b) $\alpha=1$ *and* S *is a classical* GQ;
(c) $\alpha=t+1$, $n=2$, *and* S *is a dual design in* PG(2,q) *(if* $S\neq PG(2,q)$, *then the points not in* S *constitute a maximal arc* K *of* PG(2,q) *and the lines of* S *are the non-intersecting lines of* K; *conversely, from any maximal arc* $K\neq PG(2,q)$, AG(2,q) *of* PG(2,q) *there arises a dual design* S *in* PG(2,q)).
(d) $\alpha=s$ *and* $S=H_q^n$ $(n\geq 3)$.

Remark. A maximal arc $K_{q,m}$ of PG(2,q) is a non-void set of points meeting every line in just m $(m\neq 0)$ points or in none at all [1]. Clearly $|K_{q,m}|=qm-q+m$. If m=1, then the corresponding PG is the dual of AG(2,q). For any maximal arc $K_{q,m}$ with $m\neq q+1$ there holds $m|q$. It was shown that for any m dividing $q=2^h$ there exists at least one $K_{q,m}$ in PG(2,q) [13]. But there does not exist a $K_{q,3}$ or a $K_{q,q/3}$, $q=3^h$ and $h>1$, in PG(2,q) [28]. We conjecture that for any $K_{q,m}$ of PG(2,q), with q odd, there holds $m\in\{1,q,q+1\}$.

3.3. Notice

Let S be a PG with parameters s,t,α which is fully embedded in PG(n,q). Then it is easy to show that S contains "many" partial subgeometries with parameters s,t',α fully embedded in subspaces PG(n',q) of PG(n,q), and such for any $n'\in\{2,\ldots,n-1\}$ if $\alpha>1$ and any $n'\in\{3,\ldots,n-1\}$ if $\alpha=1$.

4. SEMI PARTIAL GEOMETRIES AND $(0,\alpha)$-GEOMETRIES IN FINITE PROJECTIVE SPACES

4.1. Introduction

By (iv) in 2.3. it is not possible to show that a SPG with parameters s,t,α,μ which is fully embedded in PG(n,q), contains semi partial subgeometries with parameters s,t',α,μ' and fully embedded in subspaces PG(n',q), $n'<n$. This was the main reason to introduce the notion of $(0,\alpha)$-geometry.

4.2. Semi partial geometries fully embedded in PG(2,q) and PG(3,q)

Theorem 3. (I. Debroey and J.A. Thas [8]). *If* S=(P,B,I) *is a* SPG *fully embedded in* PG(2,q) *or* PG(3,q), *then the following*

cases may occur :
(a) S *is a* PG *and by* 3.1. *and* 3.2. S *is known*;
(b) S *is* $\overline{W(3,q)}$;
(c) S *is a Desargues configuration of* PG(3,2).

Remarks. (a) In case (c) the lines of S are the non-intersecting lines of some elliptic quadric O [18] of PG(3,2), and so $S=NQ^-(3,2)$. Here $S\cong U_{2,3}(5)$ and the point graph of S is the complement of the Petersen graph.
(b) To prove the theorem we heavily use the fact that the point graph of S is strongly regular.

4.3. (0,α)-geometries (α>1) fully embedded in PG(2,q) and PG(3,q)

Clearly a (0,α)-geometry fully embedded in PG(2,q) is a PG. Now assume that the (0,α)-geometry S=(P,B,I) is fully embedded in PG(3,q). The number of lines of S incident with a point of S is denoted by t+1.

With respect to S there are three types of planes in the space PG(3,q) :
(a) planes of type (a) containing $q\alpha-q+\alpha$ lines and $\rho_a=(q+1)(q-\frac{q}{\alpha}+1)+m$ points of S, where the constant m is the number of points of S in such a plane lying on none of the $q\alpha-q+\alpha$ lines;
(b) planes of type (b) containing exactly one line of B and $\rho_b=q+1+\frac{q(q+1)(t+1-\alpha)(\alpha-1)}{\alpha(t+1)}+m$ points of P;
(c) planes of type (c) containing no line of B and $\rho_c=\frac{(q\alpha-q+\alpha)((q+1)(t+1)-\alpha q)}{\alpha(t+1)}+m$ points of P.

We notice that $|P|=v=(q+1)\rho_c$.

A long technical argument yields the following theorem, which appears to be fundamental when handling the case n>3.

Theorem 4. (F. De Clerck and J.A. Thas [11]). *Let* S=(P,B,I) *be a* (0,α)*-geometry fully embedded in* PG(3,q).
(a) *If* m=0, *then* α=q+1 *and* S *is the design of points and lines of* PG(3,q), *or* α=q *and* $S=H_q^3$, *or* α=q=2 *and* $S=NQ^-(3,2)$ *(hence* S *is a* SPG).
(b) *If* m≠0, *then there is no plane of type* (b) *(hence* t=(α-1)(q+1)).

Corollary. *If there is a plane of type* (b), *then* $\alpha=q=2$ *and* $S=NQ^-(3,2)$.

Remark. If $m\neq 0$ and if moreover there is no plane of type (c), then $m=\frac{q}{\alpha}(q-\alpha+1)$ and $|P|=(q+1)(q^2+1)$.

Conjecture. If $m\neq 0$, then there are only two possibilities :
(a) there are no planes of type (c) and $S=\overline{W(3,q)}$ (here $q=\alpha$, $m=1$, $t=q^2-1$);
(b) there is at least one plane of type (c) and $S=NQ^+(3,2^h)$, $h\geq 2$ (here $\alpha=2^{h-1}$, $m=1$, $t+1=2^{h-1}(2^h-1)$).

Notice. If $m=1$, then it is possible to show that either $S=\overline{W(3,q)}$ or $S=NQ^+(3,2^h)$, $h\geq 2$. So to prove the conjecture it is sufficient to show that $m\in\{0,1\}$.

From Theorem 4 there easily follows

Theorem 5. (F. De Clerck and J.A. Thas [11]). *If* S *is the dual of a* SPG *with* $\alpha>1$ *and if* S *is fully embedded in* PG(n,q), $n\geq 3$, *then* $n=3$ *and* S *is the design of points and lines of* PG(3,q), *or* $S=H_q^3$, *or* $S=NQ^-(3,2)$ *(hence* S *is also a* SPG*)*.

4.4. (0,α)-geometries and semi partial geometries fully embedded in PG(n,q), n≥4 and q>2

Suppose that S=(P,B,I) is a $(0,\alpha)$-geometry fully embedded in PG(n,q), $n\geq 4$ and $q>2$. Then one shows that for any point $x\in P$ the t+1 lines of S incident with x never lie in a PG(m,q), with $m<n$. So there exist four lines of S which are incident with $x\in P$ and which span a PG(4,q). The points and lines of S which are contained in PG(4,q) and are connected to x by at least one path consisting of elements of S in PG(4,q), constitute a $(0,\alpha)$-geometry S' which is fully embedded in PG(4,q). Suppose H_x is the PG(3,q) with as point set the set of all lines of PG(4,q) through x, with as line set the set of all planes of PG(4,q) through x, and for which indicence is the natural one. The set of all lines of S' through x defines a set T(x) of t'+1 points in H_x. By Theorem 4 and its Corollary, it is possible to show that every line of H_x

intersects $T(x)$ in 0 or α points. Now by a result of G. Tallini [26] $T(x)$ is the set of all points of H_x or is the complement of a plane of H_x. Hence $\alpha\in\{q,q+1\}$. If $\alpha=q+1$, then it is easy to show that S is the design of points and lines of PG(n,q); if $\alpha=q$, then by a result of K.B.Farmer and M.P. Hale, Jr.[14], and independently G. Tallini [27], $S=\overline{W(n,2k,q)}$. Hence we have

Theorem 6. (J.A. Thas, I. Debroey and F. De Clerck [32]). *If* S *is a* $(0,\alpha)$*-geometry fully embedded in* PG(n,q), $n\geq 4$ *and* $q>2$, *then* S *is the design of points and lines of* PG(n,q) *or* $S=\overline{W(n,2k,q)}$.

And now by Theorems 3 and 6 we have

Theorem 7. (J.A. Thas, I. Debroey and F. De Clerck [32]). *The only* SPG *with parameters* $s=q,t,\alpha(>1),\mu$ *fully embedded in* PG(n,q), $n\geq 3$ *and* $q>2$, *are the design of points and lines of* PG(n,q), H_q^n, *and if* n *is odd* $\overline{W(n,q)}$.

5. $(0,\alpha)$-GEOMETRIES FULLY EMBEDDED IN PG(n,2), $n\geq 2$

If $S=(P,B,I)$ is a $(0,\alpha)$-geometry fully embedded in PG(n,2), $n\geq 2$, then clearly $\alpha\in\{2,3\}$. If $\alpha=3$, then it is easy to show that S is the design of points and lines of PG(n,2). From now on assume $\alpha=2$.

If x and y are non collinear points of a (0,2)-geometry S, with $s=2$, then we write $x\approx y$ if a point z of S is collinear with x iff it is collinear with y. Clearly $\approx$ is an equivalence relation. The (0,2)-geometry S is called reduced if all its $\approx$-classes have size 1. For $x\in P$, let x^* be the $\approx$-class containing x, and let $P^*=\{x^* \| x\in P\}$. Furthermore, for $L\in B$, let $L^*=\{x^* \| xIL\}$, $P(L)=\{x\in P \| x^*\in L^*\}$, and let $B^*=\{L^* \| L\in B\}$. Clearly $S=(P,B,I)$ is reduced iff $S\cong S^*$, with $S^*=(P^*,B^*,\in)$. The following theorem is a particular case of a result of J.I. Hall [15].

Theorem 8. *Let* $S=(P,B,I)$ *be a* (0,2)*-geometry with* $s=2$. *Then all* $\approx$*-classes of* S *have fixed size* r, *for some* $r\geq 1$. *If* x, y, z *are the points incident with* $L\in B$ *and* $x'\in x^*$, $y'\in y^*$, *then there is a point* $z'\in z^*$ *such that* x',y',z' *are the three points incident with some line* $L'\in B$. *Finally,* $S^*=(P^*,B^*,\in)$ *is a reduced* (0,2)*-geometry (with 3 points on any line) or eventually a single line.*

The following theorem was proved independently by J.I. Hall [16] and by J.A. Thas, I. Debroey and F. De Clerck [32].

Theorem 9. *Let* $S=(P,B,I)$ *be a* $(0,2)$*-geometry which is fully embedded in* $PG(n,2)$, $n\geq 2$. *Then there exist subspaces* $PG(m,2)$, $PG^{(1)}(m+1,2),\ldots,PG^{(\ell)}(m+1,2)$ *of* $PG(n,2)$ *for which* $PG^{(i)}(m+1,2)\cap\cap PG^{(j)}(m+1,2)=PG(m,2)$, *for all* $i\neq j$, *and such that* $PG^{(1)}(m+1,2)-PG(m,2),\ldots,PG^{(\ell)}(m+1,2)-PG(m,2)$ *are exactly the* $\approx$*-classes of* S. *Hence, if* $PG(n-m-1,2)$ *is a subspace of* $PG(n,2)$ *which is skew to* $PG(m,2)$, *then the points* $p_i=PG^{(i)}(m+1,2)\cap PG(n-m-1,2)$, $i=1,2,\ldots,l$, *and the lines* $PG(n-m-1,2)\cap P(L)$, *with* $L\in B$, *constitute a geometry isomorphic to* S^* *which is fully embedded in* $PG(n-m-1,2)$. *It is also clear how to apply the converse. Hence in determining all* $(0,2)$*-geometries fully embedded in a projective space of order 2, we may restrict ourselves to the reduced geometries.*

In the Cotriangle Theorem [25] E.E. Shult classified as follows all reduced $(0,2)$-geometries with $s=2$.

Theorem 10. *Every reduced* $(0,2)$*-geometry with* $s=2$ *is isomorphic to one of* $NQ^{\varepsilon}(2d-1,2)$, *for* $\varepsilon=\pm$ *and* $d\geq 2$ *but* $(d,\varepsilon)\neq(2,+)$, $\overline{W(2d-1,2)}$ *for* $d\geq 2$, *or* $U_{2,3}(m)$ *for* $m\geq 5$.

Notice that $U_{2,3}(4)$ is not reduced, $NQ^-(3,2)\cong U_{2,3}(5)$, $NQ^+(5,2)\cong\cong U_{2,3}(8)$, and $\overline{W(3,2)}\cong U_{2,3}(6)$. Theorem 10 has an easy corollary.

Corollary. *Any SPG with parameters* $s=2,t,\alpha=2,\mu$ *fully embedded in some projective space of order 2 is isomorphic to one of* $NQ^{\varepsilon}(2d-1,2)$, *for* $\varepsilon=\pm$ *and* $d\geq 2$ *but* $(d,\varepsilon)\neq(2,+)$, $\overline{W(2d-1,2)}$ *for* $d\geq 2$, H_2^n *for* $n\geq 2$, *or* $U_{2,3}(m)$ *for* $m\geq 5$.

Theorem 11. (J.I. Hall [16]). *Let* $S=(P,B,I)$ *be a* $(0,2)$*-geometry fully embedded in* $PG(n,2)$ *and assume* $S\cong NQ^{\varepsilon}(2d-1,2)$, *for* $\varepsilon=\pm$ *and* $d\geq 2$ *but* $(d,\varepsilon)\neq(2,+)$ *or* $(3,+)$. *Then* $n=2d-1$ *and* $S=NQ^{\varepsilon}(2d-1,2)$. *If* $S\cong NQ^+(5,2)$, *we have* $n\in\{5,6\}$*: if* $n=5$ *then* S *is one of two projectively different models in* $PG(5,2)$, *and if* $n=6$ S *is uniquely defined (up to a projectivity).*

Theorem 12. (J.I. Hall [16]). *Let* $S=(P,B,I)$ *be a* (0,2)*-geometry fully embedded in* PG(n,2) *and assume* $S\cong\overline{W(2d-1,2)}$ *for* $d\geq 2$. *Then* $n=2d-1$ *and* $S=W(2d-1,2)$, *or* $n=2d$ *and* $S=NQ(2d,2)$.

Let E be the vector space of all binary m - tuples containing an even number of 1's. Then E is the dimension m-1 even subcode of $GF(2)^m$. (A code is even if all its words have even weight.) Consider now $U_{2,3}(m)=(P,B,I)$, $m\geq 4$, and define $\sigma:P\rightarrow E$ as follows : $\{i,j\}^\sigma$, with $i,j\in\{1,\ldots,m\}$, is the vector of E with a 1 in coordinate position i and j and with a 0 in all other coordinate positions. Clearly σ is injective and if x,y,z are the points of a line of $U_{2,3}(m)$, then $\{x^\sigma,y^\sigma,z^\sigma,\bar{0}\}$ is a 2-dimensional subspace of E. Hence with $U_{2,3}(m)$ there corresponds a (0,2)-geometry J embedded in the projective space PG(m-2,2) deduced from E. It is easy to prove that J is fully embedded in PG(m-2,2).

Lemma (J.I. Hall [16]). *Let the* (0,2)*-geometry* $S=(P,B,I)$ *be fully embedded in* PG(n,2) *and* $S\cong U_{2,3}(m), m\geq 4$. *If* F *is the* (n+1)*-dimensional vector space deduced from* PG(n,2), *then there is a linear transformation* $T\in\mathrm{Hom}(E,F)$ *which induces an isomorphism of* J *onto* S. *Hence* $n\leq m-2$, *and if in particular* $n=m-2$ J *and* S *are projectively equivalent.*

Let the (0,2)-geometry $S=(P,B,I)$, with $S\cong U_{2,3}(m)$, be fully embedded in PG(n,2). With the notations of the preceding lemma, it is possible to show that Ker(T) is a binary, even, linear code of length m and minimal distance at least 6 [16]. Conversely, assume that C is a binary, even, linear code of length m, $m\geq 4$, and minimal distance at least 6. If x,y,z are the points of a line of $U_{2,3}(m)$, then $\{x^\sigma+C,\ y^\sigma+C,\ z^\sigma+C,\ \bar{0}+C\}$ is a 2-dimensional subspace of E/C [16]. Hence with C there corresponds a (0,2)-geometry S isomorphic to $U_{2,3}(m)$ and fully embedded in the projective space PG(n,2) deduced from E/C.

Theorem 13 (J.I. Hall [16]). *There is a one-to-one correspondence between the classes of projectively equivalent* (0,2)*-geometries* $S\cong U_{2,3}(m)$, $m\geq 4$, *fully embedded in some projective space and the equivalence classes of binary, even, linear codes of length* m *and* *minimal distance at least* 6.

Corollary (J.I. Hall [16]). *For each integer* $m \geq 4$ *there is an integer* $h(m)$ *such that there is a* (0,2)*-geometry isomorphic to* $U_{2,3}(m)$ *and fully embedded in* PG(n,2) *iff* $h(m) \leq n \leq m-2$. *The function* $h(m)$ *is a non-decreasing function of* m .

The function h(m) has been studied a great deal but is not known exactly even for relatively small values of m. The first value of h(m) not given exactly is $8 \leq h(25) \leq 9$.

Notice. For the description of all (0,2)-geometries fully embedded in PG(n,2) with $n \leq 6$ we refer to J.I. Hall [16] (and for $n \leq 4$ also to J.A. Thas, I. Debroey and F. De Clerck [32]); for the classification of all (0,2)-geometries fully embedded in PG(n,2) and possessing some additional property we refer to G. De Meur [12], J.I. Hall [16], C. Lefèvre-Percsy [20] and J.J. Seidel [24].

6. OPEN PROBLEMS

Concerning the embeddings described in the previous sections, the following problems are still open.

1° Classify all partial quadrangles fully embedded in PG(n,q), $n \geq 4$.
2° Classify all dual partial quadrangles fully embedded in PG(n,q), $n \geq 3$.

We give an example of a dual partial quadrangle S fully embedded in PG(3,q^2). Let L be a fixed line of a non-singular hermitian variety H of PG(3,q^2). Points of S are the points of H-L, and lines of S are the lines of H having no point in common with L. Then S is a dual partial quadrangle with $s=q^2$, $t=q-1$, $\mu=q^2-q$.

Another example is obtained as follows. Consider a non-singular hermitian variety H in PG(3,q^2), q odd. Let B be a set of lines on H such that any point of H is on just $\frac{q+1}{2}$ lines of B. Then the points of H together with the lines of B form a partial quadrangle with parameters $s=q^2$, $t=\frac{q-1}{2}$, $\mu=\frac{(q-1)^2}{2}$ [30]. Only for q=3 such a set B is known to exist.

3° Classify all (0,α)-geometries ($\alpha>1$) fully embedded in PG(3,q) and having no plane of type (b) (then $t=(\alpha-1)(q+1)$).

4° Determine h(m) for all $m \geq 4$, and describe all projectively inequivalent (0,2)-geometries $S \cong U_{23}(m)$ fully embedded in PG(n,2), with $h(m) \leq n \leq m-2$.

REFERENCES

[1] A. Barlotti, Sui {k;n}-archi di un piano lineare finito, *Boll. Un. Mat. Ital.* (3) 11 (1956), 553-556.

[2] R.C. Bose, Strongly regular graphs, partial geometries, and partially balanced designs, *Pac. J. Math.* 13 (1963), 389-419.

[3] F. Buekenhout and C. Lefèvre, Generalized quadrangles in projective spaces, *Arch. Math.* 25 (1974), 540-552.

[4] P.J. Cameron, Partial quadrangles, *Quart. J. Math. Oxford* (3) 25 (1974), 1-13.

[5] P.J. Cameron, J.M. Goethals, and J.J. Seidel, Strongly regular graphs having strongly regular subconstituents, *J. Algebra* 55 (1978), 257-280.

[6] I. Debroey, Semi partiële meetkunden, *Ph. D. Dissertation*, Rijksuniversiteit te Gent, 1978.

[7] I. Debroey and J.A. Thas, On semi partial geometries, *J. Comb. Th. (A)* 25 (1978), 242-250.

[8] I. Debroey and J.A. Thas, Semi partial geometries in PG(2,q) and PG(3,q), *Rend. Accad. Naz. Lincei* 64 (1978), 147-151.

[9] F. De Clerck, Partiële meetkunden, *Ph. D. Dissertation*, Rijksuniversiteit te Gent, 1978.

[10] F. De Clerck and J.A. Thas, Partial geometries in finite projective spaces, *Arch. Math.* 30 (1978), 537-540.

[11] F. De Clerck and J.A. Thas, The embedding of (0,α)-geometries in PG(n,q), Part I, *Proc. " Convegno Internazionale Geometrie Combinatorie e Loro Applicazioni"* Roma 1981, Annals of Discrete Math., to appear.

[12] G. De Meur, Espaces de Fisher hermitiens, *Ph. D. Dissertation*, Université Libre de Bruxelles, 1979.

[13] R.H.F. Denniston, Some maximal arcs in finite projective planes, *J. Comb. Th.* 6 (1969), 317-319.

[14] K.B. Farmer and M.P. Hale Jr., Dual affine geometries and alternative bilinear forms, *Lin. Alg. and Appl.* 30 (1980), 183-199.

[15] J.I. Hall, Classifying copolar spaces and graphs, to appear.

[16] J.I. Hall, Linear representations of cotriangular spaces, to appear.

[17] D.G. Higman, Partial geometries, generalized quadrangles, and strongly regular graphs, in : Barlotti A. (ed.) *Atti Convegno di Geometria e sue Applicazioni*, University Perugia , 1971.

[18] J.W.P. Hirschfeld, *Projective geometries over finite fields*, Clarendon Press-Oxford, 1979.

[19] C. Lefèvre-Percsy, Geometries with dual affine planes and symplectic quadrics, *Lin. Alg. and Appl.* 42 (1982), 31-37.

[20] C. Lefèvre-Percsy, Copolar spaces fully embedded in projective spaces, to appear.

[21] D. Olanda, Sistemi rigati immersi in uno spazio proiettivo, *Ist. Mat. Univ. Napoli*, Rel. n. 26 (1973), 1-21.

[22] D. Olanda, Sistemi rigati immersi in uno spazio proiettivo, *Rend. Accad. Naz. Lincei* 62 (1977), 489-499.

[23] S.E. Payne, Finite generalized quadrangles: a survey, *Proc. Int. Conf. Proj. Planes*, Wash. State Univ. Press (1973), 219-261.

[24] J.J. Seidel, On two-graphs and Shult's characterization of symplectic and orthogonal geometries over GF(2), *T.H.-Report* 73-WSK-02, Techn. Univ. Eindhoven 1973.

[25] E.E. Shult, Groups, polar spaces and related structures, in : M. Hall Jr. and J.H. van Lint (eds.), *Proc. of the Advanced Study Institute on Combinatorics*, Mathematical Centre Tracts no. 55, Amsterdam 1974, 130-161.

[26] G. Tallini, Problemi e resultati sulle geometrie di Galois, *Ist. Mat. Univ. Napoli*, Rel. n.30.

[27] G. Tallini, I k-insiemi di retti di PG(d,q) studiati rispetto ai fasci di rette, *Quad. Sem. Geom. Comb. n.28, Parte I, Ist. Mat. Univ. Roma* 1980, 1-17.

[28] J.A. Thas, Some results concerning {(q+1)(n-1);n}-arcs and {(q+1)(n-1)+1;n}-arcs in finite projective planes of order q, *J. Comb. Th.* 19 (1975), 228-232.

[29] J.A. Thas, Combinatorics of partial geometries and generalized quadrangles, in : Aigner M. (ed.), *Higher Combinatorics*, Reidel, Dordrecht-Holland (1977), 183-199.

[30] J.A. Thas, Ovoids and spreads of finite classical polar spaces, *Geometriae Dedicata* 10 (1981), 135-144.

[31] J.A. Thas, Combinatorics of finite generalized quadrangles : a survey, *Annals of Discrete Math.* 14 (1982), 57-76.

[32] J.A. Thas, I. Debroey and F. De Clerck, The embedding of (0,α)-geometries in PG(n,q), Part II, *Discrete Math.*, to appear.

[33] J.A. Thas and F. De Clerck, Partial geometries satisfying the axiom of Pasch, *Simon Stevin* 51 (1977), 123-137.

[34] J.A. Thas and P. De Winne, Generalized quadrangles in finite projective spaces, *J. of Geometry* 10 (1977), 126-137.

[35] J.A. Thas and S.E. Payne, Classical finite generalized quadrangles : a combinatorial study, *Ars Combinatoria* 2 (1976), 57-110.

[36] J. Tits, Sur la trialité et certains groupes qui s'en déduisent, *Publ. Math. I.H.E.S.*, Paris 2 (1959), 14-60.

[37] H. Wilbrink, private communication.

A CANONICAL FORM FOR INCIDENCE MATRICES OF FINITE PROJECTIVE PLANES AND THEIR ASSOCIATED LATIN SQUARES AND PLANAR TERNARY RINGS

STEPHEN BOURN

We refine the Paige-Wexler canonical form for incidence matrices of finite projective planes and thus obtain a simple relationship between the incidence matrix and a corresponding planar ternary ring. We also demonstrate a simple relationship between an incidence matrix and a corresponding set of mutually orthogonal latin squares.

1. INTRODUCTION

We will assume below that the reader has a basic knowledge of finite projective planes and of complete sets of mutually orthogonal latin squares. These objects are defined in many standard reference books, including for example Dénes and Keedwell [3], Hughes and Piper [5], and Kárteszi [6]. For the sake of brevity we will refer to them below as planes and CMOLS respectively.

In 1938 Bose [1] showed that the existence of a plane of order q was equivalent to the existence of a CMOLS of order q. One of the fundamental problems of finite projective geometry is to classify all planes, and although this is yet to be achieved, one useful tool at our disposal is the technique for coordinatising an arbitrary plane with a planar ternary ring, or PTR, introduced by Hall [4] in 1943. A convenient way of representing a plane is by its incidence matrix, which we will denote by N. Paige and Wexler [8] in 1953 introduced the concepts "canonical incidence matrix" and "digraph complete set of latin squares", the latter of which we will refer to as a DCLS. A DCLS of order q is a set of q-1 latin squares of order q such that in any pair of columns, m_1 and m_2, with $m_1 \neq m_2$, the q(q-1) ordered pairs (c_1, c_2), $c_1 \neq c_2$, occurring in the rows of the latin squares are all distinct. By putting N in canonical form Paige and Wexler [8] were able to show firstly that N was equivalent to a DCLS and secondly that a DCLS was equivalent to a CMOLS. Thus by combining these two results they were able to construct incidence matrices from CMOLS and vice versa. In 1968 Martin [7] defined a canonical or "normal" form for a CMOLS associated with a plane, such that a particular choice of a CMOLS in canonical form was equivalent to a particular Hall coordinatisation of the plane.

In Section 2 of the present paper we will reproduce the coordinatisation method of Hall since it is directly related to much of the following section. In Section 3 we refine the Paige-Wexler canonical form for N and thus by simply using row and column operations on N we obtain a form for N, which we will call an "ordered canonical form", which is equivalent to a Hall coordinatisation of the plane. The limited usefulness of this form in characterising planes will be demonstrated by a number of examples. The section is concluded with a brief discussion of the original motivation for this work. Finally in Section 4 we will show simple techniques to obtain any one from any other of the forms N, DCLS, or CMOLS. In particular the direct relationship between N and a CMOLS is previously unknown (see Dénes and Keedwell [3], p. 286).

2. HALL COORDINATISATION

Here we outline the coordinatisation method of Hall, using our own notation. Although no diagrams are given, the reader will find it advantageous to construct his own. It is convenient to refer to a point or line sometimes by its name, P or l say, and sometimes by its coordinate, (x,y) or [m,c] say, however this should cause no confusion. The symbols 0,1,...,q-1 together with the special symbol ∞ will be used as coordinates.

Firstly choose four points, Y, X, O and U, with no three collinear. Let O=(0,0) and U=(1,1), and let the other q-2 points of OU-YX be paired off with the q-2 coordinates (x,x) where x=2,3,...,q-1. Now let P be a point of the plane not on the line YX. Set P=(x,y) if and only if $YP \cap OU$=(x,x) and $XP \cap OU$=(y,y). A point $Q \in YX-\{Y\}$ is given the coordinate (m) if and only if $O.(1,m) \cap YX = Q$. Finally let Y=(∞).

A line l not through Y is given coordinates [m,c] if and only if l=(m).(0,c). A line s through Y other than YX is assigned the coordinate [x] if and only if it passes through (x,x). Finally let YX=[∞].

Thus we have assigned coordinates to each point and line of the plane. We may now define a ternary operation on the symbols 0,1,...,q-1 by setting T(x,m,c)=y if and only if (x,y)I[m,c], that is (x,y) lies on [m,c]. The operation T together with the elements 0,1,...,q-1 is a PTR. If we count the number of ways we could have chosen Y, X, O and U and the number of ways of assigning the coordinates (x,x), x=2,3,...,q-1, to the q-2 points of OU-{YX,O,U} then we see that for a given plane we could obtain $(q^2+q+1).(q^2+q).q^2.(q^2-2q+1).(q-2)!$ PTR's. Some of these may be identical however, and to obtain the number of distinct PTR's we must divide this number by the order of the automorphism group of the particular plane. For specific examples of these numbers see the

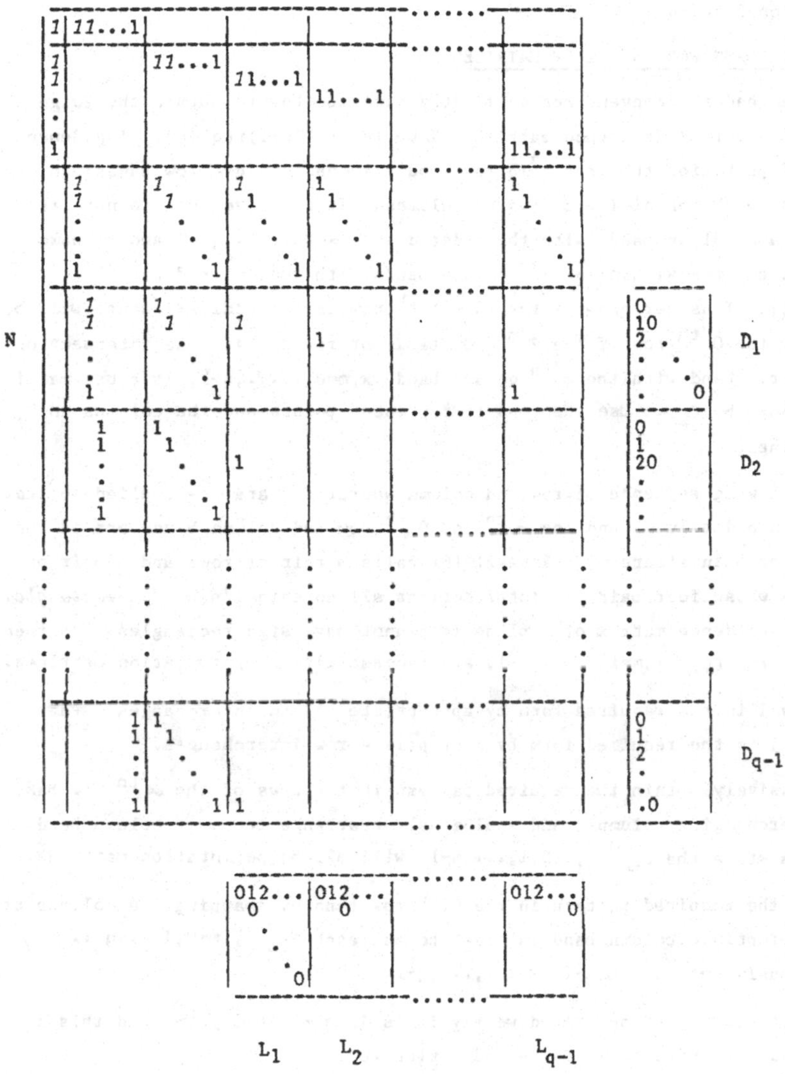

Figure 1

Note: the significance of the italics and the L_i and D_j blocks is explained at the appropriate point in the text.

end of section 3 below.

3. CANONICAL FORMS FOR INCIDENCE MATRICES

For the reader's convenience we firstly describe how to obtain the Paige and Wexler "canonical incidence matrix". Take the $(q^2+q+1)x(q^2+q+1)$ incidence matrix N and partition the last q^2+q rows and columns into q+1 row bands and column bands, each containing q rows or columns. It is convenient to name the ordered row and column bands with the ordered set $\{\infty,0,1,\ldots,q-1\}$ and to name the rows and columns within row and column bands with the ordered set $\{0,1,\ldots,q-1\}$. Thus for example the $(3q+2)'^{th}$ row in the ordinary sense will be called below the $0'^{th}$ row of the $2'^{nd}$ row band, or row $\langle 2,0\rangle$. The intersection of the x'^{th} row band with the m'^{th} column band, $x,m=0,1,\ldots,q-1$, is a qxq matrix which we denote by C_{xm}. Use the rows to represent points and the columns to represent lines.

The following sequence of row and column operations are now applied so that the first row and column, and the ∞'^{th} and $0'^{th}$ row and column bands are in the form shown for N in Figure 1. Kárteszi [6] calls a pair of rows and a pair of columns of N whose four pairwise intersections all contain 1's a *sign-rectangle*. If N is the incidence matrix of a plane it cannot have sign rectangles. It then follows that the C_{xm}, $x,m=1,2,\ldots,q-1$, are necessarily qxq permutation matrices.

(1) Put row 1 in the required form by appropriate column interchanges. Put column 1 in the required form by appropriate row interchanges.

(2) Progressively obtain the required pattern in the rows of the ∞'^{th} row band by interchanging columns, and analogously rearrange the ∞'^{th} column band. At this stage the C_{xm}, $x,m=0,1,\ldots,q-1$, will all be permutation matrices.

(3) Obtain the required pattern in the $0'^{th}$ row band by swapping the columns of the appropriate column band in order to put each $C_{0m}=I$, $m=0,1,\ldots q-1$. Analogously put each $C_{x0}=I$, $x=1,2,\ldots,q-1$.

If N is in the form just described we say it is in *canonical form*, and this is the canonical form defined by Paige and Wexler [8].

We are now ready to proceed with the new results and definitions of this paper. Coordinatise the plane in such a way that Y is row 1, X is row $\langle\infty,0\rangle$, and O is in the $0'^{th}$ row band, thus column 1 is YX and column $\langle\infty,0\rangle$ is YO.

Theorem 1. *Let N be an incidence matrix in canonical form and coordinatise the plane such that Y, X, YX and YO are positioned as described above.*

(i) *For each $x \neq \infty$, points in the x'^{th} row band all have the same first coordinate and this is also the coordinate of the line in column $\langle\infty,x\rangle$.*

(ii) *For each $m \neq \infty$, lines in the m'^{th} column band all have the same first coordinate and this is also the coordinate of the point in row $\langle\infty,m\rangle$.*

(iii) *For each $y \in \{0,1,\ldots,q-1\}$, the points in the y'^{th} rows of the last q row bands all have the same second coordinate. For each $c \in \{0,1,\ldots,q-1\}$, the lines in the c'^{th} columns of the last q column bands all have the same second coordinate. Furthermore if $y=c$ then the second coordinates of the points are the same as the second coordinates of the lines.*

Proof. Bearing in mind the structure of N the proofs follow immediately from the coordinatisation process.

(i) Points in the same row band, excluding the ∞'^{th} row band, all lie on the same line through Y.

(ii) Lines in the same column band, excluding the ∞'^{th} column band, all pass through the same point of YX.

(iii) Points in the y'^{th} rows of the last q row bands all lie on the same line l through X. Lines in the c'^{th} columns of the last q column bands all pass through the same point P on YO. Furthermore if $y=c$ then $l \cap YO=P$.

This completes the proof.

As a consequence of Theorem 1 we may define three permutations on the set $\{0,1,\ldots,q-1\}$ as follows. $S_x(x)$ is the first coordinate of the points in the x'^{th} row band, $x \neq \infty$. $S_m(m)$ is the first coordinate of the lines in the m'^{th} column band, $m \neq \infty$. $S_{yc}(y)$ is the second coordinate of a point in the y'^{th} row of any of the last q row bands. From our choice of X and YO we necessarily have $S_m(0)=S_x(0)=0$, but there are no other restrictions on the possible values of the three permutations. Thus the point in the row $\langle\infty,m\rangle$ has coordinate $(S_m(m))$, the point in row $\langle x,y\rangle$ has coordinate $(S_x(x),S_{yc}(y))$, the line in column $\langle\infty,x\rangle$ has coordinate $[S_x(x)]$ and the line in column $\langle m,c\rangle$ has coordinate $[S_m(m),S_{yc}(c)]$.

We now refine the canonical form for N by adding a fourth step. Firstly note that summing the C_{xm} across any row band or down any column band, excluding the ∞'^{th} and $0'^{th}$ row and column bands, gives E-I, where E is the qxq matrix whose entries are all 1's. This holds since N can have no sign-rectangles.

(4) Put the $1'^{st}$ row band in the form shown for N in Figure 1 by interchanging the last q-1 column bands. The ∞'^{th} row band must then be restored to its required form by interchanging its rows. Analogously put the $1'^{st}$ column band in the form shown in Figure 1, by interchanging the last q-2 row bands, and then restoring the ∞'^{th} column band by interchanging its columns.

We call this refined canonical form the *ordered canonical form*. This definition is justified by the following theorem.

<u>Theorem 2</u>. *Let N be an incidence matrix in ordered canonical form and coordinatise the plane with row* $1=Y$, *row* $\langle\infty,0\rangle=X$, *row* $\langle 0,0\rangle=0$ *and row* $\langle 1,1\rangle=U$. *Then* $S_x=S_m=S_{yc}=S_{xmyc}$ *and* $S_{xmyc}(0)=0$ *and* $S_{xmyc}(1)=1$.

<u>Proof</u>. Again bearing in mind the structure of N and the coordinatisation process the proof is straight forward. Firstly note that by our choice of Y,X,0 and U we already have that $S_x(0)=S_m(0)=S_{yc}(0)=0$ and $S_x(1)=S_m(1)=S_{yc}(1)=1$. The point in row $\langle x,x\rangle$ with coordinates $(S_x(x),S_{yc}(x))$ lies on the line OU, and so $S_x(x)=S_{yc}(x)$ for any $x=0,1,\ldots,q-1$. The line in column $\langle m,0\rangle$ with coordinates $[S_m(m),S_{yc}(0)]=[S_m(m),0]$ passes through the point in row $\langle 1,m\rangle$ with coordinates $(S_x(1),S_{yc}(m))=(1,S_{yc}(m))$, and so $S_m(m)=S_{yc}(m)$ for any $m=0,1,\ldots,q-1$. This completes the proof.

We will now outline the most general method of obtaining an ordered canonical incidence matrix for a given plane, using only row and column operations on N. The *complete reference quadrangle* consists of the four points Y, X, O and U, the six lines of this quadrangle, and the three diagonal points, which are the intersections of opposite pairs of lines. If N is in ordered canonical form with Y, X, O and U as in Theorem 2 then the 18 point and line incidences of the complete reference quadrangle are shown by *1*'s in N in Figure 1. Choose any four rows of N to be the points Y, X, O and U, provided of course that no three of them are collinear. This choice determines the complete reference quadrangle and the next step is to interchange the rows and columns of N so that the complete reference quadrangle is in its required position, which is the position it occupied in N of Theorem 2. At this stage N should have *1*'s in the positions shown in Figure 1. Now put N into ordered canonical form by following through the steps (1) to (4) in an orderly manner, and it is easily verified that at each stage the *1*'s are not moved. The row and column operations just described correspond to the choice of Y, X, O and U in the coordinatisation process.

Let R_{xmyc} be a permutation on the set $\{0,1,\ldots,q-1\}$ such that $R_{xmyc}(0)=0$ and $R_{xmyc}(1)=1$. Now apply R_{xmyc} to N to obtain a simultaneous permutation of the last q-2 row and column bands and of the last q-2 rows and columns of each row and column band. Thus for example row $\langle x,y\rangle$, $x\neq\infty$, becomes row $\langle R_{xmyc}(x),R_{xmyc}(y)\rangle$, and row $\langle\infty,m\rangle$ becomes row $\langle\infty,R_{xmyc}(m)\rangle$. By carefully considering the various parts of N it can be shown that R_{xmyc} individually fixes the *1*'s, while the 1's shown in N in Figure 1 are fixed as a set. In other words R_{xmyc} fixes our complete reference quadrangle and preserves the ordered canonical form of N. The row and column operations generated in this way by R_{xmyc} correspond to the assignment of the coordinates (x,x), x=2,3,...,q-1, in the coordinatisation process. This completes the general method of obtaining an ordered canonical N.

After choosing which rows are to be Y, X, O and U, putting the complete reference quadrangle in position, putting N in ordered canonical form and then applying an arbitrary R_{xmyc} to N, we see that the number of ordered canonical forms corresponding to a particular plane is the same as the number of PTR's, which was given in Section 2. To obtain the number of distinct ordered canonical forms we must divide this number by the order of the automorphism group of the plane, as we did to obtain the number of distinct PTR's.

We say that an ordered canonical N *corresponds naturally* to a PTR if $T(x,m,c)=y$ if and only if there is a 1 in the intersection of row $\langle x,y\rangle$ of N with column $\langle m,c\rangle$ of N. This is equivalent to saying that $S_{xmyc}=I_{xmyc}$, the identity permutation.

Theorem 3. *There is a natural one-to-one correspondence between the PTR's and the ordered canonical incidence matrices of any given plane.*

Proof. We have seen that the number of ordered canonical N's equals the number of PTR's. Let N be in ordered canonical form. Choose Y, X, O and U as in Theorem 2, and for x=2,3,...,q-1 assign the coordinate (x,x) to the point in row $\langle x,x\rangle$. On completing the coordinatisation we see that $S_{xmyc}=I_{xmyc}$ and so we have the PTR naturally corresponding to N.

Now suppose we have coordinatised the plane. Position the complete reference quadrangle and put N in ordered canonical form. Now consider for example row $\langle x,y\rangle$ with coordinates $(S_{xmyc}(x),S_{xmyc}(y))$. Apply R_{xmyc} to N to obtain N' where $R_{xmyc}^{-1}=S_{xmyc}$. Row $\langle x,y\rangle$ of N' was row $\langle R_{xmyc}(x),R_{xmyc}(y)\rangle$ of N with coordinates $(S_{xmyc}(R_{xmyc}(x)),S_{xmyc}(R_{xmyc}(y)))$, that is with coordinates (x,y). Thus N' corresponds naturally to the PTR. This completes the proof.

In particular if the PTR is linear and $T(x,m,c)=x*m+c$ then the naturally corresponding N will have a 1 in the intersection of row ⟨x,y⟩ and column ⟨m,c⟩ if and only if $x*m=y-c$, where $y-c=y+(-c)$ and $-c$ is the right additive inverse of c, that is $c+(-c)=0$. Construct the minus table, that is the table whose entry in the y'^{th} row and c'^{th} column is $y-c$. Now C_{xm} is the permutation matrix with 1's in the positions in which the element $x*m$ occurs in the minus table. Thus N can be written down immediately from the minus table and the * table. Incidentally it is easily shown that multiplication of the permutation matrices constructed from the minus table of any loop in the above manner gives an isomorphic loop.

Let $n(\pi)$ be the number of distinct ordered canonical incidence matrices corresponding to the plane π. The ordered canonical form for N uniquely characterises π if and only if $n(\pi)=1$. For a Desarguesian plane π of order $q=p^h$, p a prime, $n(\pi)=(q-2)!/h$, and this is evaluated in Table 1 for Desarguesian planes up to order 9. In Table 2 we evaluate $n(\pi)$ for the known planes of order 9. We see that the ordered canonical form only uniquely characterises the Desarguesian planes of order q, q=2, 3 or 4. The number $n(\pi)$ is related to the time it takes to identify a particular plane, given an incidence matrix corresponding to it.

We will now mention the original motivation for the work done so far. Bush [2], using a group of 9x9 permutation matrices, constructed a number of incidence matrices, in canonical form, for planes of order 9. The problem was to show that the N's constructed corresponded to the known planes of order 9. This was indeed the case as the permutation matrices used could be obtained from a minus table for GF(9), and their positioning within the N's could be obtained from a rearrangement of the multiplication tables for GF(9) or the right or left nearfields of order 9. A different, independent, solution to the problem was communicated to Bush by R. H. F. Denniston.

Bush's problem could be solved by hand, but the author has also developed a computer program which makes use of the ordered canonical form to decide whether any N for a plane of order 9 corresponds to one of the four known planes.

q	n(π)
2	1
3	1
4	1
5	6
7	120
8	240
9	2 520

Table 1

π	|aut(π)|	n(π)
Desarguesian	84 913 920	2 520
nearfield or dual nearfield	311 040	687 960
Hughes	33 696	6 350 400

Table 2

4. INCIDENCE MATRICES AND LATIN SQUARES

Let N be in canonical form. Construct q-1 latin squares $L_1,L_2,\ldots,L_{q-1}$ as follows. The entry in row x and column y of L_m is c if and only if the point in row <x,y> of N is incident with the line in column <m,c> of N, where $x,y,c=0,1,\ldots,q-1$ and $m=1,2,\ldots,q-1$. With any qxq permutation matrix C we associate the permutation $(C.(0,1,\ldots,q-1)^t)^t$. Thus we have that row x of L_m is just the permutation $(C_{xm}.(0,1,\ldots,q-1)^t)^t$. It is easy to show that the L_m are latin squares if and only if there are no sign-rectangles in N, and furthermore any pair L_{m_1}, L_{m_2} are orthogonal if and only if there are no sign-rectangles in N. Thus $\{L_1,L_2,\ldots,L_{q-1}\}$ is a CMOLS. Reversing the above process gives N from a CMOLS.

Construct q-1 latin squares $D_1,D_2,\ldots,D_{q-1}$ as follows. The entry in row y and column m of D_x is c if and only if the point in row <x,y> of N is incident with the line in column <m,c> of N, where $m,y,c=0,1,\ldots,q-1$ and $x=1,2,\ldots,q-1$. Thus column m of D_x is just $C_{xm}.(0,1,\ldots q-1)^t$. Once again it is easy to show that the D_m are latin squares if and only if there are no sign-rectangles in N, and furthermore $\{D_1,D_2,\ldots,D_{q-1}\}$ is a DCLS if and only if there are no sign-rectangles in N. Reversing this process gives N from a DCLS.

It is now apparent that row x of L_m is column m of D_x, for $x,m=1,2,\ldots,q-1$, while the first rows and columns respectively are always $(0,1,\ldots q-1)$ or its transpose, since N is in canonical form. If any set $\{L_1,L_2,\ldots,L_{q-1}\}$, with $(0,1,\ldots,q-1)$ in the first rows, is converted to a set $\{D_1,D_2,\ldots,D_{q-1}\}$, with $(0,1,\ldots,q-1)$ in the first columns, in the manner just described it is immediately obvious that the former is a CMOLS if and only if the latter is a

DCLS.

At this point we will pause to point out which parts of the preceeding paragraphs are new. The construction of the DCLS from N is exactly as described by Paige and Wexler [8]. Paige and Wexler then assumed, without loss of generality, that the CMOLS, besides having $(0,1,\ldots,q-1)$ in the first rows, have $(0,1,\ldots,q-1)^t$ in the first column of L_1, thus their L_i's are different to ours. This latter, unnecessary assumption, caused their relation between the CMOLS and the DCLS to be much more complicated then ours. Presumably this complication also prevented them from seeing the simple, direct relationship between N and the CMOLS, which is new.

The plane represented by N is equally well represented by the CMOLS or the DCLS in the obvious manner inherited from the way in which N represents the plane. The CMOLS and DCLS also inherit a canonical and an ordered canonical form from N, and this is shown in figure 1. In the case of the CMOLS the inherited ordered canonical form is the "normal" form referred to by Martin [7].

Finally we remark that the relationships shown in this section become almost self-evvident when N, the CMOLS and the DCLS are laid out as in figure 1.

REFERENCES

[1] R. C. Bose, On the application of the properties of Galois fields to the problem of construction of hyper-graeco-latin squares. *Sankhyā* 3 (1938), 323-338.

[2] K. A. Bush, "Cyclic" solutions for finite projective planes. *Ann. Disc. Math.* 18 (1983), 181-192.

[3] J. Dénes and A. D. Keedwell, *Latin Squares and Their Applications* (Akadémiai Kíado, Budapest, 1974).

[4] M. Hall, Projective planes. *Am. Math. Soc. Trans.* 54 (1943), 229-277.

[5] D. R. Hughes and F. C. Piper, *Projective Planes* (Springer Verlag, New York, Heidelberg, Berlin, 1973).

[6] F. Kárteszi, *Introduction to Finite Geometries* (North-Holland Publishing Company, Amsterdam, Oxford, 1976).

[7] G. E. Martin, Planar ternary rings and latin squares. *Matematiche (Catania)* 23 (1968), 305-318.

[8] L. J. Paige and C. Wexler, A canonical form for incidence matrices of finite projective planes and their associated latin squares. *Port. Math.* 12 (1953), 105-112.

ON CLIQUE COVERING NUMBERS OF CUBIC GRAPHS

LOUIS CACCETTA AND NORMAN J. PULLMAN

The clique covering number *of a graph is the smallest number of complete subgraphs needed to cover its edge-set. For each n, we determine the set of those integers which are clique covering numbers of connected, cubic graphs on n vertices. The analogous result for 4-regular graphs is stated.*

1. INTRODUCTION

For our purposes, graphs are finite, loopless and have no multiple edges. We call the minimum number of complete subgraphs needed to cover the edge-set of a graph G, the *clique covering number* of G, denoted cc(G).

Since the 1960's several papers have appeared concerned, in part, with the clique covering number: for example [4] to [11].

In [8] and [10] upper and lower bounds on cc(G) were determined for all k-regular graphs. This prompts the question:

Given k and n what are the possible values of cc(G) for k-regular, connected graphs G on n vertices?

In [3], the analogous question was considered for the minimum number of complete subgraphs needed to *partition* the edge set of a graph G.

Let $T_k(n)$ denote the set of values of cc(G) in question. For example $T_2(n)=\{n\}$ for all $n\geq 4$, because the only 2-regular, connected graphs are cycles. In general, $T_k(n)=\emptyset$ for all $1\leq n\leq k$ and $T_k(k+1)=\{1\}$. Moreover, $T_k(n)=\emptyset$ for all odd n when k is odd because k-regular graphs have kn/2 edges. So we can restrict our attention to $T_k(n)$ for n>k, and to only even n when k is odd.

In section 3 (Theorem 3.1) we prove that for all even n>8,

(1.1) $T_3(n)=\{m\in\mathbf{Z}:\lfloor(3n+2)/4\rfloor\leq m\leq(3n-4)/2\}\cup\{3n/2\}$

The values of $T_3(n)$ for $n\leq 8$ are also given in Theorem 3.1.

We have also determined $T_4(n)$ for all n, but (due to its length) we will publish its proof elsewhere [2]. Here we state only the result.

Theorem

If $18<n\equiv 3 \pmod 5$, *then*

$T_4(n)=\{m\in\mathbf{Z}:\lceil 3n/5\rceil+1\leq m\leq 2n\}\setminus\{2n-1\}$.

If $11\leq n$ *and* $n=13$, 18 *or* $n\not\equiv 3 \pmod 5$, *then*

$T_4(n)=\{m\in\mathbf{Z}:\lceil 3n/5\rceil\leq m\leq 2n\}\setminus\{2n-1\}$.

For $n \le 10$:

$T_4(6)=\{4\}$, $T_4(7)=\{6,7\}$, $T_4(8)=\{6,8,10,16\}$,

$T_4(9)=\{m\in \mathbf{Z}: 6\le m\le 14\}\setminus\{7,13\}$ *and*

$T_4(10)=\{m\in \mathbf{Z}: 6\le m\le 20\}\setminus\{7,18,19\}$.

2. PRELIMINARIES

Except for the term "clique" we adhere to the terminology and notation of Bondy and Murty [1].

We call a subgraph K of G a *clique* if K is a complete graph. If the clique K has r vertices we call K an *r-clique,* a clique of *order r* or a K_r. If K is a 2-clique we call it an *edge.* If K is a 3-clique we call it a *triangle.*

If G has edges, then a *clique covering* of G is a family C of cliques of G whose members *cover* E(G), the edge-set of G. That is,

$$E(G)=\cup\{E(K):K\in C\}$$

If G is 'empty' (has vertices, but no edges) then we call $\emptyset$ a clique covering of G. Among all clique coverings of G, the ones with minimum cardinality are called *minimum* clique coverings of G. Their common cardinality is called the *clique covering number* of G, denoted cc(G).

In [8] and [10] upper and lower bounds on cc(G) were determined for all k-regular graphs. For the particular case k=3 these bounds are 3n/2 and $\lfloor(3n+2)/4\rfloor$, respectively. Consequently, if $x\in T_3(n)$, then

(2.1) $\lfloor(3n+2)/4\rfloor\le x\le 3n/2$.

It was also shown in [8] and [10] that the upper and lower bounds are attained for all $n\ge 6$.

When the maximum degree of the vertices of G is 3, there is available a formula ([10], Theorem 3.1) for computing cc(G), namely

(2.2) $cc(G)=\varepsilon(G)-(3t+t_1)/2$

where t is the number of triangles in G , t_1 is the number of triangles sharing an edge with no other triangles in G and $\varepsilon(G)$ is the number of edges of G. Thus

(2.3) $(3n/2)-1\notin T_3(n)$.

3. CUBIC GRAPHS

In this section we calculate $T_3(n)$ for each $n\ge 4$. The results are summarized in the following theorem.

Theorem 3.1

(a) $T_3(4)=\{1\}$, $T_3(6)=\{5,9\}$, $T_3(8)=\{6,8,10,12\}$

(b) *If* n *is odd, then* $T_3(n)=\emptyset$

(c) *If* $n\ge 10$ *is even, then*

$T_3(n)=\{m\in\mathbf{Z}:\lfloor(3n+2)/4\rfloor\le m\le(3n-4)/2\}\cup\{3n/2\}$.

Since a cubic graph must have an even number of vertices, part (b) is immediate. Before proving parts (a) and (c) we introduce some further notation and establish two important lemmas.

Throughout this section G always denotes a cubic graph and C a minimum clique covering of G. Let T_1 be the set of K_3's in G sharing an edge with no other K_3 in G. Call such K_3's *lonely*.

Let R denote the set of non-lonely K_3's in G. By the minimality of C, $T_1 \cup R \subseteq C$. Non-lonely K_3's occur in pairs (K,K') whose intersection is an edge of G. Moreover, the subgraph $K \cup K'$ is disjoint from all other members of R. Index the partners in each pair of non-lonely triangles of G as K^i, $(K^i)'$ for $1 \le i \le |R|/2$. Then let $T_2 = \{K^i \cup (K^i)' : 1 \le i \le |R|/2\}$.
We call members of T_2, *couples*.
Let T_o be the set of K_2's in C.
Let $t_i = |T_i|$, i = o, 1, 2.
Then

(3.1) $\quad t_o + t_1 + 2t_2 = cc(G)$.

Further, since G has 3n/2 edges, and the members of C are mutually edge-disjoint we have

$$t_o + 3t_1 + 5t_2 = 3n/2.$$

Therefore

(3.2) $\quad 2t_1 + 3t_2 = 3n/2 - cc(G)$

Note that (3.2) follows from (2.2) since $t = t_1 + 2t_2$.

Lemma 3.1 *Let* $x \in T_3(8)$. *Then* $6 \le x \le 12$ *and* $x \ne 7, 9, 11$.

Proof. It follows from (2.1) and (2.3) that we need only prove that $x \ne 7, 9$. Now equations (3.1) and (3.2) become

(3.1') $\quad t_o + t_1 + 2t_2 = x$, and

(3.2') $\quad 2t_1 + 3t_2 = 12-x$.

It is clear from the above equations that if $t_2 \ne 1$, then x cannot be 7 or 9. Hence we may suppose that $t_2 = 1$. Let H be a couple in G. Necessarily $H \in C$. Consider the graph G-H obtained from G by deleting V(H), the vertex set of H, as well as all edges of G incident to a vertex of H. The graph G-H has 4 vertices and 5 edges, and hence is also a couple. Therefore $t_2 = 2$, a contradiction. □

Lemma 3.1 implies that 6, 8, 10 and 12 are the only possible candidates for $T_3(8)$. Graphs G(8,x) with 8 vertices and cc(G(8,x)) = x, x = 6, 8, 10, 12, are displayed in Fig. 3.1 below.

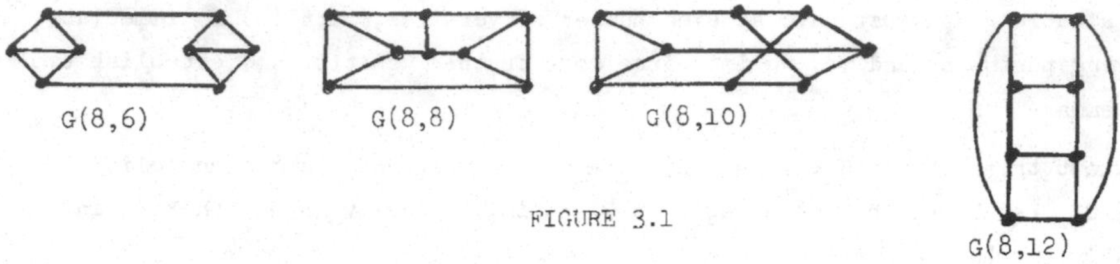

FIGURE 3.1

It thus follows that $T_3(8) = \{6, 8, 10, 12\}$.

The only cubic graphs on 6 vertices are: the complete bipartite graph $K_{3,3}$ which has clique covering number 9, and the graph pictured in Figure 3.2 which has clique covering number 5.

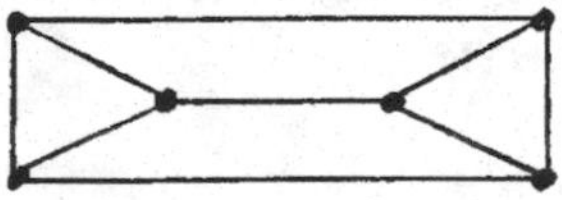

FIGURE 3.2

Consequently $T_3(6) = \{5, 9\}$. Since $T_3(4) = \{1\}$ we have established part (a) of Theorem 3.1.

Before proceeding to our next lemma we introduce some further notation and terminology.

A pair e, e' of vertex disjoint edges in G is called an *available pair* of edges if both e and e' belong to C, a minimum clique covering of G.

The following operations Φ work this way. A subgraph S of a certain type is located in G. The subgraph S is replaced by a specific graph H (given in Table 3.1 below) to form a new graph $\Phi(G)$. Each operation has the property that $\Phi(G)$ has an available pair of edges. Table 3.1 also lists the changes Δcc in cc and $\Delta\nu$ in ν the number of vertices, due to changes in G caused by operation Φ. That is

$$\Delta cc = cc(\Phi(G)) - cc(G), \text{ and}$$
$$\Delta\nu = \nu(\Phi(G)) - \nu(G).$$

Φ	S	H	$\Delta\nu$	Δcc	Restrictions on S
a	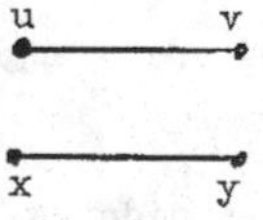	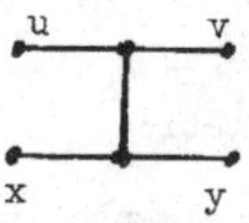	2	3	uv and xy are an available pair of edges
b	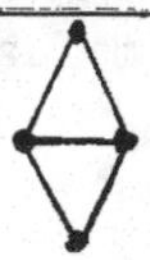		2	2	

TABLE 3.1

Availability is important for our recursive method of constructing graphs as is indicated in the following lemma.

Lemma 3.2 *Let* $G(n)$ *be the set of cubic graphs on* n *vertices and let*

$$I(n) = \{m \in \mathbb{Z}: \ \lambda(n) \le m \le (3n/2) - 2\}$$

with $\lambda(n) = \lfloor (3n+2)/4 \rfloor$.

Then for every even $n \ge 10$ *and every* $c \in I(n)$ *there exists a graph in* G *having an available pair of edges and clique covering number* c.

Proof. We prove the lemma by induction on n. That the lemma holds for n=10 follows from the graphs displayed in Figure 3.3.

As our inductive hypothesis we assume the assertion holds for $n = 2\ell \ge 10$. So there exist *(n,c)- graphs* (graphs G in *G* with $cc(G) = c$) for every $c \in I(2\ell)$, having a pair of available edges. Hence when $G(2\ell,c)$ is a $(2\ell,c)$-graph then $a(G(2\ell,c))$

is a $(2\ell+2, c+3)$ - graph, and has a pair of available edges. So $(2\ell+2,c')$ - graphs with the desired property exist for all integers

$$c' \in \{\lambda(2\ell)+3, \lambda(2\ell)+4, \ldots, \tfrac{3}{2}(2\ell+2)-2\}.$$

By definition

$$\lambda(2\ell)+3 = \begin{cases} \lambda(2\ell+2)+1, & \text{if } 2\ell \equiv 0 \pmod 4 \\ \lambda(2\ell+2)+2, & \text{otherwise} \end{cases}$$

The existence of a $(2\ell+2, \lambda(2\ell+2))$ - graph with an available pair of edges is implied by the graphs $G(4q, 3q)$ and $G(4q+2, 3q+2)$ of Figure 3.4 which were determined in [10].

Finally, if $2\ell \not\equiv 0 \pmod 4$, then the graph

$$b(G(4q+2, 3q+2))$$

with $2\ell = 4q+2$, is a $(2\ell+2, \lambda(2\ell+2)+1)$ - graph with a pair of available edges. This proves the lemma when $n = 2\ell+2$. The result now follows from the principle of mathematical induction. □

Now Lemma 3.2 together with (2.1) and (2.3) imply part (c) of Theorem 3.1.

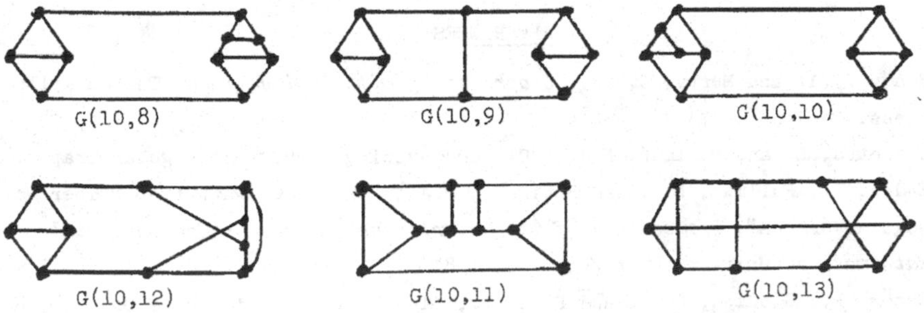

FIGURE 3.3

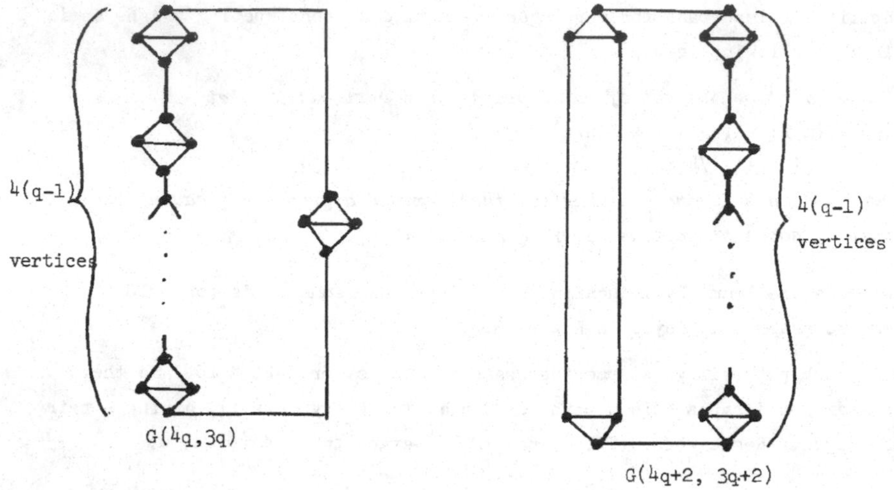

FIGURE 3.4

4. OPEN QUESTIONS

1. Determine $T_k(n)$ for all $n \geq k + 3 \geq 8$. Only even n need be considered when k is odd.
2. For $k > 4$ (n even when k is odd) is $T_k(n) \cup \{(kn/2) - 1\}$ an interval of integers for all sufficiently large n? It is when $k = 3, 4$.
3. We have $\lim_{m\to\infty} |T_3(2m)|/2m = 3/4$,

   and $\lim_{n\to\infty} |T_4(n)|/n = 7/5$. Are there analogues for $k > 4$?

ACKNOWLEDGEMENTS

This work was supported in part by grants T1821 and A4041 of the Natural Sciences and Engineering Council of Canada, and by the generosity of the School of Mathematics and Computing of the Western Australian Institute of Technology and the Faculty of Mathematics of the University of Newcastle.

REFERENCES

[1] Bondy, J.A. and Murty, U.S.R., *Graph Theory with Applications*, The MacMillan Press, London (1977).

[2] Caccetta, L. and Pullman, N.J., "Clique Covering Numbers of Regular Graphs."

[3] Eades, P., Pullman, N.J. and Robinson, P.J., "On Clique Partition Numbers of Regular Graphs", *Proceedings of The Waterloo Conference on Combinatorial Mathematics*, University of Waterloo, 1982.

[4] Erdős, P., Goodman, A.W., and Pósa, L., "The Representation of a Graph by Set Intersections", *Can. J. Math.* V.18 (1966), pp.106-112.

[5] Gregory, D.A. and Pullman, N.J., "On a Clique Covering Problem of Orlin", *Discrete Mathematics* (to appear).

[6] Lovász, L., "On covering of graphs", *Theory of Graphs*, (Proceedings of the Colloquium held at Tihany, Hungary, September 1966), (P. Erdös and G. Katona eds.) Academic Press, N.Y. (1968) - pp.231-236.

[7] Orlin, J., "Contentment in Graph Theory: Covering Graphs with Cliques", *K. Nederlandse Ak. van Wetenschappen Proc. Ser. A,V.* 80 (1977), pp.406-424.

[8] Pullman, N.J. and de Caen, D., "Clique Coverings of Graphs I: clique-partitions of regular graphs", *Utilitas Math.* V.19 (1981) pp.177-206.

[9] Pullman, N.J. and Donald, A.W., "Clique Coverings of Graphs II: complements of complete graphs", *Utilitas Math.* V.19 (1981) pp.207-214.

[10] Pullman, N.J. and de Caen, D., "Clique Coverings of Graphs III: clique-coverings of regular graphs", *Congressus Numerantium*, V.29 (1980), pp.795-808.

[11] Pullman, N.J., "Clique Coverings of Graphs IV - Algorithms", (to appear).

MODELLING COMPETITIONS BY POSET MULTIPLICATION

R. J. CLARKE

The relative merits of teams involved in a competition can be expressed by a partial ordering on the teams. The result of a round of the competition can be expressed by a multiplication of partial orderings, satisfying certain axioms. Several possible multiplications are described, and some properties of these multiplications displayed.

1. INTRODUCTION

Suppose that we have a competition between n teams, labelled $0,1,\ldots,n-1$, in which in each round certain teams play against each other in pairs. Customarily such a competition is scored by allotting two points per win and one point per draw, and splitting ties by some method, for example by goal average in the case of football. In other words, at the completion of each round we have a total ordering on the set $\{0,1,\ldots,n-1\}$ of teams. But it may be more realistic to think of a partial ordering instead.

For example, suppose that we have four teams, $0,1,2,3$, and in the first round the results are

0 defeats 1, 2 defeats 3.

We express these results by the poset

P = 

.

Here we are using the usual convention, that a downwards sloping line from a to b means $(a,b) \in P$.

If in the second round the results are

1 defeats 2, 0 defeats 3,

we express them by the poset

Q = 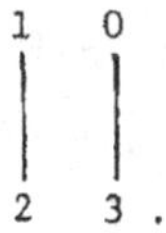

We wish to find a poset PQ which contains all of the information so far obtained. Clearly

PQ = 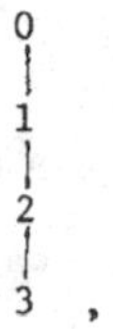

the smallest poset containing both P and Q, will do.

However, suppose that

Q' = 

represents the results of the second round. Now there is no poset which contains both P and Q'. We define

$$PQ' = Q'$$

to be the poset which represents the current positions of the teams (on the grounds that the information in Q' is more up to date than that in P). These considerations motivate the definition of a *poset multiplication*.

2. POSET MULTIPLICATIONS

Let N be the set of natural numbers. A *poset* on N is a relation on N which is irreflexive, transitive and antisymmetric. Let $\mathcal{P}$ be the set of all *finite posets* on N, i.e. all posets containing only a finite number of relations. Posets $P, Q \in \mathcal{P}$ are called *compatible*, denoted by $P \sim Q$, if there is a poset $R \in \mathcal{P}$ with $R \supseteq P \cup Q$. If $P \sim Q$ we define the poset $P+Q$ by

$$P+Q = \cap\{R \mid R \in \mathcal{P},\ R \supseteq P \cup Q\} .$$

Definition. A *poset multiplication* is a binary operation

$$\mathcal{P} \times \mathcal{P} \to \mathcal{P}$$

$$(P,Q) \to PQ$$

satisfying the following axioms.

(i) If $P \sim Q$ then $PQ = P+Q$.

(ii) $PQ = P_1 + Q$ for some $P_1 \in \mathcal{P}$, $P_1 \subseteq P$.

(iii) If σ is a permutation on N then

$$(PQ)^{\sigma} = P^{\sigma}Q^{\sigma} ,$$

where P^{σ} is defined in the obvious way.

(iv) If P^{x} denotes the dual poset to P then

$$(PQ)^{x} = P^{x}Q^{x} .$$

We point out that the following simple consequences of the axioms, the proofs of which can be left to the reader.

(1) For all P, $P^2 = P$.

(2) If $P \subseteq Q$ then $PQ = QP = Q$.

(3) For all P and Q,

$$(PQ)Q = PQ = Q(PQ).$$

(4) Suppose that there is a subset S of N such that for all $x,y \in S$, $x \neq y$, either (x,y) or $(y,x) \in Q$. (In other words, Q induces a total ordering on the elements of S.)
Suppose also that the only relations in P are between elements of S.
Then $PQ = Q$.

Proposition 1. *No poset multiplication is associative.*

Proof. Consider the posets

$$P = \begin{array}{c}0\\|\\1\end{array}, \quad Q = \begin{array}{c}1\\|\\2\end{array}, \quad R = \begin{array}{c}2\\|\\0\end{array},$$

(by which we mean $P = \{(0,1)\}$, etc.).

It can easily be seen that

$$P(QR) = QR = \begin{array}{c} 1 \\ | \\ 2 \\ | \\ 0 \end{array}$$

and

$$Q(PR) = PR = \begin{array}{c} 2 \\ | \\ 0 \\ | \\ 1 \end{array} .$$

But if the poset multiplication is associative,

$$\begin{aligned} P(QR) &= (PQ)R \\ &= (QP)R \quad \text{since} \quad PQ = QP = P+Q \\ &= Q(PR) \ , \end{aligned}$$

a contradiction.

<u>Definition</u>. A poset multiplication is called *subassociative* if for all $P,Q \in \mathcal{P}$,

$$P(PQ) = PQ.$$

Let $P,Q \in \mathcal{P}$. We define

$$S(P,Q) = \{R \mid R \in \mathcal{P},\ R \subseteq P,\ R \sim Q\}$$

and

$$M(P,Q) = \{R \mid R \text{ is maximal in } S(P,Q)\}.$$

<u>Theorem</u>. *The operation defined by*

$$P.Q = (\cap M(P,Q)) + Q \qquad \textit{for all} \quad P,Q \in \mathcal{P}$$

is a subassociative poset multiplication.

The proof of this theorem is not difficult and is omitted.

3. A LARGER MULTIPLICATION

Consider the posets

$$P = \begin{array}{c}0\\ |\\ 1\\ |\\ 2\\ |\\ 3\end{array}\ , \qquad Q = \begin{array}{c}3\\ |\\ 0\end{array}\ .$$

One would expect, given the motivation behind PQ, that

$$PQ = \begin{array}{c}1\\ |\\ 2\end{array}\quad \begin{array}{c}3\\ |\\ 0\end{array}\ .$$

However it is easy to verify that

$$P.Q = Q\ .$$

Hence we seek a product which is strictly "larger" than P.Q. Let P be a poset and let S be a subset of P. Denote by $P\backslash S$ the set theoretical difference $\{(a,b) \in P \mid (a,b) \notin S\}$. Define

$$\begin{aligned} S^* = S \cup \{(a,b) \mid \exists c \ \text{ such that } \ & (a,c) \in S,\ (a,b) \in P\backslash S, \\ & (b,c) \in P\backslash S\} \\ \cup \{(a,b) \mid \exists d \ \text{ such that } \ & (d,b) \in S,\ (a,b) \in P\backslash S, \\ & (d,a) \in P\backslash S\}\ , \end{aligned}$$

and define

$$P-S = P\backslash S^*.$$

It is easy to see that P-S is a poset and that $P-S \subseteq P\backslash S$. Now let $P,Q \in \mathcal{P}$ such that $P \not\subseteq Q$. We define PQ recursively as follows. Suppose that P'Q has been defined whenever P' is a poset containing fewer relations than P. By a P,Q *contradiction of length* r we mean a sequence

$$a_1, b_1, a_2, b_2, \ldots, a_r, b_r, a_{r+1} = a_1 \tag{1}$$

of elements of N, distinct except for the first and last members, such that

$$(a_i, b_i) \in Q, \qquad i=1,\ldots,r$$

and

$$(b_i, a_{i+1}) \in P, \quad i=1,\dots,r .$$

Let r_0 be the length of the shortest P,Q contradiction. Let S_0 be the set of all relations (b_i, a_{i+1}) occuring in any P,Q contradiction of length r_0. Then we define

$$PQ = (P-S_0)Q .$$

This effectively defines PQ by recursion, since $P-S_0$ contains fewer relations than P. We call this operation the *canonical multiplication*.

It is convenient to introduce a little more notation and terminology in connection with the above definition. It can be seen that, by successively applying the above procedure, we will obtain

$$PS = (P-S_0-S_1-\dots-S_K)Q ,$$

where the elements of S_m arise from P,Q contradictions of length $r = r_0+m$. Some S_m may of course be empty. The elements of S_m are said to *occur in* a contradiction of length r. The elements of S_m^*, where

$$P-S_0-\dots-S_m = (P-S_0-\dots-S_{m-1})\setminus S_m^* ,$$

are said to *be deleted as a result of* a contradiction of length r.

Let $(b,a) \in P$, $(b,a) \notin PQ$. Then there are two cases.

Case I. $(b,a) \in S_m$ for some m. Then there is a P,Q contradiction (1) of length $r = r_0+m$, in which

$$(b,a) = (b_r, a_1).$$

No relation $(b_i, a_{i+1}) \in S_j^*$ for any $j^* < m$.

Case II. $(b,a) \in S_m^*$, $(b,a) \notin S_m$ for some m. There are two possibilities.

(a) There is a P,Q contradiction (1) of length $r = r_0+m$, in which

$$(b,c) = (b_r, a_1)$$

for some c such that $(a,c) \in P$. Furthermore, neither (a,b), (b,a) nor any $(b_i, a_{i+1}) \in S_j^*$ for any $j^* < m$ and $(a,c) \notin S_m$, $(b,a) \notin S_m$.

(b) There is a P,Q contradiction (1) of length $r = r_0+m$, in which

$$(d,a) = (b_r, a_1)$$

for some d such that $(d,b) \in P$. This case is similar to (a) and will not be described in detail.

These cases will be referred to in the sequel. Consideration of Case II(b) will normally be omitted.

Theorem 2. *The canonical multiplication is a poset multiplication which has the following properties:*

(v) For all $P,Q \in \mathcal{P}$,

$$P.Q \subseteq PQ.$$

(vi) If $Q = Q_1 \cup Q_2$ *and* $Q_2 \subseteq P$ *then*

$$PQ = PQ_1.$$

(vii) Suppose that there is a function $f:N \to N$, *and posets* $\underline{P}$, $\underline{Q}$ *on* N, *such that for all* $a,b \in N$ *with* $f(a) \neq f(b)$,

$$(a,b) \in P \Leftrightarrow (f(a),f(b)) \in \underline{P}$$

and
$$(a,b) \in Q \Leftrightarrow (f(a),f(b)) \in \underline{Q}.$$

For each $a \in N$ *let*

$$(a) = \{b \in N | f(b) = f(a)\},$$

and let Pa *and* Qa *be the restrictions of* P *and* Q *respectively to* (a). Pa *and* Qa *may be considered as posets on* N.

Then for all $a,b \in N$,

$$\textit{if} \quad f(a) \neq f(b), \quad (a,b) \in PQ \Leftrightarrow (f(a),f(b)) \in \underline{PQ}$$

while

$$\textit{if} \quad f(a) = f(b), \quad (a,b) \in PQ \Leftrightarrow (a,b) \in PaQa .$$

Proof. It is easy to see that the canonical multiplication is a poset multiplication.

(v) Let $(b,a) \in \cap M(P,Q)$ and suppose that $(b,a) \notin PQ$. Consider the above cases.

Case I. Let $R = \{(b_i, a_{i+1}) | 1 \leq i \leq r-1\}$. Then R is a subposet of P and $R \sim Q$. Choose $M \in M(P,Q)$ such that $R \subseteq M$. Then $(b,a) \in M$, so that $R+(b,a) \subseteq M$. Hence $R+(b,a) \sim Q$, a contradiction.

Case II. (a) Let $R = \{(b_i, a_{i+1}) | 1 \leq i \leq r-1\}$. As before, R is a subposet of P with $R \sim Q$. Now $R+(a,c) \sim Q$, otherwise $(a,c) \in S_m$ or $(a,c) \in S_j^*$ for some $j < m$. As before we obtain $R+(a,c) + (b,a) \sim Q$. Thus $R+(b,c) \sim Q$, a contradiction.

Thus we have shown that

$$M(P,Q) \subseteq PQ,$$

so that

$$P.Q \subseteq PQ .$$

(vi) Let

$$a_1, b_1, \ldots, a_r, b_r, a_{r+1} = a_1 \tag{1}$$

be a P,Q contradiction of minimal length $r = r_0$. Each $(a_i, b_i) \in Q$, and $(a_i, b_i) \notin P$, otherwise we could easily find a shorter P,Q contradiction. Hence each $(a_i, b_i) \in Q_1$ and we have a P,Q_1 contradiction. Thus minimal P,Q contradictions are the same as minimal P,Q_1 contradictions. So

$$PQ = (P-S_0)Q, \quad PQ_1 = (P-S_0)Q_1 .$$

Now we claim that $Q_2 \subseteq P-S_0$. For suppose on the contrary that $(b,a) \in Q_2$, $(b,a) \notin P-S_0$. Consider the above cases.

Case I. Here, as $(b,a) \in Q_2 \subseteq Q$, $(a_r, b_1) \in Q$. Thus

$$a_r, b_1, a_2, \ldots, b_{r-1}, a_r$$

is a P,Q contradiction of length $r-1$, a contradiction.

Case II. (a) As (a_r, b) and $(b,a) \in Q$, $(a_r, a) \in Q$. But then

$$c, b_1, a_2, \ldots, a_r, a, c$$

is a P,Q contradiction of length r in which (a,c) occurs, showing that $(a,c) \in S_0$, a contradiction.

By repeating the above argument we eventually obtain

$$PQ = P_1+Q, \quad PQ_1 = P_1+Q_1,$$

where P_1 is a subposet of P such that

$$P_1 \sim Q, \quad Q_2 \subseteq P_1 .$$

Thus

$$PQ = P_1 + (Q_1+Q_2)$$

$$= P_1 + Q_1$$

$$= PQ_1 .$$

<u>(vii)</u> Let (1) above be a P,Q contradiction of minimal length $r = r_0$.

If the members of the sequence

$$f(a_1),f(b_1),\ldots,f(b_r),f(a_1) \qquad (2)$$

are distinct, except for the first and last members, then (2) is a $\underline{P},\underline{Q}$ contradiction of length r_0.

If the members of the sequence (2) are all equal then (1) is a Pa,Qa contradiction, where $a = a_1$, of length r_0.

The only other possibility can be seen to be that $f(x) = f(y)$ for two consecutive members of (2).

Suppose that

$$f(a_i) \neq f(b_i) = f(a_{i+1})$$

for some i. Then $(a_i,b_i) \in Q$, so $(a_i,a_{i+1}) \in Q$. As $(a_{i+1},b_{i+1}) \in Q$, we have $(a_i,b_{i+1}) \in Q$. Hence (1) is not a minimal P,Q contradiction.

The case that

$$f(a_i) = f(b_i) \neq f(a_{i+1})$$

for some i is treated similarly.

Now let $\underline{r}_0$ be the length of the shortest $\underline{P},\underline{Q}$ contradiction or ∞ if no such exists, and define r_{a0} similarly for Pa,Qa. Then the above argument shows that

$$r_0 = \min(\underline{r}_0, \min_{a \in N} r_{a0}) .$$

Further, let S_0 be the set of all relations from P occuring in any P,Q contradiction of length r_0, let $\underline{S}_0$ be the set of all relations from $\underline{P}$ occuring in any $\underline{P},\underline{Q}$ contradiction of length r_0, and let S_{a0} be the set of all relations from Pa occuring in any Pa,Qa contradiction of length r_0. Note that $\underline{S}$ or

S_a could be empty. Then

$$PQ = (P-S_0)Q,$$

$$\underline{PQ} = (\underline{P}-\underline{S}_0)\underline{Q}$$

and $$P_aQ_a = (P_a-S_{a0})Q_a \quad \text{for each} \quad a \in N.$$

Now we claim that if $f(a) \neq f(b)$,

$$(b,a) \in P-S_0 \Leftrightarrow (f(b),f(a)) \in \underline{P}-\underline{S}_0.$$

For suppose $(b,a) \in P$, $(b,a) \notin P-S_0$. Consider the usual three cases.

<u>Case I</u>. If $(b,a) = (b_r, a_{r+1})$ for a minimal P,Q contradiction (1) of length $r = r_0$ then

$$f(a_1), f(b_1), \ldots, f(a_1)$$

is a minimal $\underline{P},\underline{Q}$ contradiction. Hence $(f(b),f(a)) \notin \underline{P}-\underline{S}_0$.

<u>Case II</u>. (a) If $(b,c) = (b_r, a_{r+1})$ for a minimal P,Q contradiction for some c such that $(a,c) \in P$, then (b,a) and (a,c) do not occur in any P,Q contradiction of length r_0. Now if $f(a) = f(c)$ we will be able to find a contradiction of length r_0 involving (b,a), contradiction. So $f(a) \neq f(c)$. But then it is easy to see that $(f(b),f(a)) \notin \underline{P}-\underline{S}$.

Hence our claim is proven.

Now let us make the inductive hypothesis that (vii) holds for all posets P',Q' such that P' contains fewer relations than P. We may apply this hypothesis to $P-S_0$, Q to obtain

if $f(a) \neq f(b)$, $(a,b) \in (P-S_0)Q \Leftrightarrow (f(a),f(b)) \in (\underline{P}-\underline{S}_0)\underline{Q}$

and if $f(a) = f(b)$, $(a,b) \in (P-S_0)Q \Leftrightarrow (a,b) \in (P-S_0)_a Q_a$.

Now $(P-S_0)Q = PQ$ and $(\underline{P}-\underline{S}_0)\underline{Q} = \underline{PQ}$. We need only show that

$$(P-S_0)_a = P_a - S_{a0}$$

to complete the proof of (vii). However this result is clear.

Example

Let

$$P = \begin{array}{c} 0 \\ | \\ 1 \\ | \\ 2 \\ | \\ 3 \end{array} \quad , \quad Q = \begin{array}{c} 3 \\ | \\ 0 \end{array}$$

as before. Define a function $f: N \to N$ by

$$f(0) = 0, \ f(3) = 3, \ f(1) = f(2) = 1, \ f(x) = x \ \text{if} \ x > 3.$$

Define posets $\underline{P}, \underline{Q}$ on N by

$$\underline{P} = \begin{array}{c} 0 \\ | \\ 1 \\ | \\ 3 \end{array} \quad , \quad \underline{Q} = \begin{array}{c} 3 \\ | \\ 0 \end{array} \ .$$

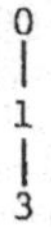

Then the conditions of property (vii) of Theorem 2 hold. Now

$$\underline{PQ} = \underline{Q} = \begin{array}{c} 3 \\ | \\ 0 \end{array} \ .$$

Also $P_1 = \begin{array}{c} 1 \\ | \\ 2 \end{array}$, $Q_1 = \phi$. So

$$P_1 Q_1 = P_1 = \begin{array}{c} 1 \\ | \\ 2 \end{array} \ .$$

It is now easy to check that

$$PQ = \begin{array}{cc} 3 & 1 \\ | & | \\ 0 & 2 \end{array}$$

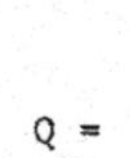

as required by our intuition.

Example

Let

$$P = \begin{array}{cccc} 0 & 2 & 4 & 6 \\ | & | & | & | \\ 1 & 3 & 5 & 7 \end{array} \ , \quad Q = \begin{array}{ccc} 1 \quad\; 3 & 5 \quad 7 \\ \bigwedge \quad | & \bigvee \\ 6 \quad 2 \quad 4 & 0 \end{array} \ .$$

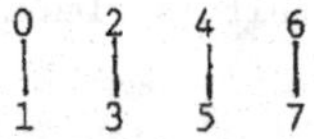

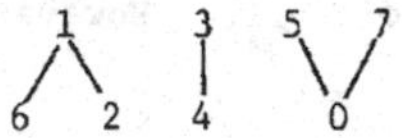

Minimal P,Q contradictions are

$$1, \ 6, \ 7, \ 0, \ 1$$

and variations of this one obtained by cyclic permutation of 1,6,7,0. Hence

$$P-S = P - \{(6,7),(0,1)\}$$

=

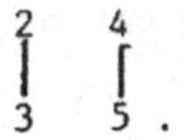

PQ = (P-S)Q =

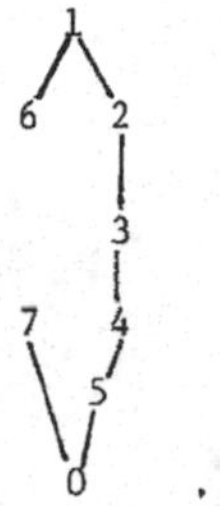

The minimal P,PQ contradiction is

1, 0, 1.

Hence

$$P - S_{PQ} = P - \{(0,1)\}$$

=

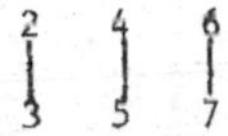

P(PQ) =

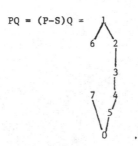

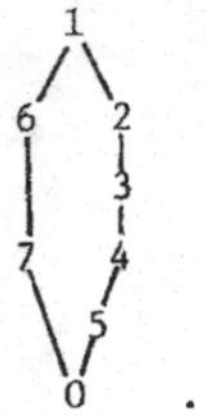

So

$$PQ \neq P(PQ)$$

and the canonical multiplication is not subassociative.

However we may define a multiplication * by

$$P{*}Q = PQ \cup P(PQ) \cup P(P(PQ)) \ldots$$

$P*Q$ is a poset because

$$PQ \subseteq P(PQ) \subseteq P(P(PQ)) \subseteq \ldots$$

It is easy to show that $*$ is a poset multiplication which satisfies (v), (vi) and (vii) and is subassociative.

DECOMPOSITION OF BLOCK DESIGNS: COMPUTATIONAL ISSUES

CHARLES J. COLBOURN AND MARLENE J. COLBOURN

Deciding whether a triple system with $\lambda = 3$ *can be decomposed, or partitioned, into triple systems with smaller* λ *is NP-complete; this contrasts with the polynomial time algorithm for decomposing triple systems with* $\lambda = 2$. *Two extensions of this result are examined here - to triple systems without repeated blocks, and to triple systems with* $\lambda = 4$.

1. INTRODUCTION

A (balanced incomplete) *block design* $B[k,\lambda;v]$ is a pair (V,B); V is a v-set of *elements* and B is a collection of k-subsets of V called *blocks*. Each unordered pair of elements of V appears in precisely λ blocks. When B contains no repeated blocks, we use the notation $NB[K,\lambda;v]$. When $k = 3$, block designs are called *triple systems*. One standard technique for constructing block designs $B[k,\lambda;v]$ is to construct systems $B[k,\lambda_1;v]$ and $B[k,\lambda_2;v]$ with $\lambda = \lambda_1 + \lambda_2$, and taking their union. Kramer [8] calls such systems *decomposable*; he demonstrates the existence of indecomposable $B[3,2;v]$ and $B[3,3;v]$. He further remarks that, for $\lambda = 2$ determining whether a design is decomposable can be carried out efficiently, i.e. in polynomial time. We have shown by way of contrast that deciding whether a $B[3,3;v]$ is decomposable is NP-complete, and hence unlikely to have any efficient solution [3]. We describe the reduction here, and extend it to $B[3,4;v]$ designs. We further eliminate the necessity for repeated blocks.

2. BACKGROUND

Given an r-regular n-vertex graph G, a *Latin background* for G, denoted $LB[G;m,s]$ is an s by s symmetric array with elements chosen from $\{1,2,\ldots,m\}$. Each diagonal entry contains the element m. In the first n rows, each position is either empty, or contains a single element from the set $\{r+1,\ldots,m\}$. In the latter $s - n$ rows, each position contains a single element of the set $\{1,2,\ldots,m\}$. Each element appears at most once in each row (and, symmetrically, each column). Finally, the pattern of empty squares forms an adjacency matrix for the graph G -- hence the term *background*.

In [1,2], Cruse's embedding technique for partial commutative Latin squares [5] is adapted to show that

Theorem 2.1: *For each* $r \geq 0$ *and each* r*-regular* n*-vertex graph* G*, there is a Latin background* LB[G;m,m] *for every even* $m \geq 2n$*. Furthermore, one can be produced in time bounded by a polynomial in* m.

Latin backgrounds are partial commutative Latin squares. In fact, a Latin background for a r-regular graph G can be completed (with no additional rows and columns) to a Latin square if and only if G is r-edge-colourable. Holyer [6] has shown that deciding whether an arbitrary cubic graph is 3-edge-colourable is NP-complete, and Levin and Galil [9] have generalized this result to r-edge-colourability for all fixed $r \geq 3$. Latin backgrounds are used to translate these graph theoretic results into the domain of combinatorial design theory.

3. DECOMPOSABILITY OF TRIPLE SYSTEMS

Our reduction in [3] is patterned closely after the standard recursive $v \to 2v+1$ construction for *Steiner triple systems*, or B[3,1;v] designs (see, for example, [11]), which employs a commutative Latin square of order $v+1$. We present the reduction here, modified to circumvent the use of repeated blocks:

Theorem 3.1: *Deciding whether an* NB[3,λ;v] *design contains a* B[3,1;v] *design is* NP*-complete.*

Proof

Membership in NP is immediate - a nondeterministically chosen sub-B[3,1;v] can easily be verified in polynomial time. To show completeness, we reduce the known NP-complete problem of r-edge-colourability of r-regular graphs to our problem. Given an arbitrary n-vertex r-regular graph G, we first determine a size for a Latin background for G. When $2n-1 \equiv 1,3 \pmod 6$, we set $v = 2n-1$; otherwise we set $v = 2n+1$. Using Theorem 2.1, we next construct a Latin background LB[G;v+1,v+1], called L, in polynomial time. We produce r disjoint Latin backgrounds $L_1,\ldots,L_r$ by repeatedly applying the permutation $(1,2,\ldots,r)(r+1,\ldots,v)(v+1)$ to the elements of L.

Using these Latin backgrounds, we construct an NB[3,3;2v+1]BD with elements $\{x_1,\ldots,x_v,y_1,\ldots,y_{v+1}\}$. The blocks of BD are as follows:

(1) On the elements $\{x_1,\ldots,x_v\}$, place r disjoint Steiner triple systems. Such systems exist (at least) for all $v > 12r$, $v \equiv 1,3 \pmod 6$

(see [12] and references therein)

(2) Let $1 \leq i < j \leq v+1$, and let the (i,j) entry of one of the Latin background be k. Include the block $\{x_k, y_i, y_j\}$.

(3) Let $1 \leq i < j \leq v+1$, and let the (i,j) entry of the Latin background L be empty. Include the blocks $\{x_1, y_i, y_j\}$, $\{x_2, y_i, y_j\}$, and $\{x_3, y_i, y_j\}$ each once.

That the set of triples so defined forms an NB[3,3;2v+1] is easily verified, and this design is constructed in polynomial time. To establish NP-completeness, then, we need only show that BD has a sub-B[3,1;2v+1] if and only if G is r-edge-colourable; further, this depends only on the triples of type (3) above.

Suppose we have an r-edge-colouring of G. To find a sub-B[3,1;2v+1], we include the triples $\{\{x_k, y_i, y_j\} | \{y_i, y_j\}$ has colour $k\}$. Together with one of each of the disjoint Steiner triple systems, and the triples arising from one of the disjoint Latin backgrounds, this constructs a B[3,1;2v+1].

In the other direction, suppose BD has a B[3,1;2v+1]. In this B[3,1;2v+1], the pairs appearing with x_1 (similarly, with x_2 and so on) form a 1-factor of G. Moreover, these 1-factors are all disjoint, and hence cover all edges of G. Since there are r disjoint 1-factors, and (rv)/2 edges in an r-regular graph, the 1-factors comprise an r-edge-colouring of G, as required.

Theorem 3.1 shows that deciding decomposability of NB[3,3;v] designs is NP-complete. However, it does not establish this for any $\lambda \geq 4$; in the case $\lambda = 4$, for example, an NB[3,4;v] may decompose into two NB[3,2;v] designs. Note that a design NB[3,λ;v] constructed by the process in Theorem 3.1 contains an NB[3;λ';v] if and only if the original λ-regular graph contains a λ'-factor. Konig [7] has shown that whenever λ is odd, there are λ-regular graphs containing no regular factors. Applying the construction in Theorem 3.1 to these graphs produces indecomposable NB[3,λ;v] designs for every odd λ. Together with the constructions in [4], this yields many infinite families of indecomposable triple systems with arbitrary odd λ.

For even λ, the construction does not have such immediate applicability. Petersen [10] has shown that every regular graph of even degree can be edge-partitioned into 2-factors; thus, all designs with even λ produced by the construction will be decomposable.

4. The Case $\lambda = 4$

The complexity of decomposition of triple systems with $\lambda = 4$, and indeed the existence of indecomposable NB[3,4;v] designs, cannot be dealt with directly using the construction in Theorem 3.1. Nonetheless, by other techniques, indecomposable NB[3,4;v] designs have been shown to exist for all $v \equiv 0,1 \pmod 3$, $v \geqq 10$ [4]. We settle the complexity question here, by showing that decomposing NB[3,4;v] designs is also NP-complete. We have not shown NP-completeness for arbitrary λ, for reasons which will become clear. Nonetheless, our result suggests that one should not expect a disparity between the complexity for even and for odd orders.

We devise a construction which transforms an NB[3,4;v] design into an NB[3,4;w] design which is decomposable if and only if it contains a B[3,1;w]. This reduces the NP-complete problem of determining the presence of a B[3,1;w] to decomposition. In this construction, we employ the observation that the unique NB[3,4;7], although decomposable, does not contain an NB[3,2;7]; thus the only decomposition can be into a B[3,1;7] and an NB[3,3;7]. Embedding this NB[3,4;7] in an NB[3,4;v] ensures that the NB[3,4;v] is decomposable if and only if it contains a B[3,1;v]. This embedding is done in the following

Theorem 4.1: *Deciding whether an* NB[3,4;v] *design is decomposable is NP-complete.*

Proof.

Let (V,B) be the unique B[3,1;7] (the Fano plane), and let (V′,B′) be an NB[3,4;v] design constructed as in the proof of Theorem 3.1. We construct an NB[3,4;7v] design which is decomposable if and only if (V′,B′) contains a B[3,1;v]. The NB[3,4;7v] design has element set $V \times V'$, and the following blocks:

1. if $\{i,j,k\} \in B$, and $\{w,x,y\} \in B'$, $\{(i,w,),(j,x),(k,y)\}$, $\{(j,w),(i,x),(k,y)\}$, $\{(i,w,(k,x),(j,y)\}$, $\{(k,w),(i,x),(j,y)\}$, $\{(j,w),(k,x),(i,y)\}$, and $\{(k,w),(j,x),(i,y)\}$ are blocks.
2. for $i \in V$ and $\{w,x,y\} \in B'$, $\{(i,w),(i,x),(i,y)\}$ is a block.
3. Let (V,B'') be the unique NB[3,4;7]. For $w \in V'$ and $\{i,j,k\} \in B''$, $\{(i,w),(j,w),(k,w)\}$ is a block.

Any B[3,1;v] in (V′,B′) immediately produces a sub-B[3,1;7v]. However, the NB[3,4;7v] cannot contain an NB[3,2;7v] since the NB[3,4;7] does not contain an NB[3,2;7]. Thus the NB[3,4;7v] is decomposable if and only if it contains a B[3,1;7v], and by Theorem

3.1, this decision problem is NP-complete.

Note that the proof of Theorem 4.1 can easily be adapted to produce many indecomposable NB[3,4;v] designs. One simply takes any 4-regular graph which is not 4-edge-colourable; produces a Latin background for it; applies the construction of Theorem 3.1 to produce an NB[3,4;v] and finally applies the construction in Theorem 4.1 to produce an indecomposable NB[3,4;7v].

This construction does not apply to larger λ, since we are lacking a "gadget" to serve the role of the NB[3,4;7] in the proof. However, we fully expect that for any fixed $\lambda > 2$, deciding decomposability of NB[3,4;v] designs is NP-complete.

ACKNOWLEDGEMENTS

Thanks are due to Derek Corneil, Eric Mendelsohn, and Alex Rosa for their valuable comments.

REFERENCES

[1] C.J. Colbourn, Some NP-complete problems on graph decompositions, *Proc. Nineteenth Allerton Conference on Communications, Control, and Computing*, (1981), 741-745.

[2] C.J. Colbourn, Embedding partial Steiner triple systems is NP-complete, *J. Combinatorial Theory A*, to appear.

[3] C.J. Colbourn and M.J. Colbourn, The computational complexity of decomposing block designs, submitted for publication.

[4] C.J. Colbourn and A. Rosa, Indecomposable triple systems with $\lambda = 4$, submitted for publication.

[5] A.B. Cruse, On embedding incomplete symmetric Latin squares, *J. Comb. Theory A*, 16 (1974), 18-22.

[6] I. Holyer, The NP-completeness of edge colouring, *SIAM J. Computing* 10 (1981), 718-720.

[7] D. König, *Theorie der Endlichen und Unendlichen Graphen*, Teubner, Leipzig, (1936).

[8] E.S. Kramer, Indecomposable triple systems, *Discrete Math.* 8 (1974) 173-180.

[9] D. Levin and Z. Galil, NP-complete problem no. 798016, submitted for publication.

[10] J. Petersen, Die Theorie der regularen Graphen, *Acta Math.* 15 (1891), 193-220.

[11] A. Rosa, Algebraic properties of designs and recursive constructions *Proc. Conf. Algebraic Aspects of Combinatorics*, Utilitas Math. Publishing Co., Winnipeg, Canada, (1975) 183-202.

[12] A. Rosa, Intersection properties of Steiner systems, *Annals of Discrete Math.* 7 (1980) 115-128.

A COMBINATORIAL PROBLEM AND THE GENERALIZED COSH

WILLIAM H. CORNISH

For $j = 1, \ldots, r$, *let* h_j *and* k_j *be integers such that* $0 \le h_j \le k_j - 1$ *and* $\omega_j = \exp\left(\frac{2\pi i}{k_j}\right)$. *Then, the number of ways of placing* $n \ge 0$ *different balls into* r *distinct cells such that, for* $j = 1, \ldots, r$, *the number of balls in the* j^{th} *cell is congruent to* h_j *modulo* k_j, *is*

$$\left(\prod_{j=1}^{r} k_j\right)^{-1} \sum_{\substack{s_1,\ldots,s_r \\ 0\le s_j\le k_j-1}} \left(\prod_{j=1}^{r} \omega_j^{-h_j s_j}\right)\left(\sum_{j=1}^{r} \omega_j^{s_j}\right)^n .$$

The proof is by means of the exponential enumerator and employs the generalized cosh: $C_k(x) = \sum_{n=0}^{\infty} \frac{x^{kn}}{(kn)!}$.

1. INTRODUCTION

A classical use of exponential generating functions is the calculation of the number of possible placements of n different balls into, say three distinct cells with regard to restrictions such as the number placed into the first cell must be even, the number placed into the second cell must be odd, while the number to be placed in the third cell is unspecified. Alternatively, we may wish to know the number of n-place sequences made up from an alphabet of A's, B's and C's so that the number of A's is even, the number of B's is odd and there are no restrictions on the number of C's. Many such problems are dealt with by Liu [5; Ch.2] and Riordan [11; Ch.5]. A variation, which is not dealt with, occurred when considering the number of ways of placing n different balls into two distinct cells so that the number in the first cell is always a multiple of three. The solution of this problem depends upon expressing the series $\sum_{n=0}^{\infty} \frac{x^{3n}}{(3n)!}$ as a linear combination of exponential functions.

Here we are concerned with a common generalization of these problems. As a consequence, we were led to the rediscovery of the so-called higher-order hyperbolic functions. Comments on these functions, together with an updated bibliography are given in Section 2 and the References.

2. THE GENERALIZED COSH

It will make no difference whether the series which are encountered are regarded as defining entire functions of a complex varaible or as formal power series over the field of complex numbers.

For any positive integer k, we define the *k-fold cosh* by

$$C_k(x) = \sum_{n=0}^{\infty} \frac{x^{kn}}{(kn)!} .$$

Define $C_k^{(0)}(x) = C_k(x)$, and $C_k^{(h)}(x)$ as the hth derivative of $C_k(x)$, for any positive integer h. Then,

$$C_k^{(1)}(x) = \sum_{n=1}^{\infty} \frac{x^{kn-1}}{(kn-1)!} = \sum_{n=0}^{\infty} \frac{x^{kn+(k-1)}}{(kn+(k-1))!} .$$

Hence, $C_k^{(2)}(x) = \sum_{n=0}^{\infty} \frac{x^{kn+(k-2)}}{(kn+(k-2))!}$, ... , $C_k^{(k-1)}(x) = \sum_{n=0}^{\infty} \frac{x^{kn+1}}{(kn+1)!}$,

and $C_k^{(k)}(x) = C_k(x)$. This accounts for the interest in these functions within the literature. Kaufman [3] gave an exhaustive bibliography on these generalized hyperbolic functions and their alterating cousins, the generalized cosine and sine functions, from Riccati (1757) through to Carlitz (1955), and listed 106 references. These functions are also discussed in the treatise of Erdélyi *et alia* [2; pp.212-217]. An additional 7 papers have been located from the Special Functions section of Mathematical Reviews; for the sake of completeness, they and their reviews, have been listed in the References. The literature is repetitive but it would seem that no paper is concerned with an application to Combinatorics.

We need the following result, wherein exp(x) denotes the exponential series $\sum_{n=0}^{\infty} \frac{x^n}{n!}$. It is well known and is a particular case of the so-called multisection of series, as given by Riordan [12; Ch.4, Section 4.3, p.131].

Lemma. *Let* h *and* k *be integers, with* $0 \le h \le k-1$, *and* $\omega = \exp\left(\frac{2\pi i}{k}\right)$. *Then,*

$$C_k^{(k-h)}(x) = \sum_{n=0}^{\infty} \frac{x^{kn+h}}{(kn+h)!} = \frac{1}{k}\sum_{s=0}^{k-1} \omega^{-hs} \exp(\omega^s x).$$

Proof. By splitting up each summand of $\sum_{s=0}^{k-1} \exp(\omega^s x)$ into the sum of k series whose n^{th} powers of x are x^{kn} through to $x^{kn+(k-1)}$, respectively, we find upon simplification that $\sum_{s=0}^{k-1} \exp(\omega^s x) = kC_k(x) + \left(\sum_{s=0}^{k-1} \omega^s\right) \sum_{r=1}^{k-1} \left(\sum_{n=0}^{\infty} \frac{x^{kn+r}}{(kn+r)!}\right)$. Hence, $C_k(x) = \frac{1}{k}\sum_{s=0}^{k-1} \exp(\omega^s x)$ and the remainder follows from differentiation.

3. COMBINATORICS

Theorem. *For* $j = 1, \ldots, r$, *let* h_j *and* k_j *be integers such that* $0 \le h_j \le k_j - 1$, *and* $\omega_j = \exp\left(\frac{2\pi i}{k_j}\right)$. *For any integer* $n \ge 0$, *let* a_n *be the number of ways of placing* n *different balls into* r *distinct cells such that, for* $j = 1, \ldots, r$, *the number of balls placed into the* j^{th} *cell is congruent to* h_j *modulo* k_j. *Then,*

$$a_n = \left(\prod_{j=1}^{r} k_j\right)^{-1} \sum_{\substack{s_1,\ldots,s_r \\ 0\le s_j\le k_j-1}} \left(\prod_{j=1}^{r} \omega_j^{-h_j s_j}\right)\left(\sum_{j=1}^{r} \omega_j^{s_j}\right)^n .$$

This is because the exponential generating function of the sequence $\{a_n\}_{n=0}^{\infty}$ *is given by*

$$\sum_{n=0}^{\infty} a_n \frac{x^n}{n!} = \prod_{j=1}^{r} C_{k_j}^{(k_j-h_j)}(x) .$$

Proof. Let $N = \{0,1,2, \ldots\}$ be the set of natural numbers and $P_n = \{(n_1, \ldots, n_r) \in N^r : n = n_1 + \ldots + n_r, n_j \equiv h_j (\mathrm{mod}\ k_j), \text{ for } j = 1, \ldots, r\}$.

For $(n_1, \ldots, n_r) \in P_n$, the number of ways of placing n different balls into the r cells so that, for $j = 1, \ldots, r$, there are n_j balls in the jth cells is

$\frac{n!}{n_1! \ldots n_r!}$.

Hence, $a_n = \sum_{(n_1,\ldots,n_r)\in P_n} \frac{n!}{n_1! \ldots n_r!}$, and, of course, $a_n = 0$ if $P_n = \emptyset$.

Thus, $\sum_{n=0}^{\infty} a_n \frac{x^n}{n!} = \sum_{n=0}^{\infty} \left(\sum_{(n_1,\ldots,n_r)\in P_n} \frac{1}{n_1! \ldots n_r!} \right) x^n = \prod_{j=1}^{r} \left(\sum_{\substack{n_j \in N \\ n_j \equiv h_j \\ (\mathrm{mod}\ k_j)}} \frac{x^{n_j}}{n_j!} \right) =$

$\prod_{j=1}^{r} \left(\sum_{n=0}^{\infty} \frac{x^{k_j n + h_j}}{(k_j n + h_j)!} \right)$. That is, $\sum_{n=0}^{\infty} a_n \frac{x^n}{n!} = \prod_{j=1}^{r} C_{k_j}^{(k_j - h_j)}(x)$.

Using the lemma of Section 2 and transforming a product of sums into a sum of products via the distributive law, we obtain

$$\sum_{n=0}^{\infty} a_n \frac{x^n}{n!} = \prod_{j=1}^{r} \left(\frac{1}{k_j} \sum_{s_j=0}^{k_j - 1} \omega_j^{-h_j s_j} \exp(\omega_j^{s_j} x) \right)$$

$$= \left(\prod_{j=1}^{r} k_j \right)^{-1} \sum_{\substack{s_1,\ldots,s_r \\ 0 \le s_j \le k_j - 1}} \left(\prod_{j=1}^{r} \omega_j^{-h_j s_j} \right) \exp\left(x \sum_{j=1}^{r} \omega_j^{s_j} \right).$$

By equating the coefficients of $\frac{x^n}{n!}$, we obtain the formula for a_n.

Sometimes the restrictions are so severe that the number of possible placements is 0. The next three corollaries are concerned with this.

<u>Corollary 1</u>. *For* $j = 1, \ldots, r$, *let* h_j, k_j *and* ω_j *be as given in the Theorem. Let* $n \ge 0$ *be a given integer. Then, there are no non-negative integers* $n_1, \ldots, n_r$ *such that* $n = n_1 + \ldots + n_r$ *and* $n_j \equiv h_j (\mathrm{mod}\ k_j)$ for each $j = 1, \ldots, r$, *if and only if*

$$\sum_{\substack{s_1,\ldots,s_r \\ 0 \le s_j \le k_j - 1}} \left(\prod_{j=1}^{r} \omega_j^{-h_j s_j} \right) \left(\sum_{j=1}^{r} \omega_j^{s_j} \right)^n = 0.$$

<u>Proof</u>. This is because $P_n = \emptyset$ if and only if $a_n = 0$.

<u>Corollary 2</u>. *For* $j = 1, \ldots, r$, *let* h_j, k_j *and* ω_j *be as given in the Theorem. Suppose the* k_j *are pairwise coprime. Then, there exists a positive integer* m *such that*

$$\sum_{\substack{s_1,\ldots,s_r \\ 0\le s_j\le k_j-1}} \left(\prod_{j=1}^{r} \omega_j^{-h_j s_j}\right)\left(\sum_{j=1}^{r} \omega_j^{s_j}\right)^m$$

is a positive integer, which is divisible by $\prod_{j=1}^{r} k_j$.

Proof. When the k_j are pairwise coprime, there is always a positive integer m such that $m \equiv h_j \pmod{k_j}$ for each $j = 1, \ldots, r$. Moreover, $m = \sum_{j=1}^{r} m_j h_j$, where the m_j are positive integers, and $m_s \equiv \delta_{sj} \pmod{k_j}$, where $\delta_{sj} = 0$ if $s \neq j$ and $\delta_{sj} = 1$ if $s = j$. This is a consequence of the Chinese Remainder Theorem, and its proof; see, for example, Niven and Zuckerman [8; Theorem 2.14, p.31]. The rest follows from the Theorem.

In the next corollary, the k_j are taken to be ≥ 2 so that the result is not trivial.

Corollary 3. *For* $j = 1, \ldots, r$, *let* h_j, k_j *and* ω_j *be as in the Theorem. Suppose* $k_j \geq 2$ *for any* $j = 1, \ldots, r$. *Then, for all* $m \geq \sum_{j=1}^{r} h_j$,

$$\sum_{\substack{s_1,\ldots,s_r \\ 0\le s_j\le k_j-1}} \left(\prod_{j=1}^{r} \omega_j^{-h_j s_j}\right)\left(1 + \sum_{j=1}^{r} \omega_j^{s_j}\right)^m$$

is a positive integer, which is divisible by $\prod_{j=1}^{r} k_j$.

Proof. Introduce an $(r+1)^{st}$ cell with the restrictions $k_{r+1} = 1$ and $h_{r+1} = 0$, whereby $\omega_{r+1} = 1$. Thus, we are considering the placement of m different balls into $(r+1)$ distinct cells so that the restrictions on the first r cells are as in the Theorem, and the number to be placed into the $(r+1)^{th}$ cell is unspecified. Provided that $m \geq \sum_{j=1}^{r} h_j$, permissible placements are always possible. The rest follows from the Theorem.

We conclude with an example. Let us take the case where $r = 2$, $k_1 = 3$, $h_1 = h$, $k_2 = 1$, $h_2 = 0$, $\omega_1 = \omega = \exp\left(\frac{2\pi i}{3}\right) = -\frac{1}{2} + \frac{\sqrt{3}}{2} i$, and $\omega_2 = 1$. Then, $1 + \omega = \rho = \exp\left(\frac{\pi i}{3}\right)$, $1 + \omega^2 = \overline{\rho}$. Hence,

$$a_n = \frac{1}{3}\sum_{s=0}^{2} \omega^{-hs}(1+\omega^s)^n$$

$$= \frac{1}{3}\left(2^n + \omega^{-h}\rho^n + \omega^{-2h}\bar{\rho}^n\right)$$

$$= \frac{1}{3}\left(2^n + \omega^{2h}\rho^n + \overline{(\omega^{2h}\rho^n)}\right)$$

$$= \frac{1}{3}\left(2^n + 2\,Re\,(\omega^{2h}\rho^n)\right)$$

$$= \frac{1}{3}\left\{2^n + 2\cos\left[\frac{(n+4h)\pi}{3}\right]\right\}$$

But also, $a_n = 0$ for $0 \le n < h$, while for $n \ge h$ it is easy to see that $a_n = \sum_{s=0}^{N}\binom{n}{3s+h}$, where $N = \left[\frac{n-h}{3}\right]$ is the integer-part of $\frac{n-h}{3}$.

We summarize this in the following proposition.

Proposition. *Let* n *and* h *be non-negative integers* with $h \le 2$. *Then, the number of ways of placing* n *different balls into two distinct cells so that the number in the first cell is always congruent to* h *modulo* 3 *is*

$$\frac{1}{3}\left\{2^n + 2\cos\left[\frac{(n+4h)\pi}{3}\right]\right\} = \begin{cases} 0 & \text{if } 0 \le n < h \\ \sum_{s=0}^{N}\binom{n}{3s+h} & \text{if } h \le n, \end{cases}$$

where $N = \left[\frac{n-h}{3}\right]$.

REFERENCES

[1] G. Battini, Su una generalizzazione delle funzioni iperboliche e delle funzioni circolari, *Riv. Mat. Univ Parma* (2) 10(1969), 39-48; *MR* 45#599.

[2] A. Erdélyi, W. Magnus, F. Oberhettinger, and F.G. Tricomi, *Higher Transcendental Functions, Vol.* 3 (McGraw-Hill, New York, 1955).

[3] H. Kaufman, A bibliographical note on higher order sine functions, *Scripta Math.* 28(1967), 29-36.

[4] H. Kaufman, A generalization of the sine function, *Amer. Math. Monthly* 64(1957), 181-183; *MR* 19-29.

[5] C.L. Liu, *Introduction to Combinatorial Mathematics* (McGraw-Hill, New York, 1968).

[6] J. Mikusiński, The trigonometry of the differential equation $x''' + x = 0$ (Polish), *Wiadom. Mat.* (2)2(1959), 207-227; *MR* 22#5749.

[7] J. Mikusiński, The trigonometry of the differential equation $x^{(4)} + x = 0$ (Polish), *Wiadom. Mat.* (2)4(1960), 73-84; *MR* 22#5750.

[8] I. Niven and H. Zuckerman, *An Introduction to the Theory of Numbers* (J. Wiley, New York, 1960).

[9] L.A. Pipes, Cyclical functions and permutation matrices, *J. Franklin Inst.* 287(1969), 285-296; *MR* 39#7148.

[10] L.A. Pipes, Cyclical functions and permutation matrices, *Matrix Tensor Quart.* 20(1970), 99-111; *MR* 42#4782.

[11] J. Riordan, *An Introduction to Combinatorial Analysis* (J. Wiley, New York, 1958).

[12] J. Riordan, *Combinatorial Identities* (J. Wiley, New York, 1968).

[13] A.G. Shannon, Arbitrary order circular functions: an extension of results of Glaisher and Lucas, *J. Natur. Sci. Math.* 19(1979), 71-76; *MR* 82(a): 33002.

GENERALISED HADAMARD MATRICES WHOSE ROWS AND COLUMNS FORM A GROUP

WARWICK DE LAUNEY

We define a group Hadamard matrix to be a generalised Hadamard matrix whose rows form a group. We show that for abelian groups G*, only group Hadamard matrices of type* p^s *for* $C_p\times\ldots\times C_p$ *exist. We also show that the matrices for* C_p *of each possible order are unique up to equivalence.*

We indicate a connection between strongly independent sets and row group Hadamard matrices.

We show that the irreducible factors of a group Hadamard matrix under tensor product are unique up to equivalence. This allows us to count and classify the group Hadamard matrices for $C_p\times C_p$*.*

1. DEFINITIONS AND NOTATION

Throughout this paper by G^n will be meant the direct product of n copies of the group G, $G^n = G\times\ldots\times G$.

A vector of length n, whose entries are taken from a group G can be regarded as an element of G^n. $\underset{\sim}{r} * \underset{\sim}{h}$ is the element of G^n obtained by multiplying $\underset{\sim}{r}$ and $\underset{\sim}{h}$ in G^n. This product is consistent with the usual Hadamard product. Two vectors $\underset{\sim}{r}$ and $\underset{\sim}{h}$ are s-orthogonal if $\underset{\sim}{r} * \underset{\sim}{h}^{-1}$ is s listings of G, $s \in \mathbb{N}$.

A (generalised) Hadamard matrix of type s *for a group* G is a square matrix whose entries come from G and whose rows are mutually s-orthogonal.

A *group Hadamard matrix* is a (generalised) Hadamard matrix whose rows form a group and whose columns form a group under the operation $*$.

Some pronumerals will only have the following meanings.

H is a Hadamard matrix

s the type of H

G the group for which H is Hadamard

h_i the ith row of H.

If H has the extra structure of a group Hadamard matrix then R is the group of rows and C the group of columns.

R_H is the group of rows of the group Hadamard matrix H.

Example 1.1. A well known generalised Hadamard matrix is Drake's matrix [1]. Given in additive form Drake's matrix is

$$D = \begin{pmatrix} 0 & \cdots & 0 \\ \vdots & & \\ \vdots & & X \\ 0 & & \end{pmatrix}$$

where

$$X = \left(x^{j-i+1 (\mathrm{mod}\ p^r - 1)}\right)$$

and x is a primitive element of $GF(p^r)$.

Note every row apart from the first contains every element of $GF(p^r)$. So if D is generalised Hadamard, then G is the additive group of $GF(p^r)$ and $s = 1$. In fact we need only check whether the rows form a group because then the rows would be 1-orthogonal. Each row has a finite order, so we need only check that the set of rows is closed under summation. The first row is the identity. If we sum the i_0th and i_1th row we obtain the vector $\left(0, x^{j-i_2+1 \pmod{p^r-1}}\right)$, where $x^{i_2} = x^{i_0} + x^{i_1}$. So D is group Hadamard.

The following lemma gives us group Hadamard matrices of type p^t for C_p^v, $t \geq v \geq 0$. We will show that for G abelian these are the only group Hadamard matrices.

Lemma 1:1 *If H is a generalised Hadamard matrix of type s for G, $\phi: G \to S$ an onto homomorphism and $\phi(H) = (\phi(h_{ij}))$ where $H = (h_{ij})$ then $\phi(H)$ is a generalised Hadamard matrix for S of type $s|G|/|S|$. Provided $S \neq \{1\}$, if H is group Hadamard then $\phi(H)$ is also group Hadamard.*

Proof: We need to show the rows of $\phi(H)$ are mutually $s|G|/|S|$-orthogonal. Now

$$\left(\phi(h_{i_0 1}), \phi(h_{i_0 2}), \ldots, \phi(h_{i_0 s|G|})\right) * \left(\phi(h_{i_1 1}), \ldots, \phi(h_{i_1 s|G|})\right)^{-1} =$$

$$= \left(\phi(h_{i_0 1} h_{i_1 1}^{-1}), \ldots, \phi(h_{i_0 s|G|} h_{i_1 s|G|}^{-1})\right)$$

This vector has each element of S appearing $s|G|/|S|$ times.

Now if the rows of H are closed under $*$ then the rows of $\phi(H)$ will be also, and if $S \neq \{1\}$ then it follows the rows are all different (otherwise they would not be orthogonal) so $\phi(H)$ is also group Hadamard. □

First we discuss the equivalence of group Hadamard matrices for an abelian group G. Two *generalised Hadamard* matrices for an *abelian* group G are said to be *equivalent* if one can be obtained from the other by *permuting* the rows and columns and *multiplying* rows or columns by *elements* of G. We write $H \sim K$.

Let P_i be a permutation matrix of suitable order and Q_i a matrix whose non-zero entries are taken from G. Q_i has the extra property that each row and column contains exactly one non-zero entry. So if $H \sim K$ then there exist Q_1, Q_2 such that $H = Q_1 K Q_2$.

Lemma 1.2: If P is a permutation matrix and H is group Hadamard then $K = HP$ and $L = PH$ are also group Hadamard.

Proof: Consider $L = PH$. Let $H = (\underset{\sim}{h}_i)_{i=1,\dots,n}$ then $L = (\underset{\sim}{h}_{\sigma(i)})_{i=1,n}$ where σ is the permutation corresponding to P. So the set of rows of L = set of rows of H, and hence the rows of L form a group.

The columns of L also form a group because

$$(h_{1j}\dots h_{nj}) * (h_{1k}\dots h_{nk}) = (h_{1i}\dots h_{ni}) \iff$$

$$(h_{\sigma(1)j},\dots,h_{\sigma(n)j}) * (H_{\sigma(1)k}\dots h_{\sigma(n)k}) = (h_{\sigma(1)i}\dots h_{\sigma(n)i}) .$$

Similarly K is group Hadamard. □

A *normalised* matrix whose entries are taken from a group G is a matrix whose entries in the *first row and column* are all *ones*, where *one* is the group identity.

Lemma 1.3: Suppose H is a normalised Hadamard matrix and L a group Hadamard matrix for an abelian group G and that there exist Q_1, Q_2 such that $H = Q_1 L Q_2$. Then H is group Hadamard and there exist P_1, P_2 such that $H = P_1 L P_2$.

Proof: Because L is group Hadamard it contains a row and column of 1's. So both H and L contain a row and column of 1's. So $Q_1 L$ must contain a column of g_0's, some $g_0 \in G$, since the columns of $Q_1 L$ are multiples of the columns of $Q_1 L Q_2$. Replace Q_1 by $g_0^{-1} Q_1$ and Q_2 by $g_0 Q_2$ then $Q_1 L$ contains a column of 1's. So multiplication by Q_1 must be equivalent to multiplication of the columns of L by the inverse of a column of L followed by a permuting of the rows. L is group Hadamard so $Q_1 L = P_3 L P_4$. Hence $Q_1 L$ is group Hadamard.

Let $K = Q_1 L$ then by a similar argument there exists P_5, P_6 such that $K Q_2 = P_5 K P_6$. So $H = K Q_2$ is group Hadamard and $H = P_5 K P_6 = P_5 P_3 L P_4 P_6 = P_1 L P_2$. □

Theorem 1.1: (a) *Let* H *and* L *be equivalent group Hadamard matrices for an abelian group* G. *Then* L *can be obtained from* H *by permuting rows and columns.*

(b) *Suppose* K *is equivalent to a group Hadamard matrix for an abelian group* G. *Further suppose* L *is obtained by normalising* K. *Then* L *is group Hadamard.*

Proof. (a) There exist Q_1, Q_2 such that $H = Q_1 L Q_2$. L is group Hadamard so applying Lemma 1.3 there exist P_1, P_2 such that $H = P_1 L P_2$.

(b) Let $K \sim H$ where H is group Hadamard then since $L \sim K$, $L \sim H$ and since L is normalised there exist P_1, P_2 such that $L = P_1 H P_2$. So L is group Hadamard. □

This theorem gives an easy test for determining whether a generalised Hadamard matrix for an abelian group, is equivalent to a group Hadamard matrix since we only need to normalise the matrix and see if the resulting matrix is group Hadamard.

2. EXISTENCE OF GROUP HADAMARD MATRICES

We now determine existence conditions for group Hadamard matrices. First we introduce a different kind of Hadamard matrix. A *row group Hadamard matrix of type* s *for a group* G is a generalised Hadamard matrix of type s for G whose *rows form a group*. To prove lemma 2.2 we need the fact that the transpose of a generalised Hadamard matrix is a generalised Hadamard matrix [2].

When not otherwise stated all Hadamard matrices will be normalised.

Lemma 2.1: *Suppose* K *(not necessarily Hadamard) is normalised, and that the rows of* K *form a group* R. *Suppose also that each row is* s *listings of a group* G, *then* K *is row group Hadamard of type* s *for* G.

Proof. Let $\underset{\sim}{k}_i$ be the ith row of K, $1 \le i \le |R|$.

We note that if $i \neq j$ then $\underset{\sim}{k}_i \neq \underset{\sim}{k}_j$ and hence $\underset{\sim}{k}_i * \underset{\sim}{k}_j^{-1} \neq (1,\ldots,1)$. So $\underset{\sim}{k}_i * \underset{\sim}{k}_j^{-1} = \underset{\sim}{k}_m \in R-\{1\}$. Since $\underset{\sim}{k}_m$ is s listings of G $\underset{\sim}{k}_i$ and $\underset{\sim}{k}_j$ are s-orthogonal. Hence K is row group Hadamard of type s for G. □

Lemma 2.2: *Suppose* H *is a row group Hadamard matrix for* G *of type* s *then the columns of* H *give* $|R|-1$ *onto homomorphisms* $\phi_i : R \to G$ *such that for all* $r \in R-\{1\}$ and $g \in G$, $\phi_i(r) = g$ *precisely* s *times.*

Conversely if we have such a set of $|R|-1$ *homomorphisms for* R *and* $G \neq \{1\}$ *with* $|R| = s|G|$ *then there exists a row group Hadamard matrix of type* s *for* G *whose rows form the group* R.

Proof. Let $H = (h_{ij}) = (\underset{\sim}{h}_i)$. Define $\phi_j : R \to G$ by $\phi_j(\underset{\sim}{h}_i) = h_{ij}$, $2 \le j \le |R|$, $i \ne 1$. The fact that R is a group ensures

$$\phi_j(\underset{\sim}{h}_{i_0} * \underset{\sim}{h}_{i_1}^{-1}) = h_{i_0 j} h_{i_1 j}^{-1} = \phi_j(\underset{\sim}{h}_{i_0})\, \phi_j^{-1}(\underset{\sim}{h}_{i_1}) .$$

Because H^T is Hadamard these homomorphisms must all be different and onto. The incidence property follows from the fact that, because H is normalised, each row of H except the first is s listings of G.

Now, for the converse, let $R = \{1, r_2, r_3, \ldots, r_{|R|}\}$ and define

$$h_{ij} = \begin{cases} 1 & ; \quad i = 1 \text{ or } j = 1 \\ \phi_j(r_i) & ; \quad 2 \le i,j \le |R| \end{cases}$$

then $\underset{\sim}{h}_i * \underset{\sim}{h}_k^{-1} = \left(\phi_j(r_i)\, \phi_j(r_k^{-1})\right)_j = \left(\phi_j(r_i r_k^{-1})\right)_j = \left(\phi_j(r_t)\right)_j = \underset{\sim}{h}_t$.

Hence the rows of H form a group, R'. Now the map $r \to \left(\phi_1(r), \phi_2(r), \ldots, \phi_n(r)\right)$ gives an isomorphism $R \approx R'$ because $|R| = s|G| = |R'|$. The matrix H is normalised and for $r \in R - \{1\}$, $\left(\phi_j(r)\right)$ is s listings of G so we may apply lemma 2.1. to see that H is a row group Hadamard matrix for G of type s whose rows form the group R. □

These conditions on R and G are very restrictive and allow us to obtain the first of our results.

Theorem 2.1: There exists a row group Hadamard matrix of type 1 *for* G *if and only if* $G = C_p \times \cdots \times C_p$.

Proof. We apply lemma 2.2 to get $|R|-1$ isomorphisms $\phi_i : R \to G$. In fact we have $|G|-1$ automorphisms such that for all $g \in G - \{1\}$, $\phi_i(g) \ne \phi_j(g)$ for all $i \ne j$. So we see that every non-identity element must be sent to every other non-identity element, since $\phi_j(g) \neq 1$ for all j. So the orders of every element apart from 1 are the same. Hence the order of every non-identity element in G must be p (where p is prime).

The only abelian groups with this property are C_p^m. Now if G is non abelian, because it is a p-group, its centre is non-trivial. But every automorphism of G fixes the centre so it cannot have the required set of automorphisms. □

To obtain the rest of our results on existence we need the following technical lemma.

Lemma 2.3: Let R *be abelian and suppose there exist* $|R|$ *homomorphisms* $\phi: R \to C_p$ *then* R *is* C_p^m.

Proof. Let $R = \langle s_i \rangle$ $i = 1,\ldots,\nu$ and let s_i have order m_i. Then $|R| = \prod_{i=1}^{\nu} m_i$. Now if $\phi: R \to C_p$ is a homomorphism and $\phi(s_i) \neq 1$, then $p|m_i$.

Let ω be the number of elements in $\{s_i | i = 1,\ldots,\nu\}$ which satisfy $p|m_i$. Now for each s_i such that $p \nmid m_i$ $\phi(s_i) = 1$, and, if $p|m_i$ then there are p possible images. Hence the number of possible homomorphisms is p^ω. So $p^\omega \geq |R|$. But $p^\omega \leq |R| = \prod_{i=1}^{\nu} m_i$, so $p^\omega = |R|$. Consequently $m_i = p$ for all $i = 1,\ldots,\nu$ and $R = C_p^\omega$. □

Let $A = \left(a^{(i-1)(j-1)}\right)$ where $C_p = \langle a \rangle$, then $A^t = A \otimes A \otimes \ldots \otimes A$ {t times} is a group Hadamard matrix of type p^t for C_p. We show that up to equivalence these are the only ones.

Theorem 2.2: Suppose H *is row group Hadamard of type* s *for* C_p *then* $s = p^{m-1}$ *and* R *is* C_p^m. *For* $s = p^{m-1}$ *the matrix exists and is unique.*

Proof. By lemma 2.2 R is a group which has $|R|-1$ homomorphisms from R onto C_p. Because C_p is abelian R is abelian, since $R \leq C_p^{|R|}$. So by lemma 2.3, $R = C_p^m$ and $s = p^{m-1}$. We need only prove uniqueness. Now there are precisely p^m homomorphisms from C_p^m to C_p. If follows that if H is row group Hadamard its columns must be the images of the complete set of homomorphisms from C_p^m to C_p. □

A *perfect group* G is a group such that $G = G'$, the *commutator subgroup* of G.

Theorem 2.3: If H *is a row group Hadamard matrix for a non-perfect group* G *then* $R = C_p^m$ *and* $G = C_p^t$, $m \geq t \geq 1$.

Proof. Let $S = G/G'$ and $\theta: G \to S$ be the coset homomorphism. Now $S \neq \{1\}$ so there exist a prime p such that $p \mid |S|$. So $S = C_{p^n} \times P (P \leq S)$. Now $C_{p^n}/C_{p^{n-1}} \approx C_p$ so we have an onto homomorphism $\psi: S \to C_p$. Composing θ and ψ we obtain an onto homomorphism $\phi: G \to C_p$.

If H is of type s for G then $\phi(H)$ is row group Hadamard of type $s|\ker \phi|$ for C_p. Now $\phi: G \to C_p$ induces a homomorphism $\psi: R_H \to R_{\phi(H)}$ where

$$\psi(\underset{\sim}{h}_i) = (\phi(h_{i1}), \phi(h_{i2}), \ldots, \phi(h_{i|R|})) .$$

Since $|R_H| = |R_{\phi(H)}|$, ψ is an isomorphism. So $R_H = R_{\phi(H)} = C_p^m$.

Now $|\ker \phi| = |G|/p$ and $s|\ker \phi| = p^{m-1}$ so $s|G| = p^m$ and $|G| = p^t$. Finally because $R = C_p^m$, G is abelian and no element of G can have order greater than p, so $G = C_p^t$, $m \geq t \geq 1$. □

We therefore have the theorem:

Theorem 2.4: *Suppose* H *is group Hadamard of type* s *for* G *then*

(i) *if* $s = 1$, $R = C = G = C_p^m$.

(ii) *if* $s > 1$ *and* G *is non-perfect then* $G = C_p^t$ *and* $R = C = C_p^m$
$m \geq t \geq 1$.

We note that by theorem 2.2 the group Hadamard matrix for C_p of type $s = p^t$ is unique up to equivalence and in fact we can say that any row group Hadamard matrix for C_p is group Hadamard. It seems likely that in general there are row group Hadamard matrices which are not group Hadamard. In the next section we will obtain algebraic equivalents for both which will indicate the difference between the two types of matrices.

3. ALGEBRAIC EQUIVALENTS FOR GROUP HADAMARD MATRICES

In view of the results in section 2 we define a *p(m,t)-Hadamard* matrix, for $m \geq t$, to be a *group Hadamard* matrix of *order* p^m *for* C_p^t.

Define a *strongly independent* set of $t \times m$ *linear maps* over GF(P) to be a set $\{A_1, A_2, \ldots, A_s\}$ such that for λ_k, not all zero, $\sum_{k=1}^{s} \lambda_k A_k$ is of *maximal rank*.

Theorem 3.1: (a) *To each* p(m,t)*-Hadamard matrix there corresponds a strongly independent set of* $t \times m$ *linear maps* $D = \{D_1, D_2, \ldots, D_m\}$.

(b) *To each* p(m,t)*-Hadamard matrix there corresponds a set of* t *strongly independent linear maps in* GL(m,p).

Conversely if we have the set in either of (a) *or* (b) *then we can construct the corresponding group Hadamard matrix.*

Proof. (a) We may regard C_p^m as a vector space of dimension m over $GF_{(p)}$. If $C_p^m = \langle a_1,\dots,a_m \rangle$ then identify $a_1^{e_1}.a_2^{e_2}\dots a_m^{e_m}$ with $(e_1,e_2,\dots,e_m)$. Then the homomorphisms in Lemma 2.2 can be represented by linear maps from a vector space of dimension m onto a vector space of dimension t.

Now if for the ith, jth and rth columns $\underset{\sim}{k}_i * \underset{\sim}{k}_j = \underset{\sim}{k}_r$ and A_i, A_j, A_r respectively represent ϕ_i, ϕ_j and ϕ_r then $A_i + A_j = A_r$. Let $L = \{A_i \mid A_i$ represents $\underset{\sim}{k}_i$, $1 \le i \le p^m\}$. Because the columns form the group C_p^m we can find m columns which multiplicatively generate the columns of H. Let $D = \{D_1,\dots,D_m\}$ be the set of matrices corresponding to these columns. We see $\Sigma\, \lambda_k D_k$, λ_k not all zero, corresponds to a column other than the column of 1's, so it is of maximal rank.

(b) Define $A_{ijk} = d_{jk}^i$, for $i = 1,\dots,m$, where

$$D_i = (d_{jk}^i). \quad \left(\text{So } D_i = (A_{ijk})\right).$$

Define $B_j = (A_{ijk})$, $\qquad j = 1,\dots,t$.

To check $\{B_j \mid j = 1,\dots,t\}$ is a set of t strongly independent maps in $GL(m,p)$ we need to show that for λ_j not all zero and $\underset{\sim}{a} \neq \underset{\sim}{0}$,

$$\left(\sum_{j=1}^{t} \lambda_j B_j\right) \underset{\sim}{a} \neq \underset{\sim}{0}.$$

Now $\displaystyle\sum_{j=1}^{t} \lambda_j B_j \underset{\sim}{a} = \sum_{k=1}^{m}\sum_{j=1}^{t} \lambda_j A_{ijk} a_k = \underset{\sim}{\lambda} D_i \underset{\sim}{a}$ where $\underset{\sim}{\lambda} = (\lambda_1,\dots,\lambda_t)$.

We wish to show the m vectors $\underset{\sim}{\lambda} D_i$ are linearly independent. Because $\underset{\sim}{\lambda} \neq \underset{\sim}{0}$ we can find an invertible matrix L with first row $\underset{\sim}{\lambda}$. We note that since L is an invertible $t \times t$ matrix $\{LD_i\}$ is strongly independent and that the vector of length m, $\underset{\sim}{\lambda} D_i$, is the first row of LD_i. So our problem is reduced to showing that if $\{C_i\}$ is a strongly independent set then the set of first rows of the matrices C_i are linearly independent. But this is immediate because $\Sigma\mu_i C_i$ is of maximal rank whenever $\mu_i \neq 0$ some $1 \le i \le m$.

So, if $\underset{\sim}{0} = \sum_{j=1}^{t} \lambda_j B_j \underset{\sim}{a}$, then $\underset{\sim}{v}\cdot\underset{\sim}{a} = 0$ for all $\underset{\sim}{v} \in V_m$ and hence $\underset{\sim}{a} = \underset{\sim}{0}$.

We note that if $\{A_i\}$ is a strongly independent set then $\{A_i^T\}$ is also. For convenience we take our strongly independent set of t linear maps to be $\{B_j^T \mid j = 1,\ldots,t\}$. So, if $B_k^T = (b_{ij}^k)$ then $b_{ij}^k = A_{jki} = d_{ki}^j$.

Before proving the converse we wish to make some observations about H and the three dimensional matrix (A_{ijk}). Let the set $\{a_1,\ldots,a_m\}$ generate R and let $\{c_1,\ldots,c_t\}$ generate G. Now arrange the matrix H so that the $(j+1)$th column is that represented by, D_j $1 \le j \le m$, and make the $(i+1)$th row that row corresponding to a_i, $1 \le i \le m$. Then we have

$$h_{i+1,j+1} = \prod_{k=1}^{t} c_k^{\delta_k} = \prod_{k=1}^{t} c_k^{\gamma_k}, \quad \delta_k = d_{ki}^j, \ \gamma_k = b_{ij}^k, \quad 1 \le i,j \le m.$$

Because the rows and columns form groups the rest of H is determined up to equivalence by this $m \times m$ square of entries.

Now if we are given a strongly independent set of t linear maps in $M_m(p)$ then we can construct an $m \times m$ square and generate by multiplying rows and columns a matrix, K, which is group Hadamard of type p^{m-t} for C_p^t. Certainly K will have its set of columns and set of rows forming a group, so all we need to check is that each row has each element of C_p^t entered exactly p^{m-t} times.

The general entry of K is $\prod_{k=1}^{t} c_k^{\varepsilon_k}$, $\varepsilon_k = \sum_{j=1}^{m} \sum_{i=1}^{m} \alpha_i b_{ij}^k \beta_j$

$\alpha_i, \beta_j \in GF(p)$. So we need only check that for fixed $(\alpha_1, \ldots, \alpha_m)$

$$(\alpha_1,\ldots,\alpha_m)\begin{pmatrix} b_{11}^k & \cdots & b_{1m}^k \\ \vdots & & \vdots \\ b_{m1}^k & \cdots & b_{mm}^k \end{pmatrix}\begin{pmatrix} \beta_1 \\ \vdots \\ \beta_m \end{pmatrix} = 0\ , \quad 1 \le k \le t\ , \tag{*}$$

exactly p^{m-t} times. But, for a strongly independent set $\{A_1, \ldots, A_t\}$ and non-zero vector $\underset{\sim}{a}$, the vectors $\underset{\sim}{a}A_1, \ldots, \underset{\sim}{a}A_{n_t}$ are all independent (provided the A_i are $t \times m$ matrices and $t \le m$). So $\underset{\sim}{\alpha} B_k^T$ are independent. Now checking (*) is the same as checking

$$\begin{pmatrix} \underset{\sim}{\alpha} B_1^T \\ \vdots \\ \underset{\sim}{\alpha} B_t^T \end{pmatrix}\begin{pmatrix} \beta_1 \\ \vdots \\ \beta_m \end{pmatrix} = \underset{\sim}{0} \quad \text{precisely } p^{m-t} \text{ times.}$$

But the matrix on the left has rank t so its null space is of dimension $m-t$. □

Example 3.1: The following set is a strongly independent set over $GF(2)$ of two 3×3 linear maps.

$$\left\{ \begin{pmatrix} 1 & 0 & 0 \\ 0 & 1 & 0 \\ 0 & 0 & 1 \end{pmatrix}, \begin{pmatrix} 0 & 1 & 0 \\ 0 & 0 & 1 \\ 1 & 1 & 0 \end{pmatrix} \right\}.$$

Form the 3×3 square, A, in fig. (i), and generate the three generating columns in fig. (ii). Then complete the matrix by generating the rest of its columns fig. (iii). Note the notation in fig. (iv).

$$A = \begin{pmatrix} a & b & 1 \\ 1 & a & b \\ b & b & a \end{pmatrix}$$

Fig. (i)

$$H = \begin{pmatrix} 1 & 1 & 1 & 1 & 1 & 1 & 1 & 1 \\ 1 & a & b & 1 & ab & a & b & ab \\ 1 & 1 & a & b & a & b & ab & ab \\ 1 & b & b & a & 1 & ab & ab & a \\ 1 & a & ab & b & b & ab & a & 1 \\ 1 & ab & 1 & a & ab & b & a & b \\ 1 & b & ab & ab & a & a & 1 & b \\ 1 & ab & a & ab & b & 1 & b & a \end{pmatrix}$$

Fig. (iii)

$$\begin{array}{ccc} 1 & 1 & 1 \\ \hline a & b & 1 \\ 1 & a & b \\ b & b & a \\ \hline a & ab & b \\ ab & 1 & a \\ b & ab & ab \\ ab & a & b \end{array}$$

Fig. (ii)

$$H = \begin{pmatrix} 1 & 1 & 1 & \cdots & 1 \\ 1 & A & \longrightarrow & & \\ 1 & \downarrow & & & \\ \cdot & & & & \\ \cdot & & & & \\ \cdot & & & & \\ 1 & & & & \end{pmatrix}$$

Fig. (iv)

We now give two corollaries of Theorem 3.1. We say the set, $D = \{\phi_1, \ldots, \phi_s\}$, of maps $\phi_i : V_m \to V_t$, $m \geq t \geq s$, is a *strongly independent set of maps* over $GF(p)$ if the following equation

$$\sum_{i=1}^{t} \lambda_i \, \phi_i \, (\underset{\sim}{v}) = \underset{\sim}{\omega}_o$$

(for fixed λ_i, not all zero, $\underset{\sim}{\omega}_o \in V_t$) has precisely p^{m-t} solutions. This agrees with our definition for linear maps. The first corollary gives two algebraic equivalents for row group Hadamard matrices. Noting the construction of the sets L and D in the proof of (i) in theorem 3.1 we have,

Corollary 3.1.1: *Every row group Hadamard matrix of type* p^{m-t} *for* C_p^t *corresponds to*

(1) *a set of* p^m $t \times m$ *linear maps* $L = \{A_i \mid 1 \le i \le p^m\}$, *such that*

(i) $\underline{O} \in L$

(ii) $A_i - A_j$, $i \ne j$, *is of maximum rank.*

(2) *a set* $D = \{\phi_1, \ldots, \phi_t\}$ *of* t *strongly independent maps (not necessarily linear),* $\phi_i : V_m \to V_t$, *over* $GF(p)$. □

Write $|[a_1, \ldots, a_m]|$ to denote the *ordered set* whose elements, in order, are $a_1, \ldots, a_m$.

Corollary 3.1.2: *Let* Σ_t *be the set of ordered* t-*sets of strongly independent linear maps in* $GL(m,p)$. *Define an action of* $GL(m,p) \times GL(m,p)$ *on* Σ_t *by*

$$(A,B)\,|[A_1, \ldots, A_t]| = |[AA_1B^{-1}, \ldots, AA_tB^{-1}]| .$$

Then to each orbit of the action of $GL(m,p) \times GL(m,p)$ *on* Σ_t *there corresponds precisely one* $p(m,t)$-*Hadamard matrix.*

Proof. We define on Σ_t the relation $\sim$. Let $X_1, X_2 \in \Sigma_t$. If the matrices obtained, as in Example 3.1, from X_1 and X_2 are equivalent then $X_1 \sim X_2$. This is an equivalence relation.

If $X_1 \sim X_2$ then the arrays formed from X_1 and X_2 can both be obtained by selecting m rows and m columns and taking the union of their pairwise intersections from one group Hadamard matrix. Because of the multiplicative structure of group Hadamard matrices there exist $A, B \in GL(m,p)$ such that $AX_1B^{-1} = X_2$. □

4. CLASSIFYING GROUP HADAMARD MATRICES

Corollary 3.1.2, of the previous section, allows us to classify the p(m,t)-Hadamard matrices by decomposing them into irreducible components. We say a *group Hadamard matrix* H is *reducible* if there exist two group Hadamard matrices of smaller order such that $H \sim H_1 \otimes H_2$; otherwise the group Hadamard matrix is said to be *irreducible*.

The corollary also allows us to count the p(m,2)-Hadamard matrices. We begin by giving a lemma which shows there are many p(m,t)-Hadamard matrices. If A is a linear map then $m_A(\lambda)$, $c_A(\lambda)$ (or just m_A, c_A) are respectively, the minimum and characteristic polynomials of A.

Lemma 4.1: The set $\{I, \Lambda, \Lambda^2, \ldots, \Lambda^{d-1}\}$ *is a strongly independent set if and only if* $d \le t$ *where* t *is the least of the degrees of the irreducible polynomials dividing* $c_\Lambda(\lambda)$.

Proof. Given a linear map M with minimum polynomial $q^{e_1}(\lambda), \ldots, q^{e_k}(\lambda)$ we may decompose it to a direct sum of maps T_i with minimum polynomial $q_i^{e_i}(\lambda)$.

$$M = T_1 \oplus \ldots \oplus T_k .$$

Now each T_i may be decomposed to a direct sum of cyclic T_i^j on their respective cyclic spaces U_i^j. For each U_i^j and T_i^j there is an element $\underset{\sim}{m}$ with the property that for all $\underset{\sim}{u} \in U_i^j$ there exists a polynomial $f(\lambda) \in GF_{(p)}[\lambda]$, $\deg f < \deg m_{T_i^j}(\lambda)$, such that $\underset{\sim}{u} = f(\lambda)\underset{\sim}{m}$.

$$M = T_1^1 \oplus \ldots \oplus T_1^{J_1} \oplus T_2^1 \oplus \ldots \oplus T_2^{J_2} \oplus \ldots \oplus T_k^1 \oplus \ldots \oplus T_k^{J_k} .$$

Now if $\underset{\sim}{a} \neq \underset{\sim}{0}$ there exist i,j such that the projection of $\underset{\sim}{a}$ onto U_i^j is not zero. Let this vector be $\underset{\sim}{b}$ and $i = I$, $j = J$.

Now suppose $\sum_{k=D}^{d-1} \lambda_k \Lambda^k \underset{\sim}{a} = q(\Lambda)\underset{\sim}{a} = \underset{\sim}{0}$, where $0 \leq$ degree $q < d$, then $q(T_I^J)\underset{\sim}{b} = \underset{\sim}{0}$ (some I, J and $\underset{\sim}{b} \in U_I^J$). Now there exists an $f(\lambda) \in GF(p)[\lambda]$ such that $\underset{\sim}{b} = f(T_I^J)\underset{\sim}{m}$, ($\underset{\sim}{m} \in U_I^J$) where degree $f <$ degree $m_{T_I}J(\lambda)$. But this means

$$q\left(T_I^J\right) f\left(T_I^J\right) \underset{\sim}{m} = \underset{\sim}{0} .$$

Hence $q\left(T_I^J\right) f\left(T_I^J\right) = \underline{\underline{0}}$, giving a contradiction since no factor of $m_{T_I}J(\lambda)$ divides q.

Now if $g(\lambda)$ is an irreducible polynomial of degree d which divides $c_\Lambda(\lambda)$ then there exist $e \geq 1$ and $i = I$, $j = J$ such that $m_{T_I}J(\lambda) = g^e(\lambda)$. Let $\underset{\sim}{a} = g^{e-1}(T_I{}^J)\underset{\sim}{m}$, (so $\underset{\sim}{a} \neq \underset{\sim}{0}$), and consider $g(T_I^J)\underset{\sim}{a}$.

$$g(T_I^J)\underset{\sim}{a} = g^e(T_I^J)\underset{\sim}{m} = \underset{\sim}{0}$$

So $g(\Lambda)$ is not invertible and hence $\{I,\Lambda,\Lambda^2,\ldots,\Lambda^d\}$ is not strongly independent. □

For $t = 2$ given any strongly independent set $\{A_1,A_2\}$ there exist $A,B \in GL(m,p)$ such that $|[AA_1B^{-1}, AA_2B^{-1}]|$ is of the form $|[I,\Lambda]|$. So we find that;

<u>*Theorem 4.1:*</u> *The number of* $p(m,2)$*-Hadamard matrices is equal to the number of similarity classes of* $GL(m,p)$ *whose corresponding characteristic polynomials have no linear divisors.*

Proof. Any orbit of the action of $GL(m,p) \times GL(m,p)$ on Σ_2 contains an ordered strongly independent 2-set of the form $|[I,\Lambda]|$. Now if $(A,B)|[I,\Lambda_1]| = |[I,\Lambda_2]|$ then $A = B$ and $A\Lambda_1 A^{-1} = \Lambda_2$. So $|[I,\Lambda_1]| \sim |[I,\Lambda_2]|$ if and only if Λ_1 is similar to Λ_2. Finally by lemma 4.1. the characteristic polynomial of Λ has no linear divisors. □

Example 4.1: We calculate the number of inequivalent $p(5,2)$-Hadamard matrices. Let $M(d)$ be the number of irreducible polynomials of degree d, then

$$\sum_{d|m} M(d) = p^m ,$$

so $M(2) = \frac{p^2-p}{2}$, $M(3) = \frac{p^3-p}{3}$, $M(5) = \frac{p^5-p}{5}$.

So the number of allowed similarity classes is

$$\frac{p^2-p}{2} \cdot \frac{p^3-p}{3} + \frac{p^5-p}{5} = \frac{1}{30} p\,(p-1)(11p^3+6p^2+p+6) .$$

The following lemmas allow us to reduce the problem of classifying group Hadamard matrices to that of classifying irreducible group Hadamard matrices. The first lemma is straightforward to check.

Lemma 4.2: If $D = \{D_1,\dots,D_t\}$ *and* $E = \{E_1,\dots,E_t\}$ *are, respectively, strongly independent sets in* $GL(m,p)$ *and* $GL(n,p)$ *then* $D \oplus E = \{D_1 \oplus E_1,\dots,D_t \oplus E_t\}$ *is a strongly independent set in* $GL(m+n,p)$.

Lemma 4.3: If $H_1,H_2,\dots,H_n$ *are group Hadamard matrices and* $X_1,X_2,\dots,X_n$ *are corresponding strongly independent sets then* $X_1 \oplus X_2 \oplus \dots \oplus X_n$ *is a strongly independent set corresponding to* $H_1 \otimes H_2 \otimes \dots \otimes H_n$.

Proof. Consider $H_1 \otimes H_2$. All we need do is find the array obtained from $X_1 \oplus X_2$ in $H_1 \otimes H_2$. Let the corresponding arrays of X_1 and X_2 be (x^1_{ij}) and (x^2_{ij}) respectively. We have (using the notation of example 3.1),

$$\begin{pmatrix} 1 \;\ldots\; 1 \;\ldots\; 1 \\ \vdots \\ 1 \;\; (x^1_{ij}) \;\; \rightarrow \\ \vdots \\ 1 \;\; \downarrow \end{pmatrix} \otimes \begin{pmatrix} 1 \;\ldots\; 1 \;\ldots\; 1 \\ \vdots \\ 1 \;\; (x^2_{ij}) \;\; \rightarrow \\ \vdots \\ 1 \;\; \downarrow \end{pmatrix} =$$

$$\left(\begin{array}{llllll}
1 \ldots 1 \ldots 1 & 1 \ldots 1 \ldots 1 & 1 \ldots 1 \ldots 1 & \cdot\;\cdot & 1 \ldots 1 \ldots 1 & 1 \ldots\ldots 1 \\
\vdots & \vdots & \vdots & & \vdots & \\
1 \;\; (x^1_{ij}) \;\; \rightarrow & 1 \;\; (x^1_{ij}) \;\; \rightarrow & 1 \;\; (x^1_{ij}) \;\; \rightarrow & & 1 \;\; (x^1_{ij}) \;\; \rightarrow & \\
\vdots & \vdots & \vdots & & \vdots & \\
1 \;\; \downarrow & 1 \;\; \downarrow & 1 \;\; \downarrow & & 1 \;\; \downarrow & \\
1 \ldots 1 \ldots 1 & x^2_{11} \ldots\ldots x^2_{11} & x^2_{12} \ldots\ldots x^2_{12} & \cdot\;\cdot & x^2_{1m} \ldots\ldots x^2_{1m} & \\
\vdots & \vdots & \vdots & & \vdots & \\
1 \;\; (x^1_{ij}) \;\; \rightarrow & \vdots\; x^2_{11}(x^1_{ij}) \;\; \rightarrow & \vdots\; x^2_{12}(x^1_{ij}) \;\; \rightarrow & & \vdots\; x^2_{1m}(x^1_{ij}) \;\; \rightarrow & \\
\vdots & \vdots & \vdots & & \vdots & \\
1 \;\; \downarrow & x^2_{11} \;\; \downarrow & x^2_{12} \;\; \downarrow & \cdot\;\cdot & x^2_{1m} \;\; \downarrow & \\
1 \ldots 1 \ldots 1 & x^2_{21} \ldots\ldots x^2_{21} & x^2_{22} \ldots\ldots x^2_{22} & \cdot\;\cdot & x^2_{2m} \ldots\ldots x^2_{2m} & \\
\vdots & \vdots & \vdots & & \vdots & \\
1 \;\; (x^1_{ij}) \;\; \rightarrow & \vdots\; x^2_{21}(x^1_{ij}) \;\; \rightarrow & \vdots\; x^2_{22}(x^1_{ij}) \;\; \rightarrow & & \vdots\; x^2_{2m}(x^1_{ij}) \;\; \rightarrow & \\
\vdots & \vdots & \vdots & & \vdots & \\
1 \;\; \downarrow & x^2_{21} \;\; \downarrow & x^2_{22} \;\; \downarrow & \cdot\;\cdot & x^2_{2m} \;\; \downarrow & \\
\cdot \quad\quad \cdot & \cdot \quad\quad \cdot & \cdot \quad\quad \cdot & & \cdot & \\
\cdot \quad\quad \cdot & \cdot \quad\quad \cdot & \cdot \quad\quad \cdot & & \cdot & \\
1 \ldots 1 \ldots 1 & x^2_{m1} \ldots\ldots x^2_{m1} & x^2_{m2} \ldots\ldots x^2_{m2} & \cdot\;\cdot & x^2_{mm} \ldots\ldots x^2_{mm} & \\
\vdots & \vdots & \vdots & & \vdots & \\
1 \;\; (x^1_{ij}) \;\; \rightarrow & \vdots\; x^2_{m1}(x^1_{ij}) \;\; \rightarrow & \vdots\; x^2_{22}(x^1_{ij}) \;\; \rightarrow & & \vdots\; x^2_{mm}(x^1_{ij}) \;\; \rightarrow & \\
\vdots & \vdots & \vdots & & \vdots & \\
1 \;\; \downarrow & x^2_{m1} \;\; \downarrow & x^2_{m2} \;\; \downarrow & & x^2_{mm} \;\; \downarrow & \\
1 & & & & & \\
\vdots & & & & & \\
1 & & & & &
\end{array}\right)$$

Thus $H_1 \otimes H_2$ contains the array

$$\begin{pmatrix} x^1_{11} & \dots & x^1_{1n} & 1 & \dots & 1 \\ \vdots & & \vdots & \vdots & & \vdots \\ x^1_{n1} & \dots & x^1_{nn} & 1 & \dots & 1 \\ 1 & \dots & 1 & x^2_{11} & \dots & x^2_{1m} \\ \vdots & & \vdots & \vdots & & \vdots \\ 1 & \dots & 1 & x^2_{m1} & \dots & x^2_{mm} \end{pmatrix}$$

which corresponds to $X_1 \oplus X_2$. □

Before proving the main result of this section we state a well known result from linear algebra [3]. Let Ω be a set of linear maps. Let U and W be invariant subspaces of Ω. Then U and W are said to be *Ω-isomorphic* if for all $A \in \Omega$ there exists $\theta \in GL(\dim U, p)$ such that $A_{|U} = \theta A_{|W}\theta^{-1}$.

<u>*Theorem 4.2:*</u> (Krull-Schmidt theorem) *If* $V = U_1 \oplus U_2 \oplus \dots \oplus U_h$ *and* $V = W_1 \oplus \dots \oplus W_k$ *are two decompositions of* V *into non-zero invariant and indecomposable subspaces relative to* Ω *then* $h = k$ *and if the* U_i *are suitably ordered* W_i *and* U_i *are Ω-isomorphic,* $1 \le i \le k$.

<u>*Theorem 4.3:*</u> *Two group Hadamard matrices are equivalent if and only if their sets of irreducible components are equal up to equivalence.*

<u>*Proof:*</u> If $H_1 = H_1^1 \otimes H_1^2 \otimes \dots \otimes H_1^n$ and $H_2 = H_2^1 \otimes H_2^2 \otimes \dots \otimes H_2^n$ and $H_1^k \sim H_2^k$ then $X_1^k \sim X_2^k$, $1 \le k \le n$. So, ensuring X_1^k and X_2^k all contain the identity of suitable order, there exists invertible A_k such that $X_1^k = A_k X_2^k A_k^{-1}$. Now by the previous lemma the corresponding strongly independent sets of H_1 and H_2 respectively, are $X_1^1 \oplus X_1^2 \oplus \dots \oplus X_1^n$ and $X_2^1 \oplus X_2^2 \oplus \dots \oplus X_2^n$. Letting

$A = A_1 \oplus A_2 \oplus \ldots \oplus A_n$ we see A is invertible and $X_1^1 \oplus \ldots \oplus X_1^n =$ $(A_1 \oplus \ldots \oplus A_n)(X_2^1 \oplus \ldots \oplus X_2^n)(A_1^{-1} \oplus \ldots \oplus A_n^{-1})$. So $H_1 \sim H_2$.

Suppose $H_1 \sim H_2$ and X_1, X_2 correspond to H_1 and H_2 respectively. Suppose further that $V = U_1^1 \oplus \ldots \oplus U_1^k$, and $V = U_2^1 \oplus \ldots \oplus U_2^h$ are respective decompositions of V with respect to X_1 and X_2. So, ensuring both X_1 and X_2 contain the identity, there exists an invertible non-linear map A such that $X_1 = A X_2 A^{-1}$.

Then $V = A(U_2^1) \oplus \ldots \oplus A(U_2^h)$ is a decomposition of V with respect to X_1 and $X_1|_{A(U_2^i)} = A X_2 A^{-1}|_{A(U_2^i)} = X_2|_{U_2^i}$. So, after reordering, by the Krull-Schmidt theorem $h = k$ and there exist θ_i such that $X_2|_{U_2^i} = \theta_i X_1|_{U_1^i} \theta_i^{-1}$, $1 \leq i \leq k$. □

We are now able to construct and classify all the $p(m,2)$-Hadamard matrices.

Theorem 4.4: *Let* $|[I,\Lambda]|$ *be a strongly independent 2-set corresponding to a* $p(m,2)$*-Hadamard matrix* H. *Let the elementary divisors of* Λ *be* $q_1(\lambda),\ldots,q_k(\lambda)$ *then*

$$H \sim H_1 \otimes H_2 \otimes \ldots \otimes H_k ,$$

where H_i *corresponds to* $|[I,C_{q_i(\lambda)}]|$ *and* $C_{q_i(\lambda)}$ *is the companion matrix of* $q_i(\lambda)$.

Any $p(m,2)$*-Hadamard matrix is of this form.*

Proof. Let H be $p(m,2)$-Hadamard then H corresponds to $|[I,\Lambda]|$ some $\Lambda \in GL(m,p)$. Suppose $H \sim H_1 \otimes H_2 \otimes \ldots \otimes H_k$ where H_i are irreducible then

$$|[I,\Lambda]| \sim |[I,D_1]| \oplus |[I,D_2]| \oplus \cdots \oplus |[I,D_k]|,$$

where each D_i is irreducible. So Λ is similar to $D_1 \oplus D_2 \oplus \ldots \oplus D_k$. By the general theory of finitely generated modules the polynomials $m_{D_1}(\lambda),\ldots,m_{D_k}(\lambda)$ are the elementary divisors of Λ. □

Example 4.2: We generate the group Hadamard matrix H for $C_p \times C_p$ corresponding to $q^e(\lambda)$ where $q(\lambda)$ is irreducible and $e \geq 1$. Let $q^e(\lambda) = \lambda^d + a_1\lambda^{d-1} + \ldots + a_d$ and set

$$\Lambda = C_{q^e(\lambda)} = \begin{pmatrix} 0 & 1 & 0 & \cdots & 0 \\ \vdots & & \ddots & \ddots & \vdots \\ \vdots & & & \ddots & 0 \\ 0 & \cdots & \cdots & 0 & 1 \\ -a_d & \cdots & \cdots & \cdots & a_1 \end{pmatrix}$$

then Λ is irreducible. Form H from $|[I,\Lambda]|$ by forming the corresponding array, shown below and generating as in Example 3.1. Because $|[I,\Lambda]|$ is not the direct sum of two smaller strongly independent sets, H is irreducible.

$$\begin{pmatrix} a & b & 1 & 1 & \cdots & 1 \\ 1 & a & b & 1 & \cdots & 1 \\ \vdots & & \ddots & \ddots & & \vdots \\ \vdots & & & \ddots & \ddots & \vdots \\ 1 & \cdots & \cdots & \cdots & a & b \\ b^{-a_d} & \cdots & \cdots & \cdots & b^{-a_2} & ab^{-a_1} \end{pmatrix}$$

These are the only irreducible $p(m,2)$-Hadamard matrices.

Theorem 4.3 reduces the problem of characterising group Hadamard matrices to that of classifying the irreducible group Hadamard matrices. This requires a classification of strongly independent m-sets. Lemma 4.1 allows us to do this for $m = 2$. The problem for $m > 2$ appears to be harder. The connection of the sets in corollary 3.1.1. and translation planes [4, p4] suggests that the following problem is worth solving:

"Characterise the strongly independent k-sets of $m \times t$ linear maps, $k \leq t \leq m$, over $GF(p)$".

Acknowledgement: I wish to thank my supervisor Dr. Jennifer Seberry for her help and encouragement.

References:

[1] D.A. Drake, "Partial λ-geometries and generalised Hadamard matrices over groups, *Canad. J. Math.* 31 (1979), 617-627.

[2] Warwick de Launey, Ph.D. Thesis, University of Sydney (in preparation).

[3] N. Jacobson, *Lectures in Abstract Algebra* Vol. 2, D. Van Nostrand Company. Inc., Princeton, 1953.

[4] Heinz Lüneburg, *Translation Planes* Springer-Verlag, Berlin, 1980.

THE ASYMPTOTIC CONNECTIVITY OF LABELLED COLOURED REGULAR BIPARTITE GRAPHS

M.N. ELLINGHAM

A labelled coloured bipartite graph, of LCBG, is a bipartite (simple) graph whose vertices have been 2-coloured and the vertices of each colour labelled independently. It is shown that for fixed $r \geq 3$ *the proportion of r-regular LCBGs on* $2n$ *vertices which are r-connected approaches* 1 *as* $n \to \infty$. *Also, fix* $r \geq 3$ *and* $q > 0$; *let* $g = \max(4, 2\{q/(2(r-2))\})$. *Then the numbers of the following types of r-regular LCBGs with* $2n$ *vertices are asymptotically equal as* $n \to \infty$: *those with girth at least* g; *those which are cyclically-q-edge-connected; and those which are cyclically-q-vertex-connected.*

1. Introduction and definitions.

This paper presents some asymptotic results on the connectivity of regular bipartite graphs. As Bollobas has remarked, almost all results in probabilistic, or asymptotic, graph theory concern *labelled* graphs. To take advantage of the bipartite nature of the graphs examined here, they must also be coloured; the vertices of each colour are then labelled independently.

The results proved here as Theorems 7 and 10 are similar to those for labelled graphs which were demonstrated by Wormald in [4], although those given here are less general.

I should like to thank B.D. McKay and N.C. Wormald for interesting me in asymptotic graph theory, and for providing the references quoted in this paper, as well as much other information.

All graphs in this paper are simple (no loops or multiple edges). A *coloured bipartite graph* (sometimes called a bicoloured graph) is a bipartite graph whose vertices have been properly 2-coloured (no two vertices of the same colour are adjacent). The "colours" used will be assumed to be black and white. A *labelled coloured bipartite graph*, or *LCBG*, is obtained from a coloured bipartite graph with b black vertices and w white vertices by labelling the black vertices with the integers $1,2,\ldots,b$, and the white vertices with $1,2,\ldots,w$. In such an LCBG u_i will denote the black vertex labelled i, and v_j the white vertex labelled j.

Two LCBGs are considered identical if there is a graph isomorphism from one to the other which maps each vertex to a vertex with the same colour and label. If no such isomorphism exists they are considered distinct.

Let $r \in \mathbb{N}$. The degree sequence of an LCBG with maximum vertex degree r can be represented by a pair (B,W), where $B = (b_0,b_1,\ldots,b_r)$ and $W = (w_0,w_1,\ldots,w_r)$ are $(r+1)$-tuples of nonnegative integers. Here b_i is the number of black, and w_i the number of white, vertices having degree i for each i, $0 \leqslant i \leqslant r$. The number of edges in such an LCBG is

$$m(B,W) = \sum_{i=0}^{r} ib_i = \sum_{i=0}^{r} iw_i \;;$$

therefore only pairs (B,W) for which $\sum_{i=0}^{r} ib_i = \sum_{i=0}^{r} iw_i$ will be considered. Let $D(r)$ be the set of all such pairs. For $(B,W) \in D(r)$ let $L(B,W)$ be the set of all LCBGs with degree sequence (B,W). Any LCBG in $L(B,W)$ will have $b = \sum_{i=0}^{r} b_i$ black vertices and $w = \sum_{i=0}^{r} w_i$ white vertices.

A *separating set* for a graph (or LCBG) G is a set $S \subset VG$ such $G - S$ is disconnected. S is called *minimal* if no subset of S is also a separating set, and *cycle-separating* if at least two components of $G - S$ contain cycles. If H is a component of $G - S$ then S is said to be *H-minimal* if every vertex of S is adjacent in G to at least one vertex of H and one vertex of $G - S - VH$; if S is minimal then it is H-minimal. G is called *s-connected* if it has no separating set of less than s elements, and *cyclically-s-(vertex)-connected* if it has no cycle-separating set of less than s elements.

A *disconnecting set* for a graph (or LCBG) G is a set $K \subseteq EG$ such that $G - K$ is disconnected. K is called *minimal* if no subset of K is also a disconnecting set, and *cycle-disconnecting* if at least two components of $G - K$ contain cycles. G is called *k-edge-connected* if it has no disconnecting set of less than k elements, and *cyclically-k-edge-connected* if it has no cycle-disconnecting set of less than k elements.

The *girth* of a graph (or LCBG) is the length of its shortest cycle.

Some special sets of LCBGs can now be defined:

$R(r,n)$ is the set of all r-regular LCBGs with $2n$ vertices.

$C(r,g,n)$ is the set of all $G \in R(r,n)$ with girth at least g, for g even, $g \geqslant 4$. Note that $C(r,4,n) = R(r,n)$.

$P(r,s,n,p)$ is the set of all $G \in R(r,n)$ which have a separating set S such that $|S| = s$ and S is H-minimal for some component H of $G - S$ having p vertices.

$Q(r,k,n,p)$ is the set of all $G \in R(r,n)$ which have a minimal disconnecting set K such that $|K| = k$ and $G - K$ has a component with p vertices.

The following terminology will be employed: "$f(x) \sim g(x)$ uniformly for $x \in X$ as $h(x) \to \infty$" means

$$\sup\{|f(x)/g(x) - 1| : x \in X,\ h(x) = n\} \to 0 \quad \text{as} \quad n \to \infty$$

(where by convention $0/0 = 1$ and the supremum of the empty set is 0).

2. Preliminary Results.

Stirling's formula.

$$n! \sim \sqrt{(2\pi)}\ n^{n+1/2} / e^n \quad \text{as} \quad n \to \infty$$

will be used frequently.

All the results in this paper depend on the following lemma, which is a consequence of the Theorem by Bender on page 218 of [1].

Lemma 1: *Suppose* r *is fixed. Then*

$$|L(B,W)| \sim \frac{m!\ b!\ w!}{\Pi\ i!^{b_i+w_i}\ \Pi\ b_i!\ \Pi\ w_i!} \exp -\left(\frac{\Sigma b_i i(i-1) \Sigma w_i i(i-1)}{2m^2}\right) \tag{1}$$

uniformly for $(B,W) \in D(r)$ *as* $m \to \infty$ *(all sums and products being taken for* $i = 0$ *to* r).

Proof: To every $G \in L(B,W)$ there corresponds a unique $b \times w$ $\{0,1\}$-matrix $[g_{jk}]$, where $g_{jk} = 1$ if $u_j v_k \in EG$, and 0 otherwise. How many such matrices are there? First, let

$$a_j = \sum_{k=1}^{w} g_{jk} \quad (= \text{degree of } u_j \text{ in } G),\ 1 \leq j \leq b$$

$$\text{and} \quad c_k = \sum_{j=1}^{b} g_{jk} \quad (= \text{degree of } v_k \text{ in } G),\ 1 \leq k \leq w.$$

Let $A = (a_1, a_2, \ldots, a_b)$ and $C = (c_1, c_2, \ldots, c_w)$. From [1], Theorem, page 218, the number of $b \times w$ $\{0,1\}$-matrices with fixed row-sum vector A and column-sum vector C approaches

$$\frac{m!}{\Pi\ a_j!\ \Pi\ c_k!} \exp -\left(\frac{\Sigma a_j(a_j-1) \Sigma c_k(c_k-1)}{2m^2}\right)$$

uniformly for all vectors A and C with entries not exceeding r, as $m = \Sigma a_j = \Sigma c_k \to \infty$ (all sums and products being taken for $j = 1$ to b or $k = 1$ to w as appropriate). Since exactly b_i a_j's and w_i c_k's are equal to i, for $0 \leqslant i \leqslant r$, this becomes

$$\frac{m!}{\Pi\, i!^{b_i}\, \Pi\, i!^{w_i}} \exp - \frac{\Sigma b_i i(i-1)\, \Sigma w_i i(i-1)}{2m^2} \qquad (2).$$

Now by elementary combinatorics the number of possible vectors A given B is just $b!/\Pi\, b_i!$. Similarly the number of possible vectors C given W is $w!/\Pi\, w_i!$. Multiplying (2) by these two factors gives the estimate (1) for $|L(B,W)|$ ●

Corollary 2: *Suppose* r *is fixed. Then*

$$|R(r,n)| \sim \frac{(rn)!}{r!^{2n}} \exp - \left(\frac{(r-1)^2}{2}\right) \quad \text{as } n \to \infty$$

and for each even $g \geqslant 4$,

$$|C(r,g,n)| \sim \frac{(rn)!}{r!^{2n}} \exp - \left(\sum_{i=1}^{g/2-1} \frac{(r-1)^{2i}}{2i}\right)$$

$$\sim |R(r,n)| \exp - \left(\sum_{i=2}^{g/2-1} \frac{(r-1)^{2i}}{2i}\right) \quad \text{as } n \to \infty.$$

Proof: For $|R(r,n)|$ substitute $m = rn$, $b_i = w_i = 0$ for $0 \leqslant i \leqslant r-1$ and $b_r = w_r = n$ in Lemma 1. The first estimate of $|C(r,g,n)|$ comes from [3], Theorem 3.12, page 169, and is equivalent to the second by simple arithmetic. Note that $|C(r,4,n)| \sim |R(r,n)|$, as expected because $C(r,4,n) = R(r,n)$ ●

For a degree sequence $(B,W) \in D(r)$ the value $r(b+w) - 2m$ indicates how far LCBGs in $L(B,W)$ depart from being r-regular. This number is in fact the sum over all vertices of the difference between r and the degree of the vertex.

Lemma 3: *Let* r *and* k *be fixed. Then for any* $(B,W) \in D(r)$ *satisfying* $r(b+w) - 2m = k$,

$$|L(B,W)| = O(1)\, n^{rn+k/2+1/2}\, r!^{-2n}\, (r/e)^{rn} \qquad (3)$$

where $n = (b+w)/2$, *and* $O(1)$ *denotes a bound depending on* r *and* k *only.*

Proof: Let $k_0 = rb - m$ (the sum over all black vertices of the difference between r and the degree of the vertex) and $k_1 = rw - m$ (the sum over all white vertices of the same); then $k_0 + k_1 = k$. Let $n = (b+w)/2$, then $k = 2rn - 2m$. Now

$$m = rn - k/2,$$
$$b = n + (b-w)/2 = n + (k_0-k_1)/(2r) = n + \alpha,$$
$$\text{and} \quad w = n - (b-w)/2 = n - (k_0-k_1)/(2r) = n - \alpha,$$

where $\alpha = (k_0-k_1)/(2r)$; note that $|\alpha| \leq k/(2r)$.

The expression (1) of Lemma 1 may now be split into factors as follows:

(a) By Stirling's formula,

$$m! = (rn-k/2)! \sim \sqrt{(2\pi)}\ (rn)^{rn-k/2+1/2} / e^{rn-k/2}$$
$$= O(1)\ n^{rn-k/2+1/2}\ (r/e)^{rn}.$$

(b) At most k_0 black vertices do not have degree r, so $b_r \geq b - k_0$ and hence

$$\frac{b!}{\Pi\, b_i!} \leq \frac{b!}{b_r!} \leq \frac{b!}{(b-k_0)!} = b(b-1)\ \ldots\ (b-k_0+1)$$
$$\leq b^{k_0} = (n+\alpha)^{k_0} = O(1)\,n^{k_0}$$

since $k_0 \leq k$ and $|\alpha| \leq k/(2r)$.

(c) By reasoning similar to (b) $w!/\Pi\, w_i! = O(1)n^{k_1}$.

(d) Since $b_r \geq b - k_0$ and $w_r \geq w - k_1$, $b_r + w_r \geq b + w - k_0 - k_1 = 2n - k$, and therefore

$$\frac{1}{\Pi\, i!^{b_i+w_i}} \leq \frac{1}{r!^{b_r+w_r}} \leq \frac{1}{r!^{2n-k}} = O(1)\,r!^{-2n}.$$

(e) $\exp - (\ldots) \leq 1$ always.

Multiplying the results of (a) - (e) and using $k_0 + k_1 = k$ gives the formula (3) •

The following lemma gives a bound for expressions of a form which will be encountered later.

Lemma 4: *Let* $t > 0$ *and* $c \in \mathbb{R}$. *For each* $a > 0$ *define*

$$f_a(x) = x^{tx+c}\ (a-x)^{t(a-x)+c}, \quad \textit{for} \quad x \in (0,a).$$

Let $y > 0$, *then there exists a constant* $K = K(t,c,y)$ *such that*

$$f_a(x) \leq K f_a(y) \quad \textit{for all} \quad a \geq 2y. \quad x \in [y,a-y].$$

Proof: Since $f_a(x)$ is symmetric about $x = a/2$ we need only show that $f_a(x) \leq Kf_a(y)$ for $x \in [y,a/2]$. Let

$$g_a(x) = \log f_a(x) = (tx+c) \log x + (t(a-x)+c) \log (a-x).$$

Suppose $c \leq 0$. Then for any $a > 0$, $g'_a(x) \leq 0$ for $x \in (0,a/2]$. Therefore $g_a(x) \leq g_a(y)$, and hence $f_a(x) \leq f_a(y)$, for $x \in [y,a/2]$. So $K = 1$ suffices.

Now suppose $c > 0$. If $a > 2c/t$ then examination of g'_a and g''_a reveals that g_a increases to a local maximum at some $x_0 \in (0,a/2)$ then decreases to a local minimum at $a/2$. If further $a > y(e^{c/(ty)}+1)$ then $g'_a(y) < 0$ and so $y > x_0$; thus $g_a(x) \leq g_a(y)$, and hence $f_a(x) \leq f_a(y)$, for $x \in [y,a/2]$. What if $a \leq a_0 = \max(2c/t, y(e^{c/(ty)}+1))$? The expression $f_a(x)/f_a(y)$ is a continuous function of (a,x) on the compact set $\{(a,x) : a \in [2y,a_0],\ x \in [y,a/2]\}$ and therefore attains a maximum M. Choosing $K = \max(1,M)$ proves the lemma •

3. Asymptotic edge-connectivity.

Lemma 5: $$|Q(r,k,n,p)| = O(1)\, n^{2n+1}\, r!^{-2n}\, (r/e)^{rn}\, g(r,k,n,p) \qquad (4)$$
$$= O(1)\, |R(r,n)|\, n^{1/2-(r-2)n}\, g(r,k,n,p)$$

where

$$g(r,k,n,p) = (p/2)^{(r-2)(p/2)+d}\, (n-p/2)^{(r-2)(n-p/2)+d}$$

$$d = (k-1)/2$$

and $O(1)$ *denotes an expression bounded by a function of* r *and* k *only.*

Proof: Let $G \in Q(r,k,n,p)$ have a minimal disconnecting set K of size k such that $G - K$ has a component H with p vertices. Let J be the union of the other components of $G - K$. H and J are coloured bipartite graphs, but are not in general LCBGs because for each graph the labels for each colour it inherits from G are not consecutive integers beginning at 1. Relabel the vertices of each colour in H in order from 1: the smallest label becomes 1, the next smallest 2 and so on; call the resulting LCBG H'. Let J' be the LCBG similarly derived from J.

Henceforth H or J as a subscript indicates to which graph a quantity refers. Notice that $k_H = r(b_H+w_H) - 2m_H$ and $k_j = r(b_j+w_j) - 2m_j$ are both equal to k.

G is now completely determined by the following factors:

(a) The degree sequence (B_H, W_H) of H (or H'). The number of these is at most $\sum_{i=0}^{k} \phi(i)\phi(k-i)$, where $\phi(i)$ is the number of partitions of i. This is $O(1)$.

(b) The degree sequence (B_J, W_J) of J (or J'): $O(1)$ possibilities.

(c) $H' \in L(B_H, W_H)$: From Lemma 3,

$$|L(B_H, W_H)| = O(1)\ n_H^{rn_H + k/2 + 1/2}\ r!^{-2n_H}\ (r/e)^{rn_H}\ .$$

(d) $J' \in L(B_J, W_J)$: Again from Lemma 3,

$$|L(B_J, W_J)| = O(1)\ n_J^{rn_J + k/2 + 1/2}\ r!^{-2n_J}\ (r/e)^{rn_J}\ .$$

(e) The distribution of the labels of the black vertices in G between H and J. There are $n!/(b_H!b_J!)$ possibilities. Applying Stirling's formula, and recalling that $b_H = n_H + \alpha_H$, $b_J = n_J + \alpha_J$ where $|\alpha_H|, |\alpha_J| \leqslant k/(2r)$, this becomes

$$\frac{O(1)\ n^{n+1/2}}{b_H^{b_H+1/2}\ \ b_J^{b_J+1/2}} = \frac{O(1)\ n^{n+1/2}}{n_H^{b_H+1/2}\ \ n_J^{b_J+1/2}}\ .$$

(f) By reasoning similar to (e) the number of distributions of the white labels in G between H and J is

$$\frac{O(1)\ n^{n+1/2}}{w_H^{w_H+1/2}\ \ w_J^{w_J+1/2}} = \frac{O(1)\ n^{n+1/2}}{n_H^{w_H+1/2}\ \ n_J^{w_J+1/2}}\ .$$

(g) The edges between H and J: At most $k!$, hence $O(1)$, possibilities.

If (B_H, W_H) and (B_J, W_J) are fixed, the number of possibilities for G is, multiplying (c) - (g) and using $n_H + n_J = n$, $b_H + w_H = n_H$ and $b_J + w_J = n_J$:

$$O(1)\ n^{2n+1}\ \frac{n_H^{(r-2)n_H + k/2 - 1/2}\ n_J^{(r-2)n_J + k/2 - 1/2}}{r!^{2n}} \left(\frac{r}{e}\right)^{rn}\ .$$

Since factors (a) and (b) are both $O(1)$, this expression also serves for $|Q(r,k,n,p)|$. Replace n_H by $p/2$ and n_J by $n - p/2$ to obtain (4); the formula in terms of $|R(r,n)|$ follows from Corollary 2 •

<u>Lemma 6</u>: *For* $r \geqslant 3$,

$$\left|\bigcup_{i=p}^{n} Q(r,k,n,i)\right| = O(1)\ n^{(k-p(r-2))/2}\ |R(r,n)|$$

where $O(1)$ *denotes a bound depending on* r, k *and* p.

Proof: By Lemma 5 there is a constant $K = K(r,k)$ such that

$$|Q(r,k,n,i)| \leq K|R(r,n)|n^{1/2-(r-2)n} g(r,k,n,i).$$

Therefore

$$\left|\bigcup_{i=p}^{n} Q(r,k,n,i)\right| \leq \sum_{i=p}^{n} |Q(r,k,n,i)|$$

$$\leq K|R(r,n)|\ n^{1/2-(r-2)n} \sum_{i=p}^{n} g(r,k,n,i) \qquad (5).$$

Now for $r \geq 3$, $g(r,k,n,p+1) = O(1)\ g(r,k,n,p)$; also $g(r,k,n,p+2) = O(n^{-1})\ g(r,k,n,p)$, and by Lemma 4 there exists a constant $L = L(r,k,p)$ such that $g(r,k,n,i) \leq L\ g(r,k,n,p+2)$ for $p+2 \leq i \leq n$. Therefore the sum in (5) is just $O(1)\ g(r,k,n,p)$, giving

$$\left|\bigcup_{i=p}^{n} Q(r,k,n,i)\right| = O(1)\ |R(r,n)|\ n^{1/2-(r-2)n}\ g(r,k,n,p)$$

$$= O(1)\ |R(r,n)|\ n^{1/2-(r-2)n}\ (n-p/2)^{(r-2)(n-p/2)+(k-1)/2}$$

$$= O(1)\ n^{(k-p(r-2))/2}\ |R(r,n)|$$

as desired •

It is now possible to prove the following asymptotic result about the cyclic-edge-connectivity of r-regular LCBGs.

Theorem 7: *Suppose* $r \geq 3$ *and* $q > 0$. *Let* $g = \max(4, 2\{q/(2(r-2))\})$ *and* $z = g(r-2) - q$ $(z \geq 0)$. *Then*

(i) *The number of cyclically-q-edge-connected* LCBGs *in* $R(r,n)$ *is*

$$(1-O(n^{-(z+1)/2}))\ |C(r,g,n)|$$

where $O()$ *denotes a bound depending on* r *and* q.

(ii) *The number of cyclically-4(r-2)-edge-connected* LCBGs *in* $R(r,n)$ *is*

$$(1-O(n^{-1/2}))\ |R(r,n)|$$

where $O()$ *denotes a bound depending on* r.

Proof: (i) For all sufficiently large n (how large depends on q), all the cyclically-q-edge-connected elements of $R(r,n)$ belong to $C(r,g,n)$. This is trivially true when $g = 4$ since $C(r,4,n) = R(r,n)$. So assume $g \geq 6$. If $G \notin C(r,g,n)$ then G has a cycle L of length $\ell \leq g - 2$. Either G is disconnected, or else the $\ell(r-2)$ edges joining L to $G - VL$ form a cycle-disconnecting set of less than q edges. In either case G is not cyclically-q-edge-connected.

Suppose $G \in C(r,g,n)$ is not cyclically-q-edge-connected. It then has a minimal cycle-disconnecting set K of k, $0 \leq k \leq q$, and since at least two components of G contain cycles, one component has p vertices where $g \leq p \leq n$. Therefore the number of such LCBGs G is at most

$$\left|\bigcup_{k=0}^{q-1} \bigcup_{p=g}^{n} Q(r,k,n,p)\right| \leq \sum_{k=0}^{q-1} O(1)\ n^{(k-g(r-2))/2}\ |R(r,n)|$$

by Lemma 6. But for each k, $k - g(r-2) \leq q - 1 - g(r-2) = -(z+1)$, and so this becomes

$$\sum_{k=0}^{q-1} O(1)\ n^{-(z+1)/2}\ |R(r,n)| = O(1)\ n^{-(z+1)/2}\ |R(r,n)|$$
$$= O(1)\ n^{-(z+1)/2}\ |C(r,g,n)|$$

since by Corollary 2 $|C(r,g,n)|$ approaches a constant proportion of $|R(r,n)|$ as $n \to \infty$.

Thus the number of cyclically-q-edge-connected LCBGs in $R(r,n)$ is $|C(r,g,n)| - O(1)\ n^{-(z+1)/2}\ |C(r,g,n)| = (1-O(n^{-(z+1)/2})\ |C(r,g,n)|$.

(ii) This follows immediately from (i) with $q = 4(r-2)$ since $C(r,4,n) = R(r,n)$ ●

4. Asymptotic vertex-connectivity

Lemma 8: $|P(r,s,n,p)| = O(1)\ n^{2n+1}\ r!^{-2n}\ (r/e)^{rn}\ f(r,s,n,p)$ (6)

$$= O(1)\ |R(r,n)|\ n^{1/2-(r-2)n}\ f(r,s,n,p)$$

where

$$f(r,s,n,p) = (p/2)^{(r-2)(p/2)+c}\ (n-s/2-p/2)^{(r-2)(n-s/2-p/2)+c}$$
$$c = (s(r-1)-1)/2$$

and $O(1)$ *denotes an expression bounded by a function of* r *and* k *only.*

Proof: Let $G \in P(r,s,n,p)$ have an H-minimal separating set S with s elements, where H is a component of $G - S$ having p vertices. Let J be the union of all the other components of $G - S$, and let F be the subgraph of G induced by S. Let H', J' and F' be the LCBGs obtained by relabelling H, J and F respectively, in the manner in the proof of Lemma 5.

Henceforth H, J and F will be used as subscripts indicating to which graph a quantity refers. Notice that $k_H = r(b_H+w_H) - 2m_H$ is the number of edges between H and F, while $k_J = r(b_J+w_J) - 2m_J$ is the number of edges between J and F. Since S is H-minimal, $k_H, k_J \geq s$; but $k_H + k_J \leq sr$, so $k_H, k_J \leq s(r-1)$.

G is now completely determined by the following factors:

(a) The possible degree sequences (B_H, W_H) for H (or H') and (B_J, W_J) for J (or J'). Altogether there are O(1) possibilities.

(b) $H' \varepsilon L(B_H, W_H)$ and $J' \varepsilon L(B_J, W_J)$. From Lemma 3 the number of these is

$$|L(B_H,W_H)|\ |L(B_J,W_J)| = O(1)\ \frac{n_H^{rn_H+k_H/2+1/2}\ n_J^{rn_J+k_J/2+1/2}}{r!^{2n_H+2n_J}} \left(\frac{r}{e}\right)^{rn_H+rn_J}$$

$$= O(1)\ \frac{n_H^{rn_H+s(r-1)/2+1/2}\ n_J^{rn_J+s(r-1)/2+1/2}}{r!^{2n}} \left(\frac{r}{e}\right)^{rn}$$

since $k_H, k_J \leqslant s(r-1)$ and $n_H + n_J = n - s/2$.

(c) The distribution of the black and white labels between H, J and F in G: the number of possibilities is

$$\frac{n!}{b_H!\ b_J!\ b_F!}\ \frac{n!}{w_H!\ w_J!\ w_F!} = O(1)\ \frac{n^{2n+1}}{n_H^{2n_H+1}\ n_J^{2n_J+1}}$$

using reasoning similar to that of (e) of the proof of Lemma 5. (Note that $|\alpha_H| \leqslant k_H/(2r) \leqslant s(r-1)/(2r)$, $|\alpha_J| \leqslant s(r-1)/(2r)$ also, and $0 \leqslant b_F$, $w_F \leqslant s$).

(d) Possible LCBGs F', and possible edges between H and F and between J and F. All these are O(1).

For fixed (B_H, W_H) and (B_J, W_J) the number of possible LCBGs can be obtained by multiplying (b), (c) and (d) to get

$$O(1)\ n^{2n+1}\ \frac{n_H^{(r-2)n_H+s(r-1)/2-1/2}\ n_J^{(r-2)n_J+s(r-1)/2-1/2}}{r!^{2n}} \left(\frac{r}{e}\right)^{rn}$$

Since factor (a) is O(1), this is also the expression for $|P(r,s,n,p)|$. Replace n_H by $p/2$ and n_J by $n - s/2 - p/2$ to get (6); the formula in terms of $|R(r,n)|$ follows from Corollary 2 •

Lemma 9: *For* $r \geqslant 3$,

$$\left|\bigcup_{i=p}^{[n-s/2]} P(r,s,n,i)\right| = O(1)\ n^{(s-p(r-2))/2}\ |R(r,n)|$$

where O(1) *denotes a bound depending on* r, s *and* p.

Proof: Use Lemma 8 in the same way as Lemma 6 uses Lemma 5 •

Theorem 10: *Suppose* $r \geq 3$, *and* $q > 0$. *Let* $g = \max(4, 2\{q/(2(r-2))\})$ *and* $z = g(r-2) - q$ $(z \geq 0)$. *Then*

(i) *The number of* r*-connected* LCBGs *in* $R(r,n)$ *is*

$$(1-O(n^{-1/2}))\ |R(r,n)| \quad \textit{if} \quad r = 3,$$

$$\textit{and} \quad (1-O(n^{(3-r)/2}))\ |R(r,n)| \ \textit{if} \ r \geq 4 \quad .$$

where $O()$ *denotes a bound depending on* r.

(ii) *The number of cyclically-*q*-vertex-connected* LCBGs *in* $R(r,n)$ *is*

$$(1-O(n^{-(z+1)/2}))\ |C(r,g,n)|$$

where $O()$ *denotes a bound depending on* r *and* q.

(iii) *The number of cyclically-*$4(r-2)$*-vertex-connected* LCBGs *in* $R(r,n)$ *is*

$$(1-O(n^{-1/2}))\ |R(r,n)|$$

where $O()$ *denotes a bound depending on* r.

Proof: (i) The result for $r = 3$ follows from Theorem 7 (ii) because every cyclically-4-edge-connected 3-regular graph is 3-connected.

Therefore assume $r \geq 4$. Any non-r-connected $G \in R(r,n)$ is in some $P(r,s,n,p)$ where $0 \leq s < r$ and $2 \leq p \leq [n-s/2]$ $(p \neq 1$ because to isolate a single vertex r vertices must be removed). The number of these is

$$\left| \bigcup_{s=0}^{r-1} \bigcup_{p=2}^{[n-s/2]} P(r,s,n,p) \right| \leq \sum_{s=0}^{r-1} O(1)\ n^{(s-p(r-2))/2}\ |R(r,n)|$$

by Lemma 9. But for each s, $s - 2(r-2) \leq (r-1) - 2(r-2) = 3 - r$, and hence this becomes

$$\sum_{s=0}^{r-1} O(1)\ n^{(3-r)/2}\ |R(r,n)| = O(1)\ n^{(3-r)/2}\ |R(r,n)|.$$

Thus the number of r-connected r-regular labelled coloured bipartite graphs is $|R(r,n)| - O(1)\ n^{(3-r)/2}\ |R(r,n)| = (1-O(n^{(3-r)/2}))\ |R(r,n)|$.

(ii) and (iii) follow from Lemma 9 in the same way as Theorem 7 (i) and (ii) follow from Lemma 6 •

5. Concluding Remarks.

The theorems of Sections 3 and 4 above deal with r-regular LCBGs only for $r \geq 3$. For 1-regular LCBGs the situation is very simple: there are $n!$ 1-regular LCBGs with $2n$ vertices, all of which are disconnected (except when $n = 1$). 2-regular LCBGs possess no finer shades of connectivity than connected or disconnected. The only connected ones are those which are coloured labelled versions of the 2n-cycle. There are $n!(n-1)!/2$ of these. Using Corollary 2 and approximating by Stirling's formula, the proportion of 2-regular LCBGs on $2n$ vertices which are connected is therefore asymptotic to $\sqrt{(\pi e/n)}/2$ as $n \to \infty$.

It may be possible to obtain asymptotic connectivity results for LCBGs with bounded degree sequences, as Wormald has done for labelled graphs in [4]. A slightly easier task would be to examine (r,s)-biregular LCBGs (all the black vertices have degree r and all the white vertices degree s).

Using some recent results of McKay and Wormald [2] it should be possible to obtain results for (unlabelled uncoloured) regular bipartite graphs from Theorems 7 and 10.

REFERENCES

[1] Edward A. Bender, The asymptotic number of non-negative integer matrices with given row and column sums, *Discrete Math.* 10, (1974), 217-223.

[2] B.D. McKay and N.C. Wormald, Automorphisms of random graphs with specified degrees, preprint.

[3] Nicholas C. Wormald, *Some problems in the Enumeration of Labelled Graphs*, Ph.D. Thesis, University of Newcastle, N.S.W. (1978).

[4] Nicholas C. Wormald, The asymptotic connectivity of labelled regular graphs, *J. Combin. Theory Ser. B*, 31 (1981), 156-167.

KRONECKER PRODUCTS OF SYSTEMS OF ORTHOGONAL DESIGNS

H.M. GASTINEAU-HILLS

The concept of a system of orthogonal designs enables many of the construction techniques of orthogonal design theory to be unified and generalized by one theorem.

1. INTRODUCTION In [2] the following concept of a system of orthogonal designs was introduced:

1.1 Definition A K-*system* of orthogonal designs, of *order* n, *genus* $(\delta_{ij})_{1 \le i < j \le K}$ (each δ_{ij} from $\{0,1\}$), *type* $(u_{11},\ldots,u_{1p_1};u_{21},\ldots,u_{2p_2};\ldots;u_{K1},\ldots,u_{Kp_K})$ (each u_{ij} a positive integer), on the distinct commuting variables $x_{11},\ldots,x_{1p_1};x_{21},\ldots,x_{2p_2};\ldots;x_{K1},\ldots,x_{Kp_K}$ is a K-tuple $(X_1,X_2,\ldots,X_K)$ of order n matrices, where for each i, X_i has entries from $\{0,\pm x_{i1},\ldots,\pm x_{ip_i}\}$, and where

$$\text{(i)} \quad X_iX_i^T = \sum_{j=1}^{p_i} u_{ij}x_{ij}^2 \, I_n \qquad (1 \le i \le K);$$

$$\text{(ii)} \quad X_jX_i^T = (-1)^{\delta_{ij}} X_iX_j^T \qquad (1 \le i < j \le K).$$

The system is called *regular* if in addition the *hadamard product* $X_i * X_j$ (defined as componentwise multiplication) is zero whenever $\delta_{ij} = 1$ $(1 \le i < j \le K)$.

Thus for each i, X_i is an orthogonal design of order n, type $(u_{i1},\ldots,u_{ip_i})$ on the variables $x_{i1},\ldots,x_{ip_i}$ ((i) is in fact the standard defining condition for such an orthogonal design); and by (ii) for each $i < j$ the pair X_i,X_j is either *amicable* ($\delta_{ij} = 0$) or *anti-amicable* ($\delta_{ij} = 1$) ($X_jX_i^T = X_iX_j^T$ is the standard defining condition for amicability, and in [2] anti-amicability was defined by the condition $X_jX_i^T = -X_iX_j^T$). A regular system has the property that those of its designs which are anti-amicable with each other nowhere "overlap".

1.2 Examples

(i) Consider a single orthogonal design on p variables: $X = \sum_{i=1}^{p} x_iA_i$, where the A_i are $\{0,\pm1\}$-matrices such that $A_i * A_j = 0$ for $i \neq j$, and $XX^T = \sum_{i=1}^{p} u_ix_i^2I_n$. This implies $A_iA_i^T = u_iI_n$ $(1 \le i \le p)$ and $A_jA_i^T = -A_iA_j^T$ $(1 \le i < j \le p)$.

X may be identified with the 1-system (X), of order n, vacuous genus, type $(u_1,\ldots,u_p)$, on $x_1,\ldots,x_p$. This system is trivially regular.

Alternatively X may be identified with the p-system $(x_1A_1,\ldots,x_pA_p)$ of order n, genus (δ_{ij}) with all $\delta_{ij} = 1$, type $(u_1;\ldots;u_p)$ on $x_1;\ldots;x_p$. This system is regular since $A_i * A_j = 0$ for $i \neq j$.

It might be suggested that the systems (X) and $(x_1A_1,\ldots,x_pA_p)$ be regarded as equivalent. The concept of equivalence will be formalized later.

(ii) A K-tuple $(X_1,\ldots,X_K)$ of order n pairwise amicable orthogonal designs is a K-system of order n, genus (δ_{ij}) with all $\delta_{ij} = 0$. Since no δ_{ij} is 1, the system is trivially regular.

(iii) A *product design* is defined in [5] by a triple of orthogonal designs, of which one pair is amicable, and each other pair has zero hadamard product and yields an orthogonal design when added: for example (X,Y,Z) where $ZY^T = YZ^T$, $X * Y = X * Z = 0$, and

$X + Y$, $X + Z$ are orthogonal designs. Using the matrix identity $(L + M)(L + M)^T = LL^T + MM^T + LM^T + ML^T$ it is easy to see that these conditions on X,Y,Z are equivalent to: $YX^T = -XY^T$, $ZX^T = -XZ^T$, $ZY^T = YZ^T$, $X * Y = X * Z = 0$. Hence a product design is a regular 3-system with genus (δ_{ij}) where $\delta_{12} = \delta_{13} = 1$, $\delta_{23} = 0$.

(iv) A *repeat design* is defined in [6] as a $(K + 2)$-tuple $(R,P_1,\ldots,P_K,H)$ of orthogonal designs such that the pairs (R,H), (P_i,H) $(i = 1,\ldots,K)$, (P_i,P_j) $(i,j = 1,\ldots,K)$ are amicable, and for each i, $R * P_i = 0$ and $R + P_i$ is an orthogonal design. These conditions are equivalent to: $P_iR^T = -RP_i^T$, $HR^T = RH^T$, $P_jP_i^T = P_iP_j^T$, $HP_i^T = P_iH^T$, $R * P_i = 0$. Hence a repeat design is a regular $(K + 2)$-system with genus (δ_{ij}) where $\delta_{12} = \ldots = \delta_{1\,K+1} = 1$ and all other $\delta_{ij} = 0$.

Non-regular systems exist - for example $\left(\begin{pmatrix} x & x \\ -x & x \end{pmatrix}, \begin{pmatrix} y & -y \\ y & y \end{pmatrix}\right)$, which is a 2-system of order 2, genus $(\delta_{12} = 1)$, type $(2;2)$ on $x;y$. However in [2] the following was shown:

1.3 Lemma *A system of type* $(1,1,\ldots,1;\ldots;1,1,\ldots,1)$ *is regular.*

In [2] it was shown that given $p_1,\ldots,p_K$ and $(\delta_{ij})_{1\le i<j\le K}$, the orders of K-systems of genus (δ_{ij}) on $p_1,\ldots,p_K$ variables are all the multiples of some power of 2 which is easily calculated using the theory of Quasi Clifford Algebras. Furthermore if n is any multiple of this power of 2, it is easy to construct such a K-system of order n and of type $(1,1,\ldots,1;\ldots;1,1,\ldots,1)$ and which is therefore regular.

In the following a theory of kronecker products of systems is developed, and many classical construction techniques of orthogonal design theory are found to be particular cases of the one simple theorem. In the past the classical construction techniques together with what was then known of the possible orders of certain simple systems (such as amicable pairs) have been used to find new designs (see [7]). So it is to be hoped that the much more general results in [2] and here will be of significance.

2. EQUIVALENCE It is natural to regard as equivalent systems which can be obtained from each other by such operations as:

2.1 (i) renaming variables (distinct variables remaining distinct); (ii) changing signs of a variable throughout; (iii) changing signs of the same row (or column) of each design of the system; (iv) applying the same permutation to the rows (or columns) of each design; (v) transposing each design; (vi) reordering the components of the system (for example obtaining (Z,X,Y) from the (X,Y,Z) of 1.3(iii)).

2.2 Remarks (a) Each of the operations 2.1 produces from a K-system a K-system of the same order. In cases (i) to (v) the genus and type are unchanged, while in (vi) the genus and type components are merely reordered.

(b) A system obtained by any of the operations 2.1 is regular if and only if the original system is regular.

In the light of the discussion 1.2(i), it is intended to extend the concept of equivalence, so that a K-system may be equivalent to certain L-systems with L not necessarily equal to K. The following construction is the key to this extension:

2.3 Lemma *Suppose* $1 \le L \le K$ *and let* $S = (X_1, \ldots, X_{L-1}, X_L, \ldots, X_K)$ *be a* K-*system of order* n, *genus* $(\delta_{ij})_{1 \le i < j \le K}$, *type* (u_{ij}) *on the variables* x_{ij} $(1 \le j \le p_i;\ 1 \le i < K)$. *Suppose that for* $j > i \ge L$, $\delta_{ij} = 1$ *and* $X_i * X_j = 0$, *and that for each* $i < L$, $\delta_{iL} = \delta_{i\ L+1} = \ldots = \delta_{iK}$. *Then*

(a) $T = (X_1, \ldots, X_{L-1}, X_L + \ldots + X_K)$ *is an* L-*system of order* n, *on the variables* $x_1, \ldots, x_{1p_1}; \ldots; x_{L-1\ 1}, \ldots, x_{L-1\ p_{L-1}};$ $x_{L1}, \ldots, x_{Lp_L}, \ldots, x_{K1}, \ldots, x_{Kp_K}$. *Its genus is* $(\delta'_{ij})_{1 \le i < j \le L}$ *where for all* i,j, $\delta'_{ij} = \delta_{ij}$. *Its type is* $(u_{11}, \ldots, u_{1p_1}; \ldots; u_{L-1\ p_{L-1}};$ $u_{L1}, \ldots, u_{Lp_L}, \ldots, u_{K1}, \ldots, u_{Kp_K})$.

(b) T *is regular if and only if* S *is regular*.

Proof The conditions $X_i * X_j = 0$ $(j > i \ge L)$ ensure that $X_L + \ldots + X_K$ is a design. It is orthogonal, of type $(u_{L1}, \ldots, u_{Lp_L}, \ldots, u_{K1}, \ldots, u_{Kp_K})$, because $X_j X_i^T + X_i X_j^T = 0$ for $j > i \ge L$ (since $\delta_{ij} = 1$) so $(X_L + \ldots + X_K)(X_L + \ldots + X_K)^T = X_L X_L^T + \ldots + X_K X_K^T = \sum_{i=L}^{K} (\sum_{j=1}^{p_i} u_{ij} x_{ij}^2) I_n$.

If $1 \le i < j < L$, $X_j X_i^T = (-1)^{\delta_{ij}} X_i X_j^T = (-1)^{\delta'_{ij}} X_i X_j^T$, and if $1 \le i < L$ we have

$$(X_L + \dots + X_K)X_i^T = X_L X_i^T + \dots + X_K X_i^T$$

$$= (-1)^{\delta_{iL}} X_i X_L^T + \dots + (-1)^{\delta_{iK}} X_i X_K^T$$

$$= (-1)^{\delta'_{iL}} X_i (X_L + \dots + X_K)^T$$

(since $\delta'_{iL} = \delta_{iL}$ and $\delta_{iL} = \dots = \delta_{iK}$).

It follows that T is an L-system of the stated genus and type. Part (b) follows from $X_i * (X_L + \dots + X_K) = X_i * X_L + \dots + X_i * X_K$.

Putting $L = 1$ in 2.3 we deduce the following:

2.4 Corollary *If* $(X_1,\dots,X_K)$ *is a regular* K-*system of order* n, *genus* (δ_{ij}) *with all* δ_{ij} *equal to* 1, *type* $(u_{11},\dots,u_{1p_1};\dots;u_{K1},\dots,u_{Kp_K})$, *then* $X_1 + \dots + X_K$ *is an orthogonal design of order* n, *type* $(u_{11},\dots,u_{1p_1},\dots,u_{K1},\dots u_{Kp_K})$.

2.5 Definition A K-system S and an L-system T are called *equivalent* if there is a sequence of systems $S = S_0, S_1,\dots,S_m = T$ such that for each $i = 1,\dots,m$, one of S_{i-1}, S_i may be obtained from the other by one of the operations 2.1, or by the construction 2.3.

Clearly equivalence is an equivalence relation on the class of systems of orthogonal designs. Also, by 2.2(b) and 2.3(b), if a system is regular so is any equivalent system. Hence we may speak of an "equivalence class" of systems, and identify any such class as being regular or not, according as all or none of its member systems is regular. Note that the two systems discussed in 1.2(i) are equivalent according to 2.5.

3. KRONECKER PRODUCTS Many constructions in orthogonal design theory have been expressed in terms of kronecker products of matrices. In order to obtain the desired unification and generalization of these constructions, we now introduce the concept of a kronecker product of systems of orthogonal designs.

3.1 Definition Let $A = (a_{ij})$, $B = (b_{k\ell})$ be $m \times m$, $n \times n$ matrices respectively, over a commutative ring R. By the *kronecker product* $A \otimes B$ we mean the $mn \times mn$ matrix whose $(i + m(k - 1), j + m(\ell - 1))$ element is $a_{ij}b_{k\ell}$.

In other words $A \otimes B$ is formed by replacing each $b_{k\ell}$ in B by the block $A\,b_{k\ell}$ (see [1]). Many writers define the kronecker product the other way (they would write $(A\,b_{k\ell})$ as $B \otimes A$), but this is of no importance.

The kronecker product is associative, and both left- and right-distributive over matrix addition. $I_m \otimes I_n = I_{mn}$, and for $r \in R$, $(rA) \otimes B = A \otimes (rB) = r(A \otimes B)$. Another property we shall need is the following:

3.2 Lemma *If for* $1 \le \ell \le r$, $A^{(\ell)}$, $B^{(\ell)}$ *are* $n^{(\ell)} \times n^{(\ell)}$ *matrices over a commutative ring, then*

$$(A^{(1)} \otimes \dots \otimes A^{(r)})(B^{(1)} \otimes \dots \otimes B^{(r)})^T$$

$$= (A^{(1)}B^{(1)\,T}) \otimes \dots \otimes (A^{(r)}B^{(r)T}).$$

Proof The case $(A^{(1)} \otimes A^{(2)})(B^{(1)} \otimes B^{(2)})^T = (A^{(1)}B^{(1)T}) \otimes (A^{(2)}B^{(2)T})$ is verified by checking corresponding elements of each side. Thus if $A^{(1)} = (a_{ij}^{(1)})$, $A^{(2)} = (A_{k\ell}^{(2)})$, $B^{(1)} = (b_{i'j'}^{(1)})$, $B^{(2)} = (b_{k'\ell'}^{(2)})$, then the $(i + n^{(1)}(k-1),\ i' + n^{(1)}(k'-1))$ element of $(A^{(1)} \otimes A^{(2)})(B^{(1)} \otimes B^{(2)})^T$ is $\sum_\ell \sum_j (a_{ij}^{(1)} a_{k\ell}^{(2)})(b_{i'j}^{(1)} b_{k'j}^{(2)})$ while the corresponding element of $(A^{(1)}B^{(1)T}) \otimes (A^{(2)}B^{(2)T})$ is $(\sum_j a_{ij}^{(1)} b_{i'j}^{(1)})(\sum_\ell a_{k\ell}^{(2)} b_{k'\ell}^{(2)})$ - clearly the same.

The general result follows by induction on r.

The kronecker product of two or more *designs* (matrices with each entry either zero or a signed variable) is not in general a design, since products of variables could appear. However it is a different matter if we take a kronecker product one factor of which is a design, and all the other factors of which are $\{0,\pm1\}$-matrices. Indeed:

3.3 Lemma *Let* $X^{(1)},\dots,X^{(r)}$ *be orthogonal designs of order* $n^{(1)},\dots,n^{(r)}$, *types* $(u_1^{(1)},\dots,u_{p^{(1)}}^{(1)}),\dots,(u_1^{(r)},\dots,u_{p^{(r)}}^{(r)})$ *on variables* $x_1^{(1)},\dots,x_{p^{(1)}}^{(1)};\dots;x_1^{(r)},\dots,x_{p^{(1)}}^{(r)}$ *respectively. Suppose we choose* k *such that* $1 \le k \le r$, *and replace all the variables of all the designs except* $X^{(k)}$ *by* 1 *(making* $\{0,\pm1\}$*-matrices), then form* $X = X^{(1)} \otimes \dots \otimes X^{(r)}$. *Then* X *is an orthogonal design of* *order* $n^{(1)}\ldots n^{(r)}$, *type* $(u^{(k)}u_1^{(k)},\ldots,u^{(k)}u_{p^{(k)}}^{(k)})$ *where*

$$u^{(k)} = \prod_{\ell\neq k}\Big(\sum_{j=1}^{p^{(\ell)}} u_j^{(\ell)}\Big),$$ *on the variables* $x_1^{(k)},\ldots,x_{p^{(k)}}^{(k)}$.

Proof After replacement we have

$$XX^T = (X^{(1)} \otimes \ldots \otimes X^{(r)})(X^{(1)} \otimes \ldots \otimes X^{(r)})^T$$

$$= (X^{(1)}X^{(1)T}) \otimes \ldots \otimes (X^{(r)}X^{(r)T}) \qquad \text{(by 3.2)}$$

$$= \Big(\sum_{j=1}^{p^{(1)}} u_j^{(1)} I_{n^{(1)}}\Big) \otimes \ldots \otimes \Big(\sum_{j=1}^{p^{(k)}} u_j^{(k)}\big(x_j^{(k)}\big)^2 I_{n^{(k)}}\Big) \otimes \ldots$$

$$\otimes \Big(\sum_{j=1}^{p^{(r)}} u_j^{(r)} I_{n^{(r)}}\Big)$$

$$= \Big(\sum_{j=1}^{p^{(k)}} u^{(k)}u_j^{(k)}\big(x_j^{(k)}\big)^2\Big) I_{n^{(1)}\ldots n^{(r)}}$$ and the result follows.

We generalize this lemma from single orthogonal designs to systems of designs.

3.4 Theorem (The kronecker product theorem) *For each* $\ell = 1,\ldots,r$ *let* $(X_1^{(\ell)},\ldots,X_K^{(\ell)})$ *be a* K-*system of order* $n^{(\ell)}$, *genus* $(\delta_{ij}^{(\ell)})$, $(1 \le i < j \le K)$, *type* $(u_{ij}^{(\ell)})$ $(1 \le j \le p_i^{(\ell)};\ 1 \le i \le K)$, *on variables* $x_{ij}^{(\ell)}$ $(1 \le j \le p_i^{(\ell)};\ 1 \le i \le K)$. *Suppose that for each* $i = 1,\ldots,K$ *we choose* k_i *such that* $1 \le k_i \le r$, *and replace all the variables of all of* $X_i^{(1)},X_i^{(2)},\ldots,X_i^{(r)}$ *except* $X_i^{(k_i)}$ *by* 1, *then form* $W = (X_1^{(1)} \otimes \ldots \otimes X_1^{(r)}, X_2^{(1)} \otimes \ldots \otimes X_2^{(r)},\ldots,X_K^{(1)} \otimes \ldots \otimes X_K^{(r)})$.

Then W *is a* K-*system of order* $n^{(1)}\dots n^{(r)}$, *genus* $(\delta_{ij})_{1\le i<j\le K}$ *where* $\delta_{ij} = \delta_{ij}^{(1)} + \dots + \delta_{ij}^{(r)}$ *reduced modulo* 2, *type* $(u_i^{(k_i)} u_{ij}^{(k_i)})$ $(1 \le j \le p_i^{(k_i)};\ 1 \le i \le K)$ *where for each* i, $u_i^{(k_i)} = \prod_{\ell\ne k}\left(\sum_j u_{ij}^{(\ell)}\right)$, *on the variables* $x_{ij}^{(k_i)}$ $(1 \le j \le p_i^{(k_i)};\ 1 \le i \le K)$.

Furthermore if each of $(X_1^{(\ell)},\dots,X_K^{(\ell)})$ *is regular* $(\ell = 1,\dots,r)$, *so is* W.

Proof From 3.3 each $X_i^{(1)} \otimes \dots \otimes X_i^{(r)}$ is an orthogonal design of order $n^{(1)}\dots n^{(r)}$, type $(u_i^{(k_i)} u_{ij}^{(k_i)})$ on the variables $x_{ij}^{(k_i)}$ $(1 \le j \le p_i^{(k_i)})$. Now for $i < j$,

$$(X_j^{(1)} \otimes \dots \otimes X_j^{(r)})(X_i^{(1)} \otimes \dots \otimes X_i^{(r)})^T$$

$$= (X_j^{(1)}X_i^{(1)T}) \otimes \dots \otimes (X_j^{(r)}X_i^{(r)T}) \qquad \text{(by 3.2)}$$

$$= ((-1)^{\delta_{ij}^{(1)}} X_i^{(1)}X_j^{(1)T}) \otimes \dots \otimes ((-1)^{\delta_{ij}^{(r)}} X_i^{(r)}X_j^{(r)T})$$

$$= (-1)^{\delta_{ij}^{(1)}+\dots+\delta_{ij}^{(r)}} (X_i^{(1)}X_j^{(1)T}) \otimes \dots \otimes (X_i^{(r)}X_j^{(r)T})$$

$$= (-1)^{\delta_{ij}}(X^{(1)} \otimes \dots \otimes X^{(r)})(X^{(1)} \otimes \dots \otimes X^{(r)})^T$$

(again by 3.2)

Hence W is a K-system of the stated genus and type. The statement on regularity follows from the facts that if $\delta_{ij} = 1$ then at least one of $\delta_{ij}^{(1)},\ldots,\delta_{ij}^{(r)} = 1$, and if any of $X_i^{(1)} * X_j^{(1)},\ldots,X_i^{(r)} * X_j^{(r)}$ is zero then so is $(X_i^{(1)} \otimes \ldots \otimes X_i^{(r)}) * (X_j^{(1)} \otimes \ldots \otimes X_j^{(r)})$.

We call a system W formed as in 3.4 a *kronecker product* of the systems $(X_1^{(\ell)},\ldots,X_K^{(\ell)})$ $(\ell = 1,\ldots,r)$. In general there are several such W for each set $\{(X_1^{(\ell)},\ldots,X_K^{(\ell)})\}$, by choice of the k_i — in other words by choice of which designs do not have their variables replaced by 1. However given such a choice of designs, by 2.1(vi) the kronecker product W is to within equivalence independent of the ordering of the set $\{(X_1^{(\ell)},\ldots,X_K^{(\ell)})\}$. (Reordering the factors merely permits rows and columns of a kronecker product of a set of matrices.)

The K-system W of 3.4 may at times be reducible to an equivalent L-system $(L < K)$ using 2.3 (possibly after reordering as in 2.1(vi)). Of special interest is the case when W is reducible to a single orthogonal design. The following definition and corollary describe this situation.

3.5 Definition A set of K-systems $\{(X_1^{(1)},\ldots,X_K^{(1)}),\ldots,(X_1^{(r)},\ldots,X_K^{(r)})\}$ is called a *compatible* set if for each $i < j$ the number of the pairs $(X_i^{(1)},X_j^{(1)}),\ldots,(X_i^{(r)},X_j^{(r)})$ which are anti-amicable is odd (in the notation of 3.4, each $\delta_{ij} = 1$).

In the case $r = 2$, we may speak of the (two) systems being compatible (but for $r > 2$, 3.5 does not mean the systems are pairwise compatible).

From 2.4 and 3.4 the following is immediate:

3.6 Corollary *If the* K*-systems* $(X_1^{(1)},\ldots,X_K^{(1)}),\ldots,(X_1^{(r)},\ldots,X_K^{(r)})$ *are regular and form a compatible set, then for any choice of the* k_i, *the construction* 3.4 *yields an orthogonal design*

$$X_1^{(1)} \otimes \ldots \otimes X_1^{(r)} + \ldots + X_K^{(1)} \otimes \ldots \otimes X_K^{(r)}$$

(Recall that "most" variables have been replaced by 1).

4. EXAMPLES We conclude by considering some well-known constructions of orthogonal design theory and showing that they are all particular cases of 3.4 (or 3.6).

4.1 Construction (Geramita, Geramita, Wallis - see [3]) *If* $X = x_1A_1 + \ldots + x_KA_K$ *is an orthogonal design, and* $(Y_1,\ldots,Y_K)$ *is an amicable* K*-tuple of orthogonal designs, then* $Y_1 \otimes A_1 + \ldots + Y_K \otimes A_K$ *is an orthogonal design.*

This is a clear case of 3.6. The K-systems $(x_1A_1,\ldots,x_KA_K)$ and $(Y_1,\ldots,Y_K)$ are regular and compatible (the genus of $(x_1A_1,\ldots,x_KA_K)$ is $(\delta_{ij}^{(1)})$ with all $\delta_{ij}^{(1)} = 1$, and the genus of $(Y_1,\ldots,Y_K)$ is $(\delta_{ij}^{(2)})$ with all $\delta_{ij}^{(2)} = 0$). Certain variables, namely the x_{ij} are replaced by 1 as required.

4.2 Construction (Robinson - see [5]) *If* S, $T = t_1A + t_2B$ *are amicable orthogonal designs, and* $(X = xC, Y, Z)$ *is a product design (as defined* 1.2(iii)) *then* $S \otimes C + A \otimes Y + B \otimes Z$ *is an orthogonal design.*

This is also a case of 3.6. The 3-systems:

(S, t_1A, t_2B) (genus: $\delta_{12}^{(1)} = 0$, $\delta_{13}^{(1)} = 0$, $\delta_{23}^{(1)} = 1$)

and (xC, Y, Z) (genus: $\delta_{12}^{(2)} = 1$, $\delta_{13}^{(2)} = 1$, $\delta_{23}^{(2)} = 0$)

are regular and compatible, and the variables chosen to be replaced by 1 are x, t_1, t_2.

4.3 Construction (Robinson, Seberry - see [6]) *If* $(X = xC,\ Y = y_1B_1 + \ldots + y_KB_K,\ Z)$ *is a product design (as defined* 1.2(iii)*), and* $(R = rA, P_1, \ldots, P_K, H)$ *is a repeat design (as defined* 1.2(iv)*), then* $A \otimes Z + P_1 \otimes B_1 + \ldots + P_K \otimes B_K + H \otimes C$ *is an orthogonal design.*

Again 3.6 applies, for the K+2-systems $(rA, P_1, \ldots, P_K, H)$ and $(Z, y_1B_1, \ldots, y_KB_K, xC)$ are regular and compatible.

4.4 Construction (Geramita, Seberry - see [4], p.273) *If* (X,Y,Z) *is a product design (as defined* 1.2(iii)*) and* $X = xA$ *is a one-variable orthogonal design, then* $\begin{pmatrix} x_1A + Y & x_2A + Z \\ -x_2A + Z & x_1A - Y \end{pmatrix}$ *is an orthogonal design.*

The 4-systems (x_1A, Y, x_2A, Z)

(genus: $\delta_{12}^{(1)} = 1$, $\delta_{13}^{(1)} = 0$, $\delta_{14}^{(1)} = 1$, $\delta_{23}^{(1)} = 1$, $\delta_{24}^{(1)} = 0$, $\delta_{34}^{(1)} = 1$)

and $t_1\begin{pmatrix} 1 & 0 \\ 0 & 1 \end{pmatrix}$, $t_2\begin{pmatrix} 1 & 0 \\ 0 & -1 \end{pmatrix}$, $t_3\begin{pmatrix} 0 & 1 \\ -1 & 0 \end{pmatrix}$, $t_4\begin{pmatrix} 0 & 1 \\ 1 & 0 \end{pmatrix}$

(genus: $\delta_{12}^{(2)} = 0$, $\delta_{13}^{(2)} = 1$, $\delta_{14}^{(2)} = 0$, $\delta_{23}^{(2)} = 0$, $\delta_{24}^{(2)} = 1$, $\delta_{34}^{(2)} = 0$)

are regular and compatible, so 3.6 with $t_1 = t_2 = t_3 = t_4 = 1$ gives an orthogonal design

$$x_1 A \otimes \begin{pmatrix} 1 & 0 \\ 0 & 1 \end{pmatrix} + Y \otimes \begin{pmatrix} 1 & 0 \\ 0 & -1 \end{pmatrix} + x_2 A \otimes \begin{pmatrix} 0 & 1 \\ -1 & 0 \end{pmatrix} + Z \otimes \begin{pmatrix} 0 & 1 \\ 1 & 0 \end{pmatrix}$$

- clearly the construction 4.4 .

4.5 Construction (Wolfe - see [7])

$X = \sum_1^s x_i A_i$, $Y = \sum_1^t y_i B_i$ *are amicable orthogonal designs, and*

$Z = \sum_1^u z_i C_i$, $W = \sum_1^v w_i D_i$ *are amicable orthogonal designs, then*

$$P = \sum_{i=1}^{u-1} z_i (B_1 \otimes C_i) + \sum_{i=1}^{s} x_i (A_i \otimes C_u),$$

$$Q = \sum_{i=1}^{v} w_i (B_1 \otimes D_i) + \sum_{i=2}^{t} y_i (B_i \otimes C_u)$$ *are amicable orthogonal*

designs.

This may be seen as an application of 3.4 to the two regular

4-systems $(bB_1, \sum_{i=1}^{s} x_i A_i, b'B_1, \sum_{i=2}^{t} y_i B_i)$

(genus: $\delta_{12}^{(1)} = 0$, $\delta_{13}^{(1)} = 0$, $\delta_{14}^{(1)} = 1$, $\delta_{23}^{(1)} = 0$, $\delta_{24}^{(1)} = 0$, $\delta_{34}^{(1)} = 1$)

and $(\sum_{i=1}^{u-1} z_i C_i, cC_u, \sum_{i=1}^{v} w_i D_i, c'C_u)$.

(genus: $\delta_{12}^{(2)} = 1$, $\delta_{13}^{(2)} = 0$, $\delta_{14}^{(2)} = 1$, $\delta_{23}^{(2)} = 0$, $\delta_{24}^{(2)} = 0$, $\delta_{34}^{(2)} = 0$),

putting $b = c = b' = c' = 1$.

These 4-systems are not compatible, but yield the regular 4-system

$$\mathcal{W} = \left(\sum_{i=1}^{u-1} z_i B_i \otimes C_i,\ \sum_{i=1}^{s} x_i A_i \otimes C_u,\ \sum_{i=1}^{v} w_i (B_1 \otimes D_i),\ \sum_{i=2}^{t} y_i (B_i \otimes C_u)\right)$$

of genus $\delta_{12} = 1$, $\delta_{13} = 0$, $\delta_{14} = 0$, $\delta_{23} = 0$, $\delta_{24} = 0$, $\delta_{34} = 1$.

Two applications of 2.3 to $\mathcal{W}$ (with components suitably reordered as in 2.1(iv)) show that the anti-amicable pairs of components of $\mathcal{W}$ (the first, second and the third, fourth) may be added giving the (equivalent) 2-system (P,Q), of genus (δ'_{ij}) given by $\delta'_{12} = 0$. The result 4.5 follows.

4.6 Construction (Robinson - see [5]) *If* $(X, Y, Z = z_1A_1 + z_2A_2)$ *is a product design (as defined* 1.2(iii)*) and* $s = sC$, $T = t_1B_1 + t_2B_2$ *are amicable orthogonal designs, then* $(z_2B_2 \otimes A_2 + C \otimes X,\ z_1B_2 \otimes A_1 + B_1 \otimes Y,\ z_1'C \otimes A_1 + z_2'B_1 \otimes A_2)$ *is a product design.*

This comes from applying 3.4 to the regular 6-systems: $(t_2B_2, sC, t_2'B_2, t_1B_1, s'C, t_1'B_1)$, $(z_2A_2, X, z_1A_1, Y, z_1'A_1, z_2'A_2)$, putting $t_2 = s = t_2' = t_1 = s' = t_1' = 1$. We get a regular 6-system but again pairs of anti-amicable designs may be added reducing the system to an equivalent 3-system which is the required product design.

4.7 Construction (Robinson - see [5]) *If* S, $X = x_1A_1 + x_2A_2$ *and* T, $Y = y_1B_1 + y_2B_2$ *are amicable pairs of orthogonal designs, then* $(zA_2 \otimes B_2,\ S \otimes B_1 + xA_1 \otimes B_2,\ A_1 \otimes T + yA_2 \otimes B_1)$ *is a product design (as defined* 1.2(iii)*).*

This comes from applying 3.4 to two regular 5-systems: $(zA_2, S, xA_1, x'A_1, z'A_2)$, $(tB_2, y_1B_1, y_2B_2, T, yB_1)$ putting $t = y_1 = y_2 = x' = z' = 1$. The resulting regular 5-system may be reduced to the required product design by adding two pairs of anti-amicable designs.

4.8 Remark The above constructions make use of 3.4 (or 3.6) for the case $r = 2$ only. However if, for example, we used 4.6 to construct a new product design, which we then used in 4.4 to construct a new orthogonal design, it could be regarded as a single application of 3.6 to a suitably chosen compatable set of three systems.

The author would like to thank Dr Jennifer Seberry for her help and encouragement.

REFERENCES

[1] M. Burrow, *Representation Theory of Finite Groups* (Academic Press, New York and London, 1965).

[2] H.M. Gastineau-Hills, *Quasi Clifford Algebras and Systems of Orthogonal Designs*, J. Aust. Math. Soc. [To appear]

[3] A.V. Geramita, J.M. Geramita, J. Seberry (Wallis), *Orthogonal Designs*, Linear and Multilinear Algebra, 3(1975/76), 281-306.

[4] A.V. Geramita, J. Seberry, *Orthogonal Designs: Quadratic Forms and Hadamard Matrices*, (Marcel Dekker, New York, 1979).

[5] P.J. Robinson, *Using Product Designs to Construct Orthogonal Designs*, Bull. Austral. Math. Soc., 16(1977), 297-305.

[6] P.J. Robinson, J. Seberry, *Orthogonal Designs in Powers of Two*, Ars Combinatoria, 4(1977), 43-57.

[7] W. Wolfe, *Orthogonal Designs - Amicable Orthogonal Designs* (Ph.D. Thesis, Queen's University Kingston, Ontario, Canada, 1975).

KRONECKER PRODUCTS OF SYSTEMS OF HIGHER DIMENSIONAL ORTHOGONAL DESIGNS

H.M. GASTINEAU-HILLS AND JOSEPH HAMMER

ABSTRACT

We generalize the results of paper [1] on systems of orthogonal designs to a theory of systems of general higher dimensional designs and present several new constructions for such designs.

Introduction: When n^2 elements are given they can be arranged in the form of a square, similarly when n^g elements ($g \geqq 3$ an integer) are given they can be arranged in the form of a g-*dimensional cube of side* n (in short a g-*cube*).

Let $X = (a_{i_1 \ldots i_g})$ be a g-cube of side n. We denote a (g-r)-dimensional layer (in short a (g-r)-layer) of X obtained by fixing the values of $i_{\alpha_1}, \ldots, i_{\alpha_r}$ at $\underline{i}_{\alpha_1}, \ldots, \underline{i}_{\alpha_r}$, respectively by

$$X(i_1, \ldots, \underline{i}_{\alpha_1}, \ldots, \underline{i}_{\alpha_2}, \ldots, \underline{i}_{\alpha_r}, \ldots, i_g)$$

where each $\underline{i}_{\alpha_1}, \ldots, \underline{i}_{\alpha_r}$ is an integer between 1, n.

For example, let X be a 5-cube $(a_{ijk\ell m})$ of side 3. Then $X(\underline{1}, j, \underline{2}, \ell, \underline{2})$ denotes the 2-layer

$$\begin{pmatrix} a_{11212} & a_{11222} & a_{11232} \\ a_{12212} & a_{12222} & a_{12232} \\ a_{13212} & a_{13222} & a_{13232} \end{pmatrix}$$

1st, 3rd, 5th subscripts fixed.

$X(\underline{i}_1,\underline{i}_2,\ldots,\underline{i}_g)$ is a 0-layer or single element (all subscripts fixed). $X(\underline{i}_1,i_2,\ldots,i_g)$ is a $(g-1)$-layer (one subscript fixed). $X(i_1,i_2,\ldots,i_g)$ is the g-layer or the whole cube X. (No subscript fixed).

(For more information on g-cubes consult [2]).

In this paper we shall generalize the results of [1] on systems of orthogonal designs to a theory of systems of general higher dimensional designs.

<u>(1.1) DEFINITION:</u> A K-system of dimension g, side n, genus $(\delta_{ij})(1 \le i < j \le k)$, type $(u_{ij})(1 \le i \le p_i;\ 1 \le j \le k)$, on variables x_{ij} $(1 \le i \le p_i;\ 1 \le j \le k)$ $(\delta_{ij} \in \{0,1\}, u_{ij}$ positive integers, p_i non-negative integers) is an ordered K-tuple of g-cubes, side n

$$(X_1,\ldots,X_K)$$

where each X_i has entries from $\{0,\pm x_1,\ldots,\pm x_{p_i}\}$ and

(i) For each $i = 1,\ldots,K$, and for each 2-layer

$$Z_i = X_i(\underline{i}_1,\ldots,\underline{i}_{\alpha-1},i_\alpha,\underline{i}_{\alpha+1},\ldots,\underline{i}_{\beta-1},i_\beta,\underline{i}_{\beta+1},\ldots,\underline{i}_g) \text{ of } X_i:$$

$$Z_i Z_i^T = \sum_{j=1}^{p_i} u_{ij} x_{ij}^2 I_n$$

(ii) For each i,j $(1 \le i < j \le k)$ and for each

$Z_i = X_i(\underline{i}_1,\ldots,\underline{i}_{\alpha-1},i_\alpha,\underline{i}_{\alpha+1},\ldots,\underline{i}_{\beta-1},i_\beta,\underline{i}_{\beta+1},\ldots,\underline{i}_g)$, a 2-layer of X_i and $Z_j = X_j(\underline{j}_1,\ldots,\underline{j}_{\alpha-1},j_\alpha,\underline{j}_{\alpha+1},\ldots,\underline{j}_{\beta-1},j_\beta,\underline{j}_{\beta+1},\ldots,\underline{j}_g)$, the *corresponding* 2-layer of X_j (so that $\underline{i}_\gamma = \underline{j}_\gamma$ for all $\gamma \ne \alpha,\beta$) $Z_j Z_i^T = (-1)^{\delta_{ij}} Z_i Z_j^T$

The system is called *regular* if in addition, for each $i < j$ for which $\delta_{ij} = 1$

$$X_i \circledH X_j = 0 \quad \text{(block of zeros)}$$

where $\circledH$ denotes the Hadamard product.

<u>Theorem</u>: Let $\{W^{(1)} = (X_1^{(1)},\dots,X_K^{(1)}),\dots, W^{(r)} = (X_1^{(r)},\dots,X_K^{(r)})\}$ be r K-systems, where for each $\ell = 1,\dots,r$, $W^{(\ell)}$ is of side $n^{(\ell)}$, genus $(\delta_{ij}^{(\ell)})$ $(1 \le i < j \le K)$, type $(u_{ij}^{(\ell)})$ $(1 \le j \le p_i^{(\ell)}, 1 \le i \le K)$ on variables $x_{ij}^{(\ell)}$ $(1 \le j \le p_i^{(\ell)}, 1 \le i \le K)$.

For each $i = 1,\dots,K$ replace all variables of all but one of $X_1^{(1)},\dots,X_i^{(r)}$, (say all but $X_i^{(k_i)}$) by 1 and form

$$W = \left(X_1^{(1)} \otimes\dots\otimes X_1^{(r)},\dots,X_K^{(1)} \otimes\dots\otimes X_K^{(r)}\right)$$

(Written $W^{(1)} \otimes\dots\otimes W^{(r)} \mid (k_1,\dots,k_K)$.) where $\otimes$ is the Kronecker product.

Then W is a K-system of side $n^{(1)}\dots n^{(r)}$, genus (δ_{ij}) $(1 \le i < j \le K)$ where $\delta_{ij} \equiv \delta_{ij}^{(1)}+\dots+\delta_{ij}^{(r)}$ mod 2, type $(u_i^{(k_i)}u_{ij}^{(k_i)})$ $(1 \le j \le p_i^{(k_i)}, 1 \le i \le K)$ where $u_i^{(k_i)} = \prod_{\ell\ne k_i} (\sum_j u_{ij}^{(\ell)})$, on the variables $x_{ij}^{(k_i)}$ $(1 \le j \le p_i, 1 \le i \le K)$.

W is <u>regular</u> if $W^{(1)},\dots,W^{(r)}$ are all regular.

To prove this theorem we need the following lemmas:

<u>Lemma (1)</u> A 2-layer $X^{(1)} \otimes\dots\otimes X^{(r)}(\underline{i}_1,\dots,\underline{i}_{\alpha-1},i_\alpha,\underline{i}_{\alpha+1},\dots,\underline{i}_{\beta-1},i_\beta,\underline{i}_{\beta+1},\dots,\underline{i}_g)$ of a kroneckerproduct $X^{(1)} \otimes\dots\otimes X^{(r)}$ is the following kroneckerproduct of 2-layers $X^{(1)}(\underline{i}_1^{(1)},\dots,\underline{i}_{\alpha-1}^{(1)},i_\alpha^{(1)},\underline{i}_{\alpha+1}^{(1)},\dots,\underline{i}_{\beta-1}^{(1)},i_\beta^{(1)},\underline{i}_{\beta+1}^{(1)},\dots,\underline{i}_g^{(1)}) \otimes\dots$

$\dots \otimes X^{(r)}(\underline{i}_1^{(r)},\dots,\underline{i}_{\alpha-1}^{(r)},i_\alpha^{(r)},\underline{i}_{\alpha+1}^{(r)},\dots,\underline{i}_{\beta-1}^{(r)},i_\beta^{(r)},\underline{i}_{\beta+1}^{(r)},\dots,\underline{i}_g^{(r)})$

where for each $\gamma \ne \alpha,\beta$

$$\underline{i}_\gamma-1 = (\underline{i}_\gamma^{(1)}-1) + n^{(1)}(\underline{i}_\gamma^{(2)}-1) + n^{(1)}n^{(2)}(\underline{i}_\gamma^{(3)}-1) + \dots$$

$$\dots + n^{(1)}n^{(2)}\dots n^{(r-1)}(\underline{i}_\gamma^{(r)}-1). \quad (1 \le \underline{i}_\gamma^{(\ell)} \le n^{(\ell)}).$$

(Clearly $\underline{i}_\gamma^{(1)}-1$ is the residue of $\underline{i}_\gamma-1$ modulo $n^{(1)}$, $i_\gamma^{(2)}-1$ is the residue of $\left((i_\gamma-1) - (i_\gamma^{(1)}-1)\right)/n^{(1)}$ modulo $n^{(2)}$ and so on.)

For example the 9th row of the Kronecker product

$$A \otimes E \otimes R = \begin{pmatrix} a & b \\ c & d \end{pmatrix} \otimes \begin{pmatrix} e & f \\ g & h \end{pmatrix} \otimes \begin{pmatrix} r & s & t \\ u & v & w \\ x & y & z \end{pmatrix} \quad \text{is the row}$$

$$a\,e\,x \quad b\,e\,x \quad a\,e\,y \quad b\,e\,y \quad a\,f\,x \quad b\,f\,x \quad \ldots$$

using rows 1, 1, 3 of A, E, R respectively

$$\left(9-1 = 1 - 1 + 2(1-1) + 2 \times 2(3-1)\right).$$

We define the kronecker product of cubes $H = (h_{ij})$ by $K = (k_{ij}\ldots)$ to be the cube obtained by replacing the $ij\ldots$ entry of K by the cube $H\ k_{ij}\ldots$.

<u>Lemma (2)</u> Suppose that for $1 \le i \le r$, $A^{(i)}$, $B^{(i)}$ are $n^{(i)} \times n^{(i)}$ matrices over a commutative ring. Then

$$(A^{(1)} \otimes \ldots \otimes A^{(r)})(B^{(1)} \otimes \ldots \otimes B^{(r)})^T = (A^{(1)}B^{(1)T}) \otimes \ldots \otimes (A^{(r)}B^{(r)T})$$

This is proved in [2]

<u>Lemma (3)</u> Suppose that for $1 \le i \le r$, $A^{(i)}$, $B^{(i)}$ are $n^{(i)} \times n^{(i)}$ matrices over a commutative ring. Then

$$(A^{(1)} \otimes \ldots \otimes A^{(r)}) \textcircled{H} (B^{(1)} \otimes \ldots \otimes B^{(r)}) = (A^{(1)} \textcircled{H} B^{(1)}) \otimes \ldots \otimes (A^{(r)} \textcircled{H} B^{(r)}).$$

This is obvious since an element of either expression is obtained by taking certain elements of $A^{(1)}, \ldots, A^{(r)}$ and *corresponding* elements of $B^{(1)}, \ldots, B^{(r)}$ respectively, and forming their product.

<u>Proof of Theorem:</u>

(1) W is an ordered K-tuple of g-cubes of side $n^{(1)} \ldots n^{(r)}$, since a kronecker product of r g-cubes of sides $n^{(1)}, \ldots, n^{(r)}$ is a g-cube of side $n^{(1)} \ldots n^{(r)}$. The ith g-cube $X_i = X_i^{(1)} \otimes \ldots \otimes X_i^{(r)}$ of W is a kronecker

product of r-1 $\{0,\pm 1\}$ g-cubes and the g-cube $X_i^{(k_i)}$ whose entries are from $\{0,\pm x_{i_1}^{(k_i)},\ldots,\pm x_{i_{p_i}}^{(k_i)}\}$ so x_i itself has entries from $\{0,\pm x_{i_1}^{(k_i)},\ldots,\pm x_{i_{p_i}}^{(k_i)}\}$.

(2) Let $i \in \{1,\ldots,K\}$, and let $Z_i = Y_i(\underline{i}_1,\ldots,i_\alpha,\ldots,i_\beta,\ldots,\underline{i}_g)$ be a 2-layer of the ith component of W.

For $\gamma \neq \alpha$ or β, let $\underline{i}_\gamma^{(1)},\ldots,\underline{i}_\gamma^{(r)}$ be the numbers determined by $\underline{i}_\gamma$ as in lemma (1)

(i.e. $\underline{i}_\gamma - 1 = (\underline{i}_\gamma^{(1)}-1) + n^{(1)}(\underline{i}_\gamma^{(2)}-1) + \ldots + n^{(1)}\ldots n^{(r-1)}(\underline{i}_\gamma^{(r)}-1))$ and write $Z_i^{(\ell)} = X_i^{(\ell)}(\underline{i}_1^{(\ell)},\ldots,i_\alpha^{(\ell)},\ldots,i_\beta^{(\ell)},\ldots,\underline{i}_g^{(\ell)})$. Then

$$Z_i Z_i^T = (Z_i^{(1)} \otimes \ldots \otimes Z_i^{(r)})(Z_i^{(1)} \otimes \ldots \otimes Z_i^{(r)})^T \quad \text{by lemma (1)}$$

$$= Z_i^{(1)} Z_i^{(1)T} \otimes \ldots \otimes Z_i^{(r)} Z_i^{(r)T} \quad \text{by lemma (2)}$$

$$= \sum_{j=1}^{p_i^{(1)}} u_{ij}^{(1)} I_{n^{(1)}} \otimes \ldots \otimes \sum_{j=1}^{p_i^{(k_i)}} u_{ij}^{(k_i)} x_{ij}^{(k_i)} I_{n^{(k_i)}} \otimes \ldots \otimes \sum_{j=1}^{p_i^{(r)}} u_{ij}^{(r)} I_{n^{(r)}}$$

by definition (1.1)(i)

(since each $Z_i^{(\ell)}$ is an orthogonal design with variables replaced by 1 for all $\ell \neq k_i$.)

$$= \Big(\sum_{j=1}^{p_i^{(k_i)}} u_i^{(k_i)} u_{ij}^{(k_i)} x_{ij}^{(k_i)} \Big) I_{n^{(1)}\ldots n^{(r)}}$$

So each 2-layer of the ith component of W is an orthogonal design of type $(u_i^{(k_i)} u_{ij}^{(k_i)})$ $(1 \le j \le p_i^{(k_i)})$.

(3) Let $i,j \in \{1,\ldots,K\}$, $i < j$, and let $Z_i = Y_i(\underline{i}_1,\ldots,i_\alpha,\ldots,i_\beta,\ldots,\underline{i}_g)$ be a 2-layer of the ith component of W, and $Z_j = Y_j(\underline{j}_1,\ldots,j_\alpha,\ldots,j_\beta,\ldots,\underline{j}_g)$ the *corresponding* 2-layer of the jth component of W (so that $\underline{i}_\gamma = \underline{j}_\gamma$ for all $\gamma \neq \alpha,\beta$).

For $\gamma \neq \alpha,\beta$ let $\underline{i}_\gamma^{(1)} = \underline{j}_\gamma^{(1)},\dots,\underline{i}_\gamma^{(r)} = \underline{j}^{(r)}$ be the integers determined by the $\underline{i}_\gamma = \underline{j}_\gamma$ as in lemma (1)

(That is $\underline{i}_\gamma - 1 = (\underline{i}_1^{(1)}-1) + n^{(1)}(\underline{i}_\gamma^{(2)}-1) + n^{(1)}n^{(2)}(i_\gamma^{(3)}-1) + \dots$

$$+ n^{(1)}n^{(2)}\dots n^{(r-1)}(i_\gamma^{(r)}-1)$$

and $1 \le j_\gamma^{(\ell)} \le n^{(\ell)}$ for all $\ell = 1,\dots,r$.) .

Write $Z_i^{(\ell)} = X_i^{(\ell)}(\underline{i}_1^{(\ell)},\dots,i_\alpha^{(\ell)},\dots,i_\beta^{(\ell)},\dots,\underline{i}_g^{(\ell)})$

and $Z_j^{(\ell)} = X_j^{(\ell)}(\underline{j}_1^{(\ell)},\dots,j_\alpha^{(\ell)},\dots,j_\beta^{(\ell)},\dots,\underline{j}_g^{\ell})$ for each $\ell = 1,\dots,r$.

Then

$$Z_j Z_i^T = (Z_j^{(1)}\otimes\dots\otimes Z_j^{(r)})(Z_i^{(1)}\otimes\dots\otimes Z_i^{(r)})^T \qquad \text{by lemma (1)}$$

$$= Z_j^{(1)}Z_i^{(1)T}\otimes\dots\otimes Z_j^{(r)}Z_i^{(r)T} \qquad \text{by lemma (2)}$$

$$= (-1)^{\delta_{ij}^{(1)}} Z_i^{(1)}Z_j^{(1)T}\otimes\dots\otimes(-1)^{\delta_{ij}^{(r)}} Z_i^{(r)}Z_j^{(r)T} \qquad \text{(definition (1.1)(ii)}$$

$$= (-1)^{\delta_{ij}^{(1)}+\dots+\delta_{ij}^{(r)}} Z_i^{(1)}Z_j^{(1)T}\otimes\dots\otimes Z_i^{(r)}Z_j^{(r)T}$$

$$= (-1)^{\delta_{ij}}(Z_i^{(1)}\otimes\dots\otimes Z_i^{(r)})(Z_j^{(1)}\otimes\dots\otimes Z_j^{(r)})^T \qquad \text{by lemma (1)}$$

$$= (-1)^{\delta_{ij}} Z_i Z_j^T \qquad \text{by lemma (2)}$$

(4) Finally suppose $W^{(1)},\dots,W^{(r)}$ all regular. Let $i, j \in \{1,\dots,K\}$, $i < j$. Suppose $\delta_{ij} = 1$. We have to show that

$$Y_i \,(H)\, Y_j = 0$$

(Y_i, Y_j the i, j terms of W).

$$Y_i \text{ (H) } Y_j = (X_i^{(1)} \otimes \ldots \otimes X_i^{(r)}) \text{ (H) } (X_j^{(1)} \otimes \ldots \otimes X_j^{(r)})$$

$$= (X_i^{(1)} \text{ (H) } X_j^{(1)}) \otimes \ldots \otimes (X_i^{(r)} \text{ (H) } X_j^{(r)}) \qquad \text{by Lemma (3)}$$

However $\delta_{ij} = 1$ and $\delta_{ij} \equiv \delta_{ij}^{(1)} + \ldots + \delta_{ij}^{(r)} \pmod 2$.

So $\delta_{ij}^{(\ell)} = 1$ for at least one $\ell \in \{1,\ldots,r\}$. So since $W^{(\ell)}$ is regular we have one of the factors in the last kronecker product is zero. So $Y_i \text{ (H) } Y_j = 0$.

<u>(1.2) DEFINITION:</u> A set $W^{(1)},\ldots,W^{(r)}$ of K-systems iscalled compatible if δ_{ij} $(=\delta_{ij}^{(1)}+\ldots+\delta_{ij}^{(r)}$ reduced mod 2) is 1 for all $i < j$.

<u>Corollary</u>: A kronecker product of a compatible set of regular K-systems $\{W^{(1)},\ldots,W^{(r)}\}$ produces an orthogonal g-cube

$$X_1^{(1)} \otimes \ldots \otimes X_1^{(r)} + \ldots + X_K^{(1)} \otimes \ldots \otimes X_K^{(r)}$$

(since all the pairs are anti-amicable) and any set of pairwise anti-amicable members of a regular system may be added to give an orthogonal design). Next we present a few constructions using the theorem or the corollary.

<u>Construction 1</u>: *Let* X, Y, Z *be g-dimensional orthogonal designs of side* n, *such that*

(i) $X = Y + Z$,

(ii) $Y \text{ (H) } Z = 0$.

Then Y, Z *are anti-amicable, that is for each* 2-*layer* Y_i *of* Y *and corresponding* 2-*layer* Z_i *of* Z,

$$Z_i Y_i^T = - Y_i Z_i^T .$$

Proof. In [1] this was proved in the case $g = 2$.

Now, let Y_i, Z_i be corresponding 2-layers of Y, Z respectively. Then Y_i+Z_i is the corresponding 2-layer of X, and Y_i+Z_i, Y_i, Z_i are 2-dimensional orthogonal designs of side n, such that $Y_i \textcircled{H} Z_i = 0$. It follows that $Z_i Y_i^T = -Y_i Z_i^T$.

Figure 2 is an anti-amicable pair obtained from the orthogonal design of Figure 1.

Conversely, given g-*dimensional designs* Y, Z *of side* n, *such that*

(i) $Y \textcircled{H} Z = 0$

(ii) Y, Z *anti-amicable, then* $Y+Z$ *is an orthogonal design.*

Construction 2: *Let* $W^{(1)} = (X_1^{(1)}, X_2^{(1)})$ *be a 2-system, side 2, genus* (0), *type* (1,1) *on* a; b, *and let* $W^{(2)} = (X_1^{(2)}, X_2^{(2)})$ *be a 2-system, side 2, genus* (1), *type* (1,1) *on* c; d. *Let us form* $W = W^{(1)} \otimes W^{(2)}$ *replacing* c, d *by* 1.

We get

$$W = (X_1^{(1)} \otimes X_1^{(2)}, X_2^{(1)} \otimes X_2^{(2)})$$

which is a 2-system of genus (1).

This is a consequence of the Theorem.

Figure 5 provides such a 2-system obtained from the 2-systems of figures 3 and 4.

Construction 3: *Let* $A = x_1A_1 + \ldots + x_KA_K$ *be a* g-*dimensional orthogonal design and* $(Y_1, \ldots, Y_K)$ *be an amicable* K-*tuple then* $B = Y_1 \otimes A_1 + \ldots + Y_K \otimes A_K$ *is a* g-*dimensional orthogonal design.*

This is an obvious consequence of the Corollary.

Figure 6 provides an example for this construction.

Remark. We have proved the Theorem for systems in which each g -cube is of propriety (2,2,...,2), that is each 2-dimensional layer in every direction is an orthogonal design (as defined in [2]). The proof will carry through for systems in which the cubes are of propriety $(d_1,...,d_g)(2 \le d_i \le g)$.

References

1. H.M. Gastineau-Hills, Kronecker Products of Systems of Orthogonal Designs, (same Proceedings).
2. Joseph Hammer and Jennifer Seberry, Higher Dimensional Orthogonal Desgins and Applications, IEEE Transactions on Information Theory, Volume IT-27, p. 772-779.

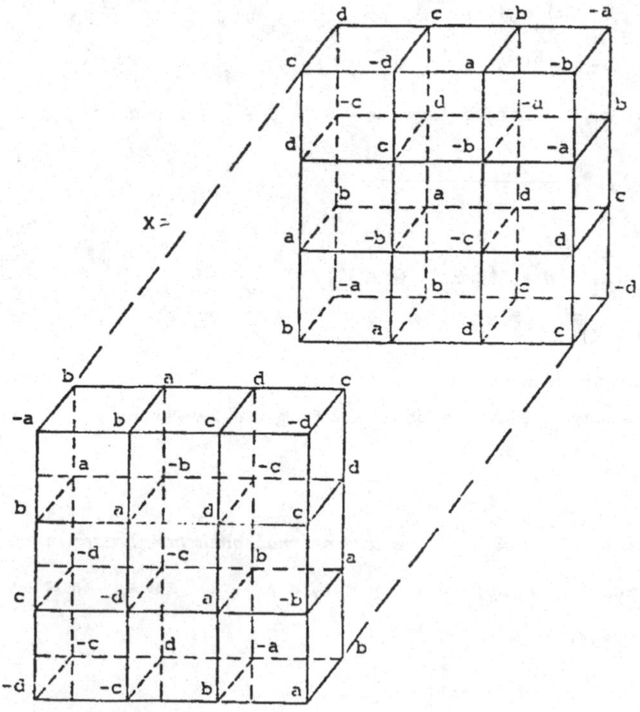

Figure 1

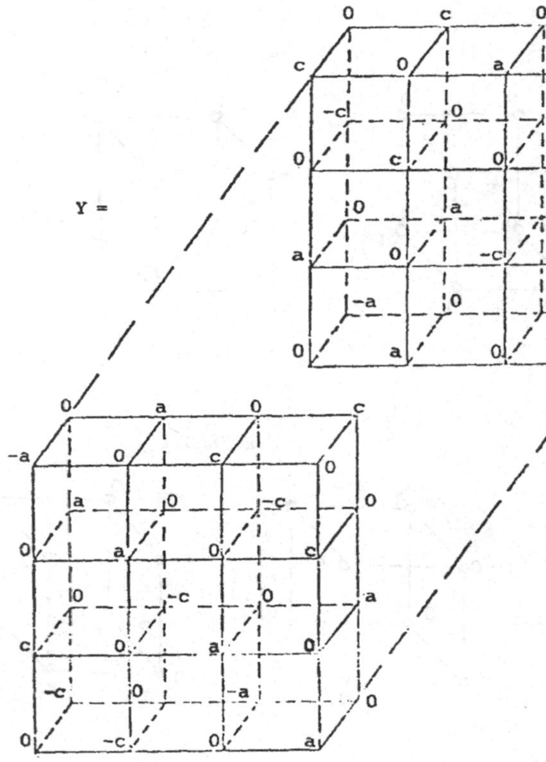

Figure 2a

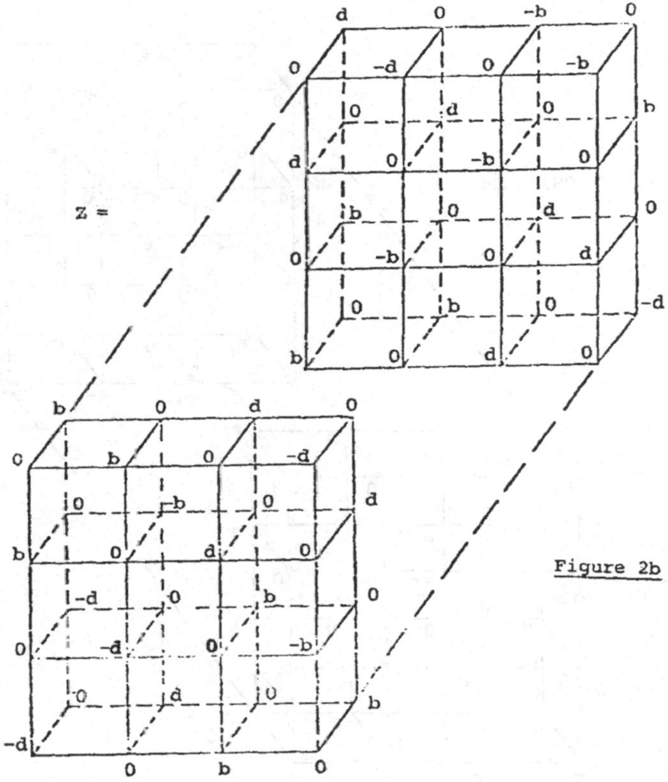

Figure 2b

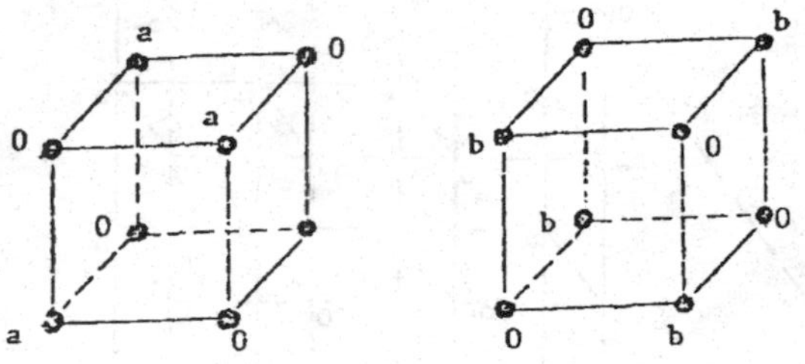

FIGURE 3

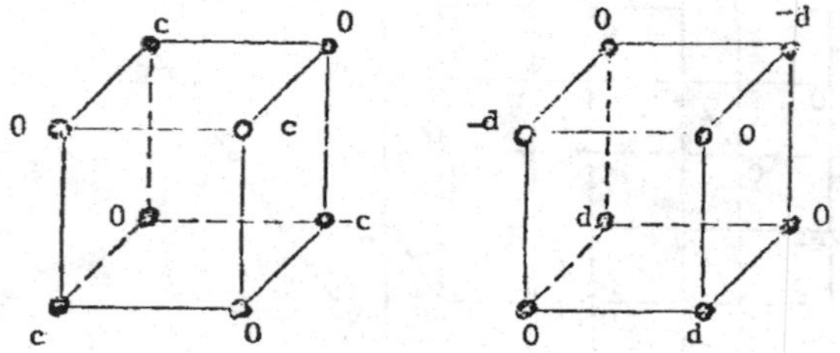

FIGURE 4

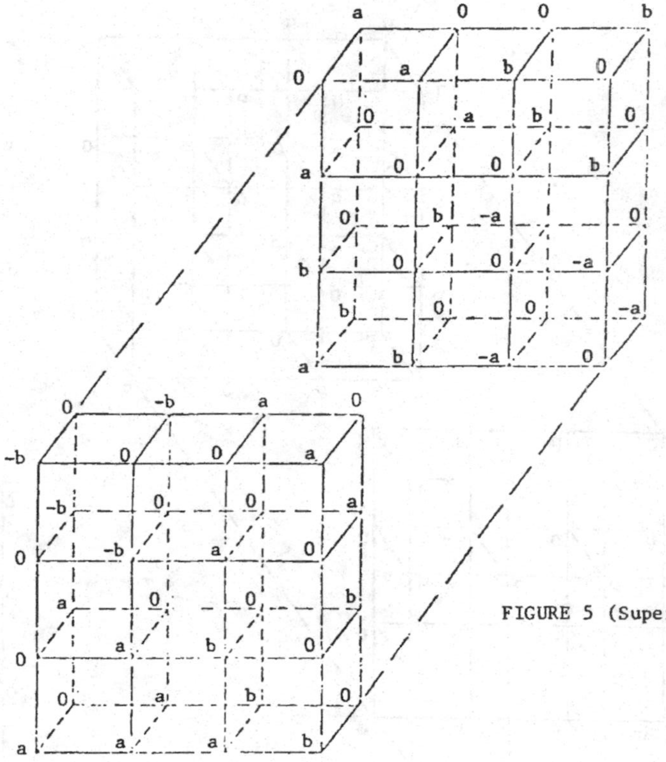

FIGURE 5 (Superimposed)

TWO NEW SEQUENCES OF OVALS IN FINITE DESARGUESIAN PLANES OF EVEN ORDER

DAVID G. GLYNN

According to the definition of the famous Italian geometer, Beniamino Segre, an oval of a finite projective plane is a maximal sized set of points, no three of which are collinear. One of the most influential theorems of finite geometry has been Segre's 1954 result that every oval of a finite Desarguesian plane of odd order is an irreducible conic. In 1957 and 1962 he showed that the even order case is more complicated by constructing two infinite sequences of non-conic ovals in Desarguesian planes of even order. This paper doubles the number of known sequences of ovals in these planes from two to four. Also a short survey of results about ovals is given.

1. INTRODUCTION

Let $GF(q)$ denote the Galois field of q elements, where $q = p^h$, p is a prime and h is a positive integer. $PG(2,q)$, the Desarguesian plane of order q, is constructed as follows. The points of $PG(2,q)$ are the homogeneous triples (x_0,x_1,x_2), $x_i \in GF(q)$, $\forall i$, and the lines are the homogeneous triples $[y_0,y_1,y_2]$, $y_i \in GF(q)$, $\forall i$. A point (x_0,x_1,x_2) is on a line $[y_0,y_1,y_2]$ in $PG(2,q)$ if and only if $x_0y_0+x_1y_1+x_2y_2 = 0$. (For further details about $PG(2,q)$ see [2].) Now the maximum size of a set of points of $PG(2,q)$, no three of which are collinear, is $q+1$ if q is odd, or $q+2$ if q is even. (See [16].) Hence, according to the definition of Beniamino Segre, (the one used here), an oval of $PG(2,q)$ is a set of points no three collinear, of cardinality $q+1$ if q is odd, or $q+2$ if q is even. (Note that for some authors, an "oval" is a set of $q+1$ points, no three collinear, while their term "complete oval" means the same as Segre's oval.) In this paper, some of the progress made towards the goal of classifying all ovals of $PG(2,q)$ is described. Section two contains Segre's complete classification of ovals in $PG(2,q)$, q odd, and Sections three to five contain results about the more complicated case of q even. In particular, Section three gives some results about Galois fields and then uses these to obtain a simple numerical condition that a certain type of set $D(k)$ is an oval. (The author used this condition to find by computer all ovals of type $D(k)$ up to $q = 2^{19}$.) The two new sequences of ovals suggested by this evidence are proved in Section four to be actually infinite. In Section five all the known existence results about ovals are collected together and finally Section six draws some conclusions about the problem of ovals of $PG(2,q)$.

2. THE OVALS OF $PG(2,q)$, q ODD

The biggest impetus to the study of finite geometry was probably the discovery by B. Segre in 1954 [14], that every oval of $PG(2,q)$, q odd, is the set of points of an irreducible conic. Thus, it is the set of points (x_0, x_1, x_2) satisfying an equation

$$\sum_{i=0}^{2} \sum_{j=0}^{2} a_{ij} x_i x_j = 0,$$

where $A = (a_{ij})$ is a symmetric non-singular 3×3 matrix over $GF(q)$. Conversely every irreducible conic is an oval and also all conics are homographically equivalent. Since any five points of the plane, no three collinear, determine a unique conic, $(q > 3)$, it is easily calculated that there are $q^5 - q^2$ conic-ovals of $PG(2,q)$. This is also the number of symmetric non-singular homogeneous 3×3 matrices over $GF(q)$. See [16] for further details.

3. RESULTS ABOUT $GF(q)$ AND OVALS OF TYPE $D(k)$

Various results and definitions about $GF(q)$ useful for the study of ovals in $PG(2,q)$ are now listed.

Definition 1. *A permutation polynomial of* $GF(q)$ *is a polynomial* $f(t)$ *over* $GF(q)$ *such that* $f(a) = f(b)$, $(a,b \in GF(q))$, $\Rightarrow a = b$.

Result 1. (Dickson's criterion) [3]. If $f(a) = 0 \Leftrightarrow a = 0$, then $f(t)$ is a permutation polynomial of $GF(q) \Leftrightarrow [f(t)]^r$, reduced modulo $t^q - t$, has zero coefficient for t^{q-1}, $(\forall r,\ 1 \le r \le q-2)$.

Result 2. [3]. Let g be any function from $GF(q)$ to itself. Then g corresponds to a unique polynomial f over $GF(q)$ and of degree less than or equal to $q-1$. In fact, if $f(\lambda) = g(\lambda)$, $(\forall\lambda \in GF(q))$, then $f(t) = \sum_{i=0}^{q-1} a_i t^i$, $(a_i \in GF(q))$, where

$$a_0 = g(0), \quad a_r = -\sum_{\lambda \in GF(q)^*} g(\lambda)\lambda^{-r}, \qquad (1 \le r \le q-2),$$

and $a_{q-1} = -\sum_{\lambda\in GF(q)} g(\lambda)$. ($GF(q)^*$ denotes $GF(q) \setminus \{0\}$.)

From now on in this paper, q *is assumed to be equal to a power of two. That is,* $q = 2^h$, *(h a positive integer).*

Definition 2. *The partial ordering* $\lesssim$ *on the set of integers* $N_q = \{n \mid 0 \le n \le q-1\}$ *is defined as follows:*
if $a,b \in N_q$, *then* $a \lesssim b \Leftrightarrow a = \sum_{i=0}^{h-1} a_i 2^i$, $b = \sum_{i=0}^{h-1} b_i 2^i$, $(a_i, b_i \in \{0,1\}, \forall i)$, *and* $a_i \le b_i$, $(\forall i)$.

Result 3. If $k \in Z$, (the set of integers), then the expansion of $(1+t)^k \pmod{t^q-t}$ over $GF(q)$, may be calculated as follows. Firstly, if $k=0$, then $(1+t)^k = 1$. Secondly, if $k \ne 0$, then reduce $k \pmod{q-1}$ so that $1 \le k \le q-1$. Then

$$(1+t)^k = \sum_{\substack{c \lesssim k \\ c \in N_q}} t^c \pmod{t^q-t}.$$

For example, $(1+t)^{2^i} = 1+t^{2^i}$, $(\forall i \in Z,\ 1 \le i)$, and $(1+t)^{(q-1)i} = \sum_{c\in N_q} t^c$, $(\forall i \in Z,\ i \ne 0)$.

Definition 3. *Let* $g \in GF(q)$, *then we say* g *is first category* $\Leftrightarrow tr(g) = \sum_{i=0}^{h-1} g^{2^i} = 0$, *and* g *is second category* $\Leftrightarrow tr(g) = 1$. *(Note that* $tr(g) = 0$ *or* 1 *as* $[tr(g)]^2+tr(g) = 0$. *Also the first category elements of* $GF(q)$ *form a subgroup of index two in the additive group of* $GF(q)$, *while the second category elements are the complement of this subgroup in* $GF(q)$. *See [3] or [16].)*

Result 4. [3], [16].

(a) 0 is first category in $GF(q)$,

(b) 1 is first category in $GF(q) \Leftrightarrow h$ is even,

(c) Let $g \in GF(q)$, then $\exists x \in GF(q)$ with $g = x^2+x \Leftrightarrow g$ is first category in $GF(q)$,

(d) g is first category in $GF(q) \Leftrightarrow g^{2^i}$ is first category, $(\forall i \in Z,\ i \ge 0)$.

Result 5. Let $n \in Z$, $(n,h) = 1$, $n \ge 1$, and $a,b \in GF(q)$, then

$$(*)\ x^{2^n}+ax+b = 0 \text{ has } \begin{cases} \text{no solution} \Leftrightarrow a \ne 0 \text{ and } tr(b/a^{g+1}) = 1 \\ \text{one solution} \Leftrightarrow a = 0 \\ \text{two solutions} \Leftrightarrow a \ne 0 \text{ and } tr(b/a^{g+1}) = 0, \end{cases}$$

where $g \equiv (2^n-1)^{-1} \pmod{q-1}$.

Proof. Note that $(2^n-1, q-1) = (2^n-1, 2^h-1) = 2^{(n,h)}-1 = 2-1 = 1$ and so $g \equiv (2^n-1)^{-1} \pmod{q-1}$ is properly defined.

(1) Suppose $a \ne 0$. Then dividing $(*)$ by a^{g+1} gives the equation

$(x/a^g)^{2^n}+(x/a^g)+b/a^{g+1} = 0$, (since $g.2^n \equiv g+1 \pmod{q-1}$). Let $y = x/a^g$, then $y^{2^n}+y+b/a^{g+1} = 0$ ...(**).

We show that:

(a) if $tr(b/a^{g+1}) = 0$, then (**) has at least two solutions,

(b) if (**) has two solutions it has precisely two solutions,

(c) if (**) has a solution then $tr(b/a^{g+1}) = 0$.

These results imply that if $a \neq 0$, then (**) and therefore (*) has 2 or 0 solutions as $tr(b/a^{g+1}) = 0$ or 1 respectively.

(a) Let $tr(b/a^{g+1}) = 0$. Then $b/a^{g+1} = z^2+z$, for some $z \in GF(q)$, (by Result 4(c)). Since $(n,h) = 1$, $\exists m \in \{1,\dots,h\}$ with $mn \equiv 1 \pmod h$. Then $y = z+z^{2^n}+z^{2^{2n}}+\dots+z^{2^{(m-1)n}}$ and $y = z+z^{2^n}+z^{2^{2n}}+\dots+z^{2^{(m-1)n}}+1$ are solutions of (**).

(b) Suppose (**) has two solutions s_1 and s_2.

Then $s_1^{2^n}+s_1+b/a^{g+1} = s_2^{2^n}+s_2+b/a^{g+1} = 0$

$\Rightarrow (s_1+s_2)^{2^n} = s_1+s_2$

$\Rightarrow (s_1+s_2)^{2^n-1} = 1$

$\Rightarrow s_1+s_2 = 1.$

If a third solution s_3 existed then there would hold

$$(s_1+s_2)+(s_1+s_3)+(s_2+s_3) = 2s_1+2s_2+2s_3 = 0 = 1+1+1 = 1,$$

which would be a contradiction.

(c) Suppose $s^{2^n}+s+b/a^{g+1} = 0$, $s \in GF(q)$.

Then $tr(s^{2^n}+s+b/a^{g+1}) = tr(0) = 0$

$= tr(s^{2^n})+tr(s)+tr(b/a^{g+1})$

$= tr(s)+tr(s)+tr(b/a^{g+1})$

$= tr(b/a^{g+1})$.

Hence $tr(b/a^{g+1}) = 0$.

(2) Suppose $a = 0$. Then $x^{2^n}+b = 0 \Rightarrow x = b^{2^{-n}}$ and so (*) has a unique solution in this case.

<u>Result 6</u>. [18]. Suppose K is an oval of $PG(2,q)$ through the two points $(0,1,0)$ and $(0,0,1)$. Then $K = \{(0,1,0),(0,0,1)\} \cup \{(1,\lambda,f(\lambda)) \mid \lambda \in GF(q)\}$, where f is a permutation polynomial over $GF(q)$ such that $f(\lambda) = \sum_{i=0}^{q-2} a_i\lambda^i$, $\forall\lambda \in GF(q)$, and $a_i = 0$ if i is odd.

This result shows that the problem of finding ovals in $PG(2,q)$ is the same as finding certain permutation polynomials f. The simplest polynomials are the monomials, hence the following definition is a natural one.

<u>Definition 4</u>. *Let* $D(k)$, $(k \not\equiv 0 \pmod{q-1}$, $k \in \mathbb{Z})$, *be the set of points of* $PG(2,q)$ $\{(0,1,0),(0,0,1)\} \cup \{(1,\lambda,\lambda^k) \mid \lambda \in GF(q)\}$. *Since*

$\lambda^k = \lambda^\ell$, $\forall\lambda$, *if* $k \equiv \ell \pmod{q-1}$, k *is always assumed to satisfy* $1 \le k \le q-2$. *(Thus, by Result 6,* $D(k)$ *is an oval only if* k *is even.)*

Result 7. [7]. $D(k)$ is an oval of $PG(2,q)$ if and only if $(k,q-1) = (k-1,q-1) = 1$, and

$$f_k(t) = \frac{(t+1)^k+1}{t} \text{ is a permutation polynomial of } GF(q).$$

Result 8. If $(k,q-1) = (k-1,q-1) = 1$, then

$$D(k) \cong D(1-k) \cong D(k^{-1}) \cong D(1-k^{-1}) \cong D((1-k)^{-1}) \cong D(k(k-1)^{-1}),$$

where $\cong$ denotes homographical equivalence in $PG(2,q)$.

Proof. Note that k, $1-k$, k^{-1}, $1-k^{-1}$, $(1-k)^{-1}$ and $k(k-1)^{-1}$ are all considered to be integers modulo $q-1$, and that k^{-1} and $1-k^{-1}$ are well-defined as $(k,q-1) = 1$, and also $(1-k)^{-1}$ and $k(k-1)^{-1}$ are well-defined as $(k-1,q-1) = 1$. Let A,B,C and D be the points $(1,0,0),(0,1,0),(0,0,1)$ and $(1,1,1)$ respectively. Then the homographies (linear collineations) of $PG(2,q)$ fixing D and taking A,B,C in turn to $B,A,C; A,C,B; C,A,B; B,C,A;$ and $C,B,A;$ also take $D(k)$ to $D(1-k)$; $D(k^{-1})$; $D(1-k^{-1})$; $D((1-k)^{-1})$; and $D(k(k-1)^{-1})$ respectively. Thus there is a group of six homographies of $PG(2,q)$ permuting the above sets of points.

A corollary to Results 6 and 8 is that $D(k)$ is an oval only if $(k,q-1) = (k-1,q-1) = 1$, and k, $1-k$, k^{-1}, $1-k^{-1}$, $(1-k)^{-1}$ and $k(k-1)^{-1}$ are all even when reduced modulo $q-1$ to lie between 1 and $q-2$. This corollary was used to classify ovals of type $D(k)$ for small values of q in [7] and [8].

Theorem A. $D(k)$ *is an oval of* $PG(2,q)$ *if and only if* $d \not\le kd$, $\forall d \in \{1,2,\ldots,q-2\}$, *where* dk *is reduced modulo* $q-1$ *to lie in* $N_q = \{0,1,\ldots,q-1\}$. *(As usual* 0 *is reduced to* 0 *and all the other multiples of* $q-1$ *are reduced to* $q-1$. *See Definition 2.)*

Proof. Suppose $k \in \{1,2,\ldots,q-2\}$ and $(k,q-1) = (k-1,q-1) = 1$. Then $D(k)$ is an oval of $PG(2,q)$

$\Leftrightarrow f_k(t) = \dfrac{(t+1)^k+1}{t}$ is a permutation polynomial, (Result 7)

$\Leftrightarrow [f_k(t)]^r \pmod{t^q-t}$ has zero coefficient for t^{q-1} $\forall r \in \{1,\ldots,q-2\}$. (See Result 1. $f_k(a) = 0 \Leftrightarrow (a+1)^k+1 = 0$ $\Leftrightarrow a+1 = 1^{k^{-1}} \Leftrightarrow a+1 = 1 \Leftrightarrow a = 0$.)

$\Leftrightarrow \displaystyle\sum_{\lambda\in GF(q)^*} [((\lambda+1)^k+1)\lambda^{-1}]^r = 0$, $\forall r \in \{1,\ldots,q-2\}$, (Result 2)

$\Leftrightarrow$ the coefficient of t^r in $[(t+1)^k+1]^r$ is zero, $\forall r \in \{1,\ldots,q-2\}$, (by Result 2 again)

$\Leftrightarrow \sum_{d\lesssim r}(t+1)^{kd}$ has zero coefficient for t^r, $\forall r \in \{1,\dots,q-2\}$, (by Result 3)

$\Leftrightarrow \sum_{d\lesssim r}\ \sum_{e\lesssim kd} t^e$ has zero coefficient for t^r, $\forall r \in \{1,\dots,q-2\}$, (by Result 3 again)

$\Leftrightarrow \sum_{d\lesssim r\lesssim kd} t^r = 0, \quad \forall r \in \{1,\dots,q-2\}$

$\Leftrightarrow |\{d|d \lesssim r \lesssim kd\}| \equiv 0 \pmod 2$, $\forall r \in \{1,\dots,q-2\}$

$\Leftrightarrow d \not\lesssim kd$, $\forall d \in \{1,\dots,q-2\}$.

This last step needs some explanation. If $|\{d|d \lesssim r \lesssim kd\}| \equiv 0 \pmod 2$, $\forall r \in \{1,\dots,q-2\}$, and if there is a minimal d', (with respect to $\lesssim$), with $d' \lesssim kd'$, $d' \in \{1,\dots,q-2\}$, then putting $r = d'$ implies that $\{d|d \lesssim d' \lesssim kd\} = \{d'\}$. Since $|\{d'\}| = 1 \not\equiv 0 \pmod 2$ this shows that no such d' exists and so $d \not\lesssim kd$, $\forall d \in \{1,\dots,q-2\}$. Conversely, if $d \not\lesssim kd$, $\forall d \in \{1,\dots,q-2\}$, then $\{d|d \lesssim r \lesssim kd\} = \phi$, $\forall r \in \{1,\dots,q-2\}$ and so $|\{d|d \lesssim r \lesssim kd\}| \equiv 0 \pmod 2$, $\forall r \in \{1,\dots,q-2\}$.

Now $d \not\lesssim kd$, $\forall d \in \{1,\dots,q-2\}$ actually implies that $(k,q-1) = (k-1,q-1) = 1$, for if $(k,q-1) = \ell \neq 1$, then $m = (q-1)/\ell$ satisfies $m \in \{1,\dots,q-2\}$ and $mk \equiv q-1 \pmod{q-1}$. Hence $m \lesssim mk \pmod{q-1}$. Similarly, if $(k-1,q-1) = n \neq 1$, then $p = (q-1)/n$ satisfies $p(k-1) \equiv 0 \pmod{q-1}$. Hence $p \equiv pk \pmod{q-1}$, and so $p \lesssim pk$, $p \in \{1,\dots,q-2\}$. Theorem A has now been proved.

Computer Result. Using Theorem A, all the ovals of type $D(k)$ were found by computer for $q = 2^h$, $h = 1,\dots,19$. These ovals all belonged to the infinite sequences described in Section 5. In fact, this computer evidence gave the inspiration for the two constructions of the next section. The condition of Theorem A does not involve $GF(q)$: it only involves multiplying modulo $q-1$ and checking binary expansions. Thus a very fast computer algorithm was easily developed. If the reader has a fast computer, the ovals of type $D(k)$ could almost certainly be classified for some larger values of h. (See Section 6.)

4. THE TWO NEW SEQUENCES OF OVALS

In this section $q = 2^h$, h odd. Also let $\sigma = 2^{h+1/2}$, (thus $\sigma \equiv \sqrt{2} \pmod{q-1}$), and

$$\gamma = \begin{cases} 2^n & \text{if } h = 4n-1 \\ 2^{3n+1} & \text{if } h = 4n+1 \end{cases}, \quad (\text{thus } \gamma \equiv \sqrt[4]{2} \pmod{q-1}).$$

Hence $\gamma^2 \equiv \sigma$, and $\gamma^4 \equiv \sigma^2 \equiv 2 \pmod{q-1}$.

Theorem B. $D(\sigma+\gamma)$ *is an oval of* $PG(2,q)$, $q = 2^h$, h *odd*, $h \geq 3$.

Proof. Note that the following is only the first proof of the author and it is to be expected that in the future a much simpler one could be found. The idea of the proof came from that of $D(6)$, found in [6]. First, a lemma that is used in the proof is given.

Lemma. *Let* $S = s+t$, $T = st$, $Y = TS^{-2}$, *and* $\beta_a = (s^a+t^a)YS^{-a}$, *where* $s,t \in GF(q)$, $s \neq 0$, $t \neq 0$, *and* $s \neq t$. *Also let* $K = \beta_{\gamma-1}$. *Then*

(1) $\beta_0 = 0$

(2) $\beta_{2^m} = Y$, $(\forall m \in Z,\ m \geq 0)$

(3) $\beta_a = Y^{-1}\beta_p\beta_{a-p+1}+\beta_{p-1}\beta_{a-p}$, $(\forall p \in \{1,\ldots,a\},\ a \geq 1)$

(4) $\beta_{2^m-1} = \sum_{i=1}^{m} Y^{2^{i-1}}$, $(\forall m \in Z,\ m \geq 1)$

(5) $\beta_{q-1} = 0$, $Y = K+K^\gamma+K^\sigma+K^{\sigma\gamma}$ and $tr(Y) = tr(K) = 0$

(6) $\beta_{\sigma-1} = K+K^\gamma$

(7) $\beta_{\sigma+\gamma-1} = K+K^2+K^\gamma+K^{\gamma+1}+K^\sigma+K^{\sigma\gamma}$.

Proof.

(1) $\beta_0 = (s^0+t^0)YS^0 = (1+1)Y = 0$

(2) $\beta_{2^m} = (s^{2^m}+t^{2^m})YS^{-2^m} = (s+t)^{2^m}YS^{-2^m} = Y$

(3)
$$\begin{aligned}
Y^{-1}&\beta_p\beta_{a-p+1}+\beta_{p-1}\beta_{a-p}\\
&= Y^{-1}(s^p+t^p)YS^{-p}.(s^{a-p+1}+t^{a-p+1})YS^{-(a-p+1)}\\
&+ (s^{p-1}+t^{p-1})YS^{-(p-1)}(s^{a-p}+t^{a-p})YS^{-(a-p)}\\
&= YS^{-a-1}(s^{a+1}+s^pt^{a-p+1}+s^{a-p+1}t^p+t^{a+1})\\
&+ Y^2S^{-a+1}(s^{a-1}+s^{p-1}t^{a-p}+s^{a-p}t^{p-1}+t^{a-1})\\
&= YS^{-a-1}(s^{a+1}+s^pt^{a-p+1}+s^{a-p+1}t^p+t^{a+1}+s^at+s^pt^{a-p+1}+s^{a-p+1}t^p+st^a),\\
&\qquad (\text{as } YS^2 = T = st)\\
&= YS^{-a-1}(s^{a+1}+s^at+st^a+t^{a+1})\\
&= YS^{-a-1}(s^a+t^a)S\\
&= (s^a+t^a)YS^{-a}\\
&= \beta_a
\end{aligned}$$

(4) Use induction on m. It is true for $m=1$, as $\beta_{2-1} = \beta_1 = Y$, (by (2) above). Suppose it is true for $m=p$. Then

$$\begin{aligned}
\beta_{2^{p+1}-1} &= Y^{-1}\beta_{2^p}\beta_{2^p}+\beta_{2^p-1}\beta_{2^p-1}, &&\text{(by (3) above)}\\
&= Y^{-1}YY + \Big(\sum_{i=1}^{p} Y^{2^{i-1}}\Big)^2, &&\text{(by (2) and the inductive hypothesis)}
\end{aligned}$$

$$= Y + \sum_{i=1}^{p} Y^{2^i} = \sum_{i=1}^{p+1} Y^{2^{i-1}} .$$

Hence it is true for $m = p+1$, and so for all $m \geq 1$.

(5) By (4) above, $\beta_{q-1} = \beta_{2^h-1} = \sum_{i=1}^{h} Y^{2^{i-1}} = tr(Y)$.

But $Y = \frac{st}{s^2+t^2} = \frac{s}{s+t} + (\frac{s}{s+t})^2$, hence $tr(Y) = 0$.

Since $K = \beta_{\gamma-1}$ = a sum of powers of two of Y, (by (4)), then $tr(K) = 0$ also. Now, if $h = 4n-1$, then

$$K+K^{\gamma}+K^{\sigma}+K^{\sigma\gamma}$$

$$= \sum_{i=1}^{n} Y^{2^{i-1}} + \sum_{i=1}^{n} Y^{2^{i-1}.2^n} + \sum_{i=1}^{n} Y^{2^{i-1}.2^{2n}} + \sum_{i=1}^{n} Y^{2^{i-1}.2^{3n}}$$

$$= \sum_{i=1}^{4n-1} Y^{2^{i-1}} + Y^{2^{4n-1}} = tr(Y)+Y = Y.$$

And if $h = 4n+1$, then

$$K+K^{\gamma}+K^{\sigma}+K^{\sigma\gamma}$$

$$= \sum_{i=1}^{3n+1} Y^{2^{i-1}} + \sum_{i=1}^{3n+1} Y^{2^{i-1}.2^{3n+1}} + \sum_{i=1}^{3n+1} Y^{2^{i-1}.2^{2n+1}} + \sum_{i=1}^{3n+1} Y^{2^{i-1}.2^{n+1}}$$

$$= \sum_{i=1}^{4n+1} Y^{2^{i-1}} + \sum_{i=1}^{4n+1} Y^{2^{i-1}} + \sum_{i=1}^{4n+1} Y^{2^{i-1}} + Y^{2^{4n+1}} = 3tr(Y)+Y = Y.$$

(6) If $h = 4n-1$, then $\beta_{\sigma-1} = \sum_{i=1}^{2n} Y^{2^{i-1}} = \sum_{i=1}^{n} Y^{2^{i-1}} + (\sum_{i=1}^{n} Y^{2^{i-1}})^{2^n}$,

and so $\beta_{\sigma-1} = K+K^{\gamma}$. If $h = 4n+1$, then

$$\beta_{\sigma-1} = tr(Y) + \beta_{\sigma-1} = \sum_{i=1}^{4n+1} Y^{2^{i-1}} + \sum_{i=1}^{2n+1} Y^{2^{i-1}}$$

$$= \sum_{i=1}^{3n+1} Y^{2^{i-1}} + (\sum_{i=1}^{3n+1} Y^{2^{i-1}})^{2^{3n+1}} = K+K^{\gamma}.$$

(7) $\beta_{\sigma+\gamma-1} = Y^{-1}\beta_{\sigma}\beta_{\gamma}+\beta_{\sigma-1}\beta_{\gamma-1}$, (by (3) above)

$= Y^{-1}YY+(K+K^{\gamma})K$, (by (2) and (6))

$= K+K^{\gamma}+K^{\sigma}+K^{\sigma\gamma}+K^2+K^{\gamma+1}$, (by (5))

$= K+K^2+K^{\gamma}+K^{\gamma+1}+K^{\sigma}+K^{\sigma\gamma}$.

Now Result 7 is used to prove that $D(\sigma+\gamma)$ is an oval. Firstly, $(\sigma+\gamma,q-1) = 1$ as $(\sigma+\gamma)^{-1} \equiv -\gamma^{-1}+\sigma-\gamma+1 \pmod{q-1}$, and $(\sigma+\gamma-1,q-1) = 1$ as $(\sigma+\gamma-1)^{-1} \equiv 3^{-1}(\sigma\gamma+\gamma-1) \pmod{q-1}$. (Note that 3^{-1} is defined $(\bmod\ q-1)$ as $(3,q-1) = (2^2-1,2^h-1) = 2^{(2,h)}-1 = 1$.)
It remains to show that $f_{\sigma+\gamma}(t) = ((t+1)^{\sigma+\gamma}+1)t^{-1}$ is a permutation polynomial of $GF(q)$. Consider $f_{\sigma+\gamma}(s) = f_{\sigma+\gamma}(t)$, where $s \neq t$, $(s,t \in GF(q))$. Then we may assume that $s \neq 0$ and $t \neq 0$ because

$f_{\sigma+\gamma}(s) = 0$

$$\Leftrightarrow \quad (s+1)^{\sigma+\gamma} + 1 = 0$$

$$\Leftrightarrow \quad s+1 = 1 \Leftrightarrow s = 0.$$

Now

$$f_{\sigma+\gamma}(s) + f_{\sigma+\gamma}(t) = 0$$

$$\Rightarrow \quad s^{\sigma+\gamma-1}+s^{\sigma-1}+s^{\gamma-1}+t^{\sigma+\gamma-1}+t^{\sigma-1}+t^{\gamma-1} = 0$$

$$\Rightarrow \quad (s^{\sigma+\gamma-1}+t^{\sigma+\gamma-1})YS^{-(\sigma+\gamma-1)}+(s^{\sigma-1}+t^{\sigma-1})YS^{-(\sigma+\gamma-1)} +(s^{\gamma-1}+t^{\gamma-1})YS^{-(\sigma+\gamma-1)} = 0$$

$$\Rightarrow \quad \beta_{\sigma+\gamma-1}+\beta_{\sigma-1}S^{-\gamma}+\beta_{\gamma-1}S^{-\sigma} = 0$$

$$\Rightarrow \quad Kx^{\gamma}+(K+K^{\gamma})x+K+K^2+K^{\gamma}+K^{\gamma+1}+K^{\sigma}+K^{\sigma\gamma} = 0,$$

where $x = S^{-\gamma}$, (using the above Lemma).
Now $K \neq 0$ as $K = 0 \Rightarrow s^{\gamma-1}+t^{\gamma-1} = 0 \Rightarrow s = t$. Thus

$$x^{\gamma} + (1+K^{\gamma-1})x + 1 + K + K^{\gamma-1} + K^{\gamma} + K^{\sigma-1} + K^{\sigma\gamma-1} = 0.$$

The coefficient of x in this equation, $1+K^{\gamma-1}$, is non-zero as $K^{\gamma-1} + 1 = 0 \Rightarrow K = 1$, but $tr(K) = 0 \neq tr(1) = 1$. Hence the equation has zero or two solutions. (See Result 5.) Let $A = 1+K^{\gamma-1}$. Thus $K = (A+1)^{g}$, where $g \equiv (\gamma-1)^{-1} \equiv \sigma\gamma+\sigma+\gamma+1 \pmod{q-1}$. Then the equation becomes $x^{\gamma} + AX + B = 0$, where

$$B = A + KA + K^{\sigma-1}A^{\sigma}$$
$$= A + (A+1)^{g}A + (A+1)^{g(\sigma-1)}A^{\sigma}.$$

Using Result 5, there holds

$$tr\left(\frac{B}{A^{g+1}}\right) = 0$$

$$\Rightarrow \quad tr(A^{-g}(1+(A+1)^{g}+(A+1)^{g(\sigma-1)}A^{\sigma-1})) = 0$$

$$\Rightarrow \quad tr(A^{-g}(1+(A+1)^{\sigma\gamma+\sigma+\gamma+1}+(A+1)^{\gamma+1}A^{\sigma-1})) = 0$$

$$\Rightarrow \quad tr(A^{-g}(1+A^{\sigma\gamma+\sigma+\gamma+1}+A^{\sigma\gamma+\sigma+\gamma}+A^{\sigma\gamma+\sigma+1}+A^{\sigma\gamma+\sigma}+A^{\sigma\gamma+\gamma+1}+A^{\sigma\gamma+\gamma} +A^{\sigma\gamma+1}+A^{\sigma\gamma}+A^{\sigma+\gamma+1}+A^{\sigma+\gamma}+A^{\sigma+1}+A^{\sigma}+A^{\gamma+1} +A^{\gamma}+A+1+A^{\sigma+\gamma}+A^{\sigma+\gamma-1}+A^{\sigma}+A^{\sigma-1})) = 0$$

$$\Rightarrow \quad tr(1+A^{-1}+A^{-\gamma}+A^{-\gamma-1}+A^{-\sigma}+A^{-\sigma-1}+A^{-\sigma-\gamma}+A^{-\sigma-\gamma-1}+A^{-\sigma\gamma} +A^{-\sigma\gamma-\gamma}+A^{-\sigma\gamma-\sigma}+A^{-\sigma\gamma-\sigma-1}+A^{-\sigma\gamma-\sigma-\gamma}+A^{-\sigma\gamma-2}+A^{-\sigma\gamma-\gamma-2}) = 0$$

$$\Rightarrow \quad tr(1)+tr(A^{-1}+A^{-\gamma})+tr(A^{-\gamma-1}+A^{-\sigma\gamma-\sigma})+tr(A^{-\sigma}+A^{-\sigma\gamma}) +tr(A^{-\sigma-1}+A^{-\sigma\gamma-\gamma})+tr(A^{-\sigma-\gamma}+A^{-\sigma\gamma-2})+tr(A^{-\sigma-\gamma-1}+A^{-\sigma\gamma-\sigma-\gamma}) +tr(A^{-\sigma\gamma-\sigma-1}+A^{-\sigma\gamma-\gamma-2}) = 0$$

$$\Rightarrow \quad 1+tr(A^{-1})+tr(A^{-1})^{\gamma}+tr(A^{-\gamma-1})+tr(A^{-\gamma-1})^{\sigma}+tr(A^{-\sigma})+tr(A^{-\sigma})^{\gamma} +tr(A^{-\sigma-1})+tr(A^{-\sigma-1})^{\gamma}+tr(A^{-\sigma-\gamma})+tr(A^{-\sigma-\gamma})^{\sigma}$$

$$+\mathrm{tr}(A^{-\sigma-\gamma-1})+\mathrm{tr}(A^{-\sigma-\gamma-1})^{\gamma}+\mathrm{tr}(A^{-\sigma\gamma-\sigma-1})+\mathrm{tr}(A^{-\sigma\gamma-\sigma-1})^{\gamma} = 0$$

$\Rightarrow \quad 1 = 0$, a contradiction.

Hence $f_{\sigma+\gamma}(s) = f_{\sigma+\gamma}(t)$, $(s,t \in GF(q))$

$\Rightarrow \quad s = t$.

Thus $f_{\sigma+\gamma}(t)$ is a permutation polynomial of $GF(q)$, and so $D(\sigma+\gamma)$ is an oval of $PG(2,q)$.

Theorem C. $D(3\sigma+4)$ *is an oval of* $PG(2,q)$, $q = 2^h$, h *odd*, $h \geq 3$.

Proof. By Result 8, we consider $D(\frac{\sigma+2}{3})$ instead of $D(3\sigma+4)$, since $(3\sigma+4)(3\sigma+4-1)^{-1} \equiv (3\sigma+4)(\frac{\sigma-1}{3}) \equiv \frac{\sigma+2}{3}$ (mod $q-1$). The method of proof will be to show that any line of $PG(2,q)$ intersects $D(\frac{\sigma+2}{3})$ in less than three points. Firstly, any lines through $(0,1,0)$ or $(0,0,1)$ intersect $D(\frac{\sigma+2}{3})$ in precisely two points as $t^{\frac{\sigma+2}{3}}$ is a permutation polynomial. Consider a general line $[\ell,k,1]$, $(\ell,k \in GF(q),\ k \neq 0)$, not through $(0,1,0)$ or $(0,0,1)$. Its equation of intersection with $D(\frac{\sigma+2}{3})$ is $x^{\frac{\sigma+2}{3}} + kx + \ell = 0$. Let $x = k^{3(\sigma+1)}X^3$. (This is a 1-1 substitution since $(3,q-1) = 1$.) Thus

(1)...$X^{\sigma+2} + X^3 + m = 0$, (where $m = \ell k^{-3\sigma-4}$).

Assume this equation has two distinct solutions: $X = \alpha,\ \beta \in GF(q)$. (We shall show there is no other.) Then $(X+\alpha)(X+\beta) = X^2+(\alpha+\beta)X+\alpha\beta = 0$, (for $X = \alpha$ or β), and so $X^{\sigma}+aX+b = 0$, (for some a, b, functions of α and β). Hence $X^{\sigma^2}+(aX)^{\sigma}+b^{\sigma} = X^2+a^{\sigma}X^{\sigma}+b^{\sigma} = X^2+a^{\sigma+1}X+a^{\sigma}b+b^{\sigma} = 0$. Thus $a^{\sigma+1} = \alpha+\beta$ and $a^{\sigma}b+b^{\sigma} = \alpha\beta$. We now substitute

(2)...$X^{\sigma} = aX + b$, and

(3)...$X^2 = a^{\sigma+1}X + a^{\sigma}b + b^{\sigma}$

into (1) above. Now (1) $\Rightarrow X^2(X^{\sigma}+X)+m = 0$

$\Rightarrow \quad (a^{\sigma+1}X+a^{\sigma}b+b^{\sigma})((a+1)X+b)+m = 0$

$\Rightarrow \quad a^{\sigma+1}(a+1)(a^{\sigma+1}X+a^{\sigma}b+b^{\sigma})+(a^{\sigma+1}b+(a^{\sigma}b+b^{\sigma})(a+1))X+(a^{\sigma}b+b^{\sigma})b+m = 0$

$\Rightarrow \quad (a^{2\sigma+2}(a+1)+(a^{\sigma}b+b^{\sigma})(a+1)+a^{\sigma+1}b)X+a^{\sigma+1}(a+1)(a^{\sigma}b+b^{\sigma})+(a^{\sigma}b+b^{\sigma})b+m = 0.$

Since this equation is satisfied for $X = \alpha$ and $X = \beta$, then it is satisfied for all X. In particular, the coefficient of X in the equation is zero, and so

(4)...$a^{2\sigma+2}(a+1) = a^{\sigma}b+(a+1)b^{\sigma}$.

Now $a \neq 0$, (otherwise (2) would have only one solution), and so multiplying (4) by $(a+1)^{\sigma+1}a^{-2\sigma-2}$ gives

(5)...$(a+1)^{\sigma+2} = (a+1)^{\sigma+1}ba^{-\sigma-2}+((a+1)^{\sigma+1}ba^{-\sigma-2})^{\sigma}$.

Hence $(a+1)^{\sigma+2}$ is first category in $GF(q)$. Thus

$$\mathrm{tr}((a+1)^{\sigma+2}) = \mathrm{tr}(a^{\sigma+2}+a^{\sigma}+a^2+1)$$

$$= tr(a^{\sigma+2})+tr(a^{\sigma})+tr(a^2)+tr(1)$$
$$= tr(a^{\sigma+2})+2tr(a^2)+1$$
$$= tr(a^{\sigma+2})+1 = 0$$

$\Rightarrow \quad tr(a^{\sigma+2}) = 1$

$\Rightarrow \quad tr((a^{\sigma+2})^{\sigma/2}) = 1$

$\Rightarrow \quad tr(a^{\sigma+1}) = 1$

$\Rightarrow \quad tr(\alpha+\beta) = 1.$

Now if a third solution, say $\delta \in GF(q)$, to (1) existed, then $tr(\alpha+\delta)$ and $tr(\beta+\delta)$ would also equal 1. Hence

$$tr(\alpha+\beta)+tr(\alpha+\delta)+tr(\beta+\delta) = 1+1+1 = 1$$
$$= tr(2\alpha+2\beta+2\delta) = tr(0) = 0, \quad \text{a contradiction.}$$

Thus δ does not exist, and so $[\ell,k,1]$ intersects $D(\frac{\sigma+2}{3})$ in at most two points. Hence $D(\frac{\sigma+2}{3})$ is an oval of PG(2,q) and by Result 8, so is $D(3\sigma+4)$.

5. THE KNOWN OVALS OF PG(2,q), q EVEN

Here all the classes of ovals of PG(2,q) known in 1982 are listed. Let $q = 2^h$.

(a) $D(2^n)$, $(n,h) = 1$ (Discovered by Segre in 1957 [15].)

These are called the translation ovals. See [7] and [12] for further details. Note that $D(2)$ gives the sub-class of irreducible conics.

(b) $D(6)$, h odd (See Segre 1962 [17].)

Further references are [6] and [7].

(c) $D(\sigma+\gamma)$, h odd, $h \geq 3$ ($\sigma\equiv\sqrt{2}$, $\gamma\equiv\sqrt[4]{2}$ (mod q-1). See Theorem B of Section 4.)

Here is the list of ovals $D(k)$, equivalent to $D(\sigma+\gamma)$ by Result 8.

k	$1-k$	k^{-1}	$1-k^{-1}$	$(1-k)^{-1}$	$k(k-1)^{-1}$
$\sigma+\gamma$	$1-\sigma-\gamma$	$-\gamma^{-1}+\sigma-\gamma+1$	$\gamma^{-1}-\sigma+\gamma$	$\frac{1}{3}(-2\gamma^{-1}-\gamma+1)$	$\frac{1}{3}(2\gamma^{-1}+\gamma+2)$

(d) $D(3\sigma+4)$, h odd, $h \geq 3$ (See Theorem C of Section 4.)

Again, here is the list of ovals $D(k)$, equivalent to $D(3\sigma+4)$.

k	$1-k$	k^{-1}	$1-k^{-1}$	$(1-k)^{-1}$	$k(k-1)^{-1}$
$3\sigma+4$	$-3\sigma-3$	$\frac{3}{2}\sigma-2$	$-\frac{3}{2}\sigma+3$	$\frac{1-\sigma}{3}$	$\frac{\sigma+2}{3}$

(e) An irregular oval in PG(2,16) (Lunelli, Sce 1958 [11]).

This oval and the irreducible conic (plus its nucleus) are the only ovals of PG(2,16) [5]. For the characterization of this irregular oval in terms of having a point-transitive group, see [9]. The author has calculated that this oval is projectively equivalent to $C \cup C''\setminus(C \cap C'')$, where C and C'' are the non-singular cubic curves

$$C : x^3+y^3+z^3+dxyz = 0, \text{ and}$$

$$C'': x^3+y^3+z^3+d^4xyz = 0, \text{ where } d \in GF(16),\ d^5 = 1,\ d \neq 1.$$

Note that the classes (a) to (d) above sometimes overlap for small values of q, but they are distinct for large values. The more complicated question of isomorphism between various classes is left to the interested reader. However, see [8] for some ideas on this.

6. CONCLUSION

In this paper we have looked mainly at the ovals of type D(k) and constructed two infinite sequences of these. (A discussion of the many applications of ovals has been left to another survey.) The computer evidence given in Section three leads naturally to the following two conjectures. (As with all conjectures, it is quite possible that they are false.)

Conjecture A. The only ovals of type D(k) in PG(2,q), ($q = 2^h$, h even), are the translation ovals, $D(2^n)$, $(n,h) = 1$.

Conjecture B. The only ovals of type D(k) are those given by the four sequences (a) to (d) of Section five.

It would also be interesting to know if the numerical condition of Theorem A could be of direct applicability, (without using a computer). Finally, the possible existence of further ovals (also not of type D(k)), is of utmost interest in the theory of finite Desarguesian planes.

REFERENCES

[1] U. Bartocci, *Considerazioni sulla teoria delle ovali*, (Thesis, University of Rome, 1967).

[2] P. Dembowski, *Finite Geometries* (Springer-Verlag, Berlin-Heidelberg - New York, 1968).

[3] L.E. Dickson, *Linear Groups, with an Exposition of the Galois Field Theory*, (Teubner, Leipzig, 1901).

[4] D.G. Glynn, *Finite Projective Planes and Related Combinatorial Systems*, (Ph.D. Thesis, Adelaide, 1978).

[5] M. Hall Jr., Ovals in the Desarguesian plane of order 16, *Ann. Mat. Pura Appl.*, (4), 102, (1975), 159-176.

[6] J.W.P. Hirschfeld, Rational curves on quadrics over finite fields of characteristic two, *Rend. Mat. Appl.*, 34, (1971), 773-795.

[7] J.W.P. Hirschfeld, Ovals in Desarguesian planes of even order, *Ann. Mat. Pura Appl.*, (4), 102, (1975), 79-89.

[8] J.W.P. Hirschfeld, *Projective Geometries over Finite Fields* (Oxford Mathematical Monographs, 1979).

[9] G. Korchmàros, Gruppi di collineazioni transitivi sui punti di una ovale [(q+2)-arco] di $S_{2,q}$, q pari, *Atti. Sem. Mat. Fis. Univ. Modena*, 27, (1978), 89-105.

[10] B. Larato, Sull'esistenza delle ovali del tipo $D(x^k)$. *Atti Sem. Mat. Fis. Univ. Modena*, 29, (1980), 345-348.

[11] L. Lunelli and M. Sce, *K-archi completi nei piani proiettivi desarguesiani di rango 8 e 16*, Centro Calcoli Numerici, Politecnico di Milano, (1958).

[12] S.E. Payne, A complete determination of translation ovoids in finite Desarguesian planes, *Atti Accad. Naz. Lincei Rend.* (8), 51, (1971), 328-331.

[13] L.A. Rosati, Sulle ovali dei piani desarguesiani finiti d'ordine pari, *Annals of Discrete Mathematics*, 18, (1983), 713-720.

[14] B. Segre, Ovals in a finite projective plane, *Can. J. Math.*, 7, (1955), 414-416.

[15] B. Segre, Sui k-archi nei piani finiti di caratteristica 2, *Revue de Math. Pures Appl.*, 2, (1957), 289-300.

[16] B. Segre, *Lectures on Modern Geometry* (Cremonese, Rome, 1961).

[17] B. Segre, Ovali e curve σ nei piani di Galois di caratteristica due, *Atti Accad. Naz. Lincei Rend.*, (8), 32, (1962), 785-790.

[18] B. Segre and U. Bartocci, Ovali ed altre curve nei piani di Galois di caratteristica due, *Acta Arith.*, 18, (1971), 423-449.

STOCHASTIC PROCESSES AND COMBINATORIC IDENTITIES

W. HENDERSON, R.W. KENNINGTON AND C.E.M. PEARCE

Despite the existence of such salient successes as the use of ergodic theory to establish Szemerédi's theorem, dynamical systems and stochastic processes have not been widely employed in the derivation of combinatoric results. This paper offers instances of what the authors believe should be a more widespread tool in the routine analysis of combinatoric problems.

1. INTRODUCTION

Takács [24] has observed that many results in the field of stochastic processes are essentially combinatoric in nature, which in view of the genesis of probability in elementary combinatorics is not entirely surprising. However this century probability theory has developed into a branch of analysis and results concerning stochastic processes have often been established both by analytical and by combinatoric arguments which may, at first sight, look quite unrelated. A good example is provided by fluctuation theory, the foundations of which were laid in the 1950's independently by Andersen [1], [2], [3] employing combinatoric means and by Pollaczek [17], Spitzer [23] and others analytically. A unified and very readable account is given by Port [19]. See also P. Erdös and J. Spencer [6].

To date the traffic in ideas between stochastic processes and combinatorics has been largely one-way, combinatoric arguments being used to simplify or clarify the analytical and probabilistic treatment of particular processes. The possibility of a reverse traffic, with formal process interpretations of combinatoric structures being employed as a source both of results and of methods, appears at least as attractive. There is an appreciable corpus of theory available for broad classes of stochastic processes together with the chance of physical 'feel' to augment more formal intuition. The intuitive apprehension of a process is especially useful when one is faced with problems of a 'to find' rather than a 'to verify' character.

An illustration of how effective this reverse traffic can be is the use of the multiple Poincaré recurrence theorem of ergodic theory to derive Szemerédi's theorem of combinatoric number theory. The proof makes use of special cases of rather more general results, suggesting that stronger conclusions may be available. Furstenberg [9] gives a fine exposition of the ideas involved. In the same vein, though with less overkill in the proof, a structurally simpler version of Schmidt's result [22] on normal numbers has been provided by utilising results from ergodic theory and probability theory (Pearce and Keane [16]). A simple and elegant, though little known, exemplar of the 'process' motif within the field of stochastic processes itself is the 'method of marks' (see Runnenburg [21]), which ascribes a physical interpretation as a probability to the carrier variable in probability generating functions.

Nevertheless, while the utilisation of stochastic processes to establish combinatoric results is not new, it is not common either. Our present purpose is to advocate an increased use of 'process orientation' as part of the combinatorialist's routine repertoire. A good introduction is offered by Feller [7] and Karlin [12]. In the following sections we offer two simple examples.

Before proceeding to treat these, we note that stochastic processes have a useful tendency to possess characteristics insensitive to variations in the particular assumptions prescribing the situation concerned. The most long standing and celebrated instances are the various probability limit theorems, the central limit theorem, the law of the iterated logarithm, etc.

2. PATH FORMALISMS

In this section we elaborate the 'path formalism' interpretation of recurrence relations. Recent applications of this idea have been made by Antippa and Toan [4], [5] in the study of systems of equations. This approach has proved useful for bringing combinatorial ideas into play to improve solution procedures for certain classes of equations, particularly when symmetries are involved. The underlying notion is as follows. Consider a weighted directed graph with vertices numbered $0,1,2,3,\ldots$ and let w_{ij} be the weight of the edge from i to j, if present, and 0 otherwise. The weight of a (finite) path is the product of the weights of its edges, and the work in [4], [5] involves calculation of the sum $H(P(i,j))$ of the weights of all paths from i to j. This, in turn, depends on the $S(i,j)$'s, where $S(i,j)$ is the sum of the weights of all paths from i to j which do not pass through j en route. The idea of the example to be considered is that if $W = [w_{ij}]$ is a stochastic matrix then the whole situation can be viewed in terms of a Markov chain and the calculation of $S(i,j)$ becomes much easier. In our example we evaluate $S = S(0,0)$. Consider the following prescription.

Take $1 \in C \subset Z^+ = \{1,2,\ldots\}$ and let $0 < a_j < 1$ for $j \in Z^+$. Set $D = \{a_j: j > 0\}$, $D^- = \{1-a_j: j > 0\}$ and define

$$[j] = \begin{cases} \max\{\ell: \ell \in C, \ell < j\} & , \; j \notin C \\ 0 & , \; j \in C \end{cases} .$$

Suppose we are interested in the sum

$$S = \Sigma \prod_{i=1}^{n} b_i \tag{1}$$

where the summation is over all products with $n \in Z^+$ satisfying the conditions

(i) each $b_i \in D \cup D^-$

(ii) $b_1 = a_1$ or $1-a_1$

(iii) if $b_i = a_k$, then $b_{i+1} = a_{k+1}$ or $1-a_{k+1}$

(iv) if $b_i = 1-a_k$, $k \notin C$, then $b_{i+1} = a_{[k]}$ or $1-a_{[k]}$

(v) if $b_i = 1-a_k$, $k \in C$, then $i = n$.

The form of (1) is reminiscent of expressions for first passage and absorption probabilities in discrete-state Markov chains (see, for example, Karlin [12]) and we shall exploit this form to extend the path formalism behind (1) and give a 'look see' interpretation and analysis in terms of two Markov chains.

We set up an irreducible Markov chain on the state space $\{0\} \cup Z^+$ and ascribe a special role to the states labelled by C. We take $a_0=1$ and define our chain by the one-step transition matrix (P_{ij}) given by

$$P_{ij} = \begin{cases} a_i & , \; j = i+1 \\ 1-a_i & , \; j = [i] \end{cases} . \tag{2}$$

Each sample path for the process M thus defined, for the origin as initial state, is then a walk on the non-negative integers built up as follows. A step to the right occurs by moving to the immediately adjacent state and a step to the left from state i by dropping to the nearest leftmost state of the subset C when $i \notin C$ and by dropping to the origin when $i \in C$. By inserting a leading term $b_0 = a_0 = 1$ at the beginning of each product of (1), we see that there is a one to one correspondence between the products $b_0 b_1 \ldots b_n$ and sample paths of the Markov chain beginning at the origin and making a first return there on the last step. Furthermore, the value of each such product is simply the probability associated with the corresponding sample path. The sum S may thus be interpreted as the probability that the process will ever return to the origin, given that it commences there.

one-step state transitions in the chain M

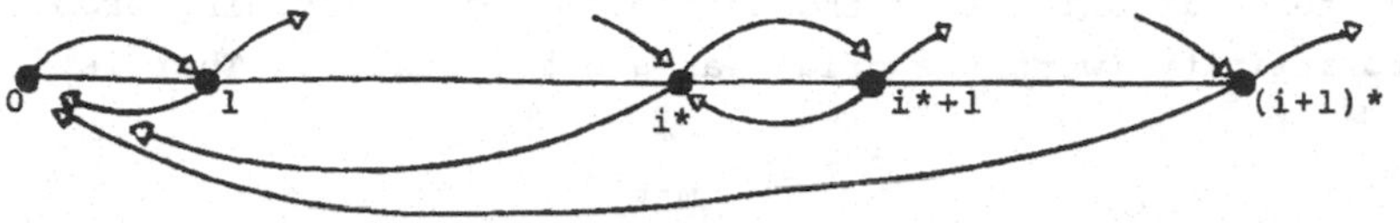

Suppose we label the states of C as $1^* < 2^* < \ldots$. When $|C| = N < \infty$ we set $(N+1)^* = \infty$. If the process M enters the state i^*, it either proceeds to $(i+1)^*$ (with probability $a_{i^*}a_{i^*+1}\ldots a_{(i+1)^*-1}$) by a sequence of steps to the right, or proceeds directly to the origin (with probability $1 - a_{i^*}$) in a single step, or executes a loop $i^* \to i^* + 1 \to \ldots \to j - 1 \to j \to i^*$ for some $j \in (i^*,(i+1)^*)$. In the lattermost event the process is then faced with the same possibilities afresh. The probability of looping back to i^* infinitely often is zero and either the transition $i^* \to (i+1)^*$ (with 0 as a taboo state) or the transition $i^* \to 0$ (with $(i+1)^*$ as a taboo state) will eventually occur, with probability one. The two possibilities occur with probabilities

$$c_i = \frac{\prod_{r=i^*}^{(i+1)^*-1} a_r}{1 - a_i^* + \prod_{r=i^*}^{(i+1)^*-1} a_r} \qquad (3)$$

and $1 - c_i$ respectively.

This suggest that we may conceptualise the process more simply by amalgamating the collection of states $[i^*,(i+1)^*)$ into a single state. More precisely, define a new Markov chain M^* with states $0,1,2,\ldots$ (possibly only a finite set) and one-step transition matrix (Q_{ij}) given by

$$Q_{ij} = \begin{matrix} c_i & , & j = i + 1 \\ 1 - c_i & , & j = 0 \end{matrix} \quad , \qquad (4)$$

where we take $c_0 = 1$. We think of the new process M^* as being in state i when the process M is in the set of states $[i^*,(i+1)^*)$ (state 0 when M is in state 0). The transition instants for M^* are the instants at which M moves from one such set of states to another. The probability that M^* will make a return to the origin, conditional on commencing there, is the same as the corresponding probability S for M. The process M* has a particularly simple structure: it continues stepping to the right one unit at a time until it makes a leftward step, when it returns at once to the origin. The probability that the process ever returns to the origin, conditional on its starting there, is unity less the probability of proceeding from the origin to infinity (with the origin as a taboo state). That is

$$S = 1 - \prod_{m=0}^{\infty} c_m . \tag{5}$$

one-step state transitions in the chain M*

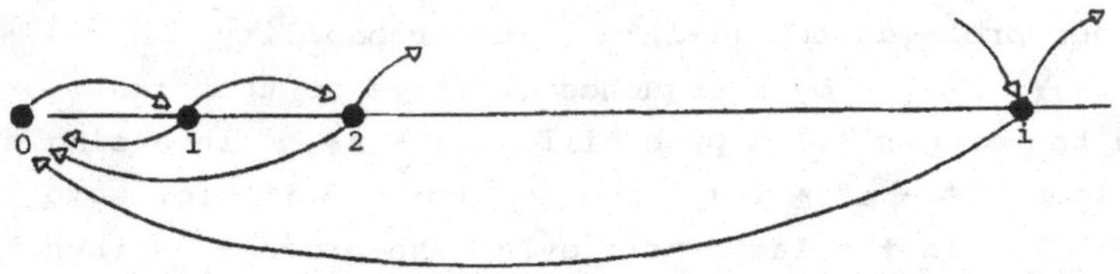

Observations.

This evaluation of S is, of course, available directly by combinatoric means but would entail algebraic detail together with a limiting procedure. The present 'look see' development is entirely structural and involves virtually no algebra and no limiting operations. The argument extends trivially to more involved sequencing of product terms $b_1 b_2 \ldots b_n$ corresponding to a nested hierarchy $Z^* \supset C_1 \supset C_2 \supset \ldots$ of special subsets of labels.

It is clear from (5) that unless the sequence (c_m) converges to unity fairly rapidly we will have $S = 1$. This illustrates the insensitivity phenomenon alluded to in the introductory section. We observe parenthetically that the right hand side of the standard equivalence

$$\prod_{m=1}^{\infty} c_m > 0 \quad \leftrightarrow \quad \sum_{m=1}^{\infty} (1-c_m) < \infty$$

suggests the Borel-Cantelli lemmas, again inviting physical interpretation of the circumstances when $S < 1$.

3. PROCESS ON RANDOM SETS

A seminal paper of Polya [18] treats the following allocation problem. Particles occur in successive complexes of size t. The particles of each complex in turn are assigned to $N > t$ cells at random, subject to the constraint that no two particles of any one complex may be assigned to the same cell. The question of initial interest was the mean number of complexes required before each cell had at least one particle assigned to it. The corresponding question for $t = 1$ and each cell having at least m particles was considered by Newman and Shepp [15]. The exact expressions involved turn out to be awkward computationally and interest centred on their asymptotic forms as $m,N \to \infty$. Newman and Shepp show that the mean number of particles $E_m(N)$ required for each cell to have at least m particles is given asymptotically for N large and m fixed by

$$E_m(N) = N[\log N+(m-1)\log\log N + C_m + 0(1)].$$

That is, although 'first completion' requires $\approx N \log N$ particles, interestingly each additional completion requires only a further $\approx N \log \log N$ particles.

Subsequent developments have been numerous and mostly posit an independent distribution (not necessarily uniform) over cells $1,2,\ldots,N$ for each particle of a given complex. Successive complexes are allocated independently, possibly with different distributions. An account of the literature to 1978 is given by Kolchin, Sevast'yanov and Chistyakov [14]. This area has independent combinatoric interest in that many of its results may be interpreted in terms of graphs associated with random mappings (see the review paper of Kolchin and Chistyakov [13]).

We revert to the original Polya scheme and consider a single complex for the case $t = N$. We endow the complex with some stochastic structure by regarding the cells to which successive particles are allocated as the successive states visited by a random process and singling out for attention two subsets G,A of the N cells. As a stochastic process, the model admits the following natural formulation.

Suppose a process with a discrete state space moves in succession into states $x_1,x_2,\ldots$ at times $t_1 < t_2 < \ldots$ and that $x_i \neq x_j$ for $1 \leq i \neq j \leq N$, that is, the process has not entered any state twice by time t_N. This occurs typically in the study of a process generated by a single orbit. For notational convenience we set $x_j = j$ and $t_j = j$ for $1 \leq j \leq N$, so that up to time N the process is in state j at time j. Denote by S the set of states $\{1,2,\ldots,N\}$. Let $G \subset A \subset S$ and denote $m = |A \setminus G| > 0$, $\rho = |G| > 0$. We imagine that G,A are chosen randomly by uniform sampling from S without replacement. Equivalently we may regard G and A as fixed and the orbit as random, in which case the process picks out a sequence of the elements of S by choosing an integer from S, without replacement, at each time point. Suppose that g is the last state of G visited by the process. We pose the following question:

What is the distribution of the first passage time T_m (completion time for the set A), possibly zero, from the epoch at which the process visits state g to the first epoch when it has visited all the states of A?

We may rephrase the latter version with a more combinatoric slant as follows. For fixed N call a sequence of m elements of A/G, ρ of G and $N - m - \rho$ of δ/A an (N,m,ρ) sequence. For verbal convenience we refer to this loosely as a sequence of m a's, ρ g's and $N - m - \rho$ δ's. The event $T_m = k$ occurs if and only if the last a or g is exactly k places after the last g; and for $k \geqslant 1$ this means the last a is k places after the last g. We require the probability that this occurs when a sequence is chosen at random from the $\frac{N!}{m!\rho!(N-m-\rho)!}$ distinguishable (N,m,ρ) sequences.

Before proceeding to examine this question, we note that the case $\rho = 1$ admits an application in applied combinatorics to a problem of queueing in lanes (see Hauer and Templeton [9]). N cars in a parking lot are so situated that a given car of interest cannot move until m others have moved. The N drivers reach their respective cars in a random sequence. Once the driver of the given car is ready, for how many further drivers does he have to wait before he can move his car? A generalisation of the Hauer and Templeton model with general ρ and more than one set A is treated by the authors [10].

It is not our intention to detail probability limit theorems which arise with our specific formulation. Rather we show that results of combinatoric interest arise in a natural way even within the very circumscribed ambit imposed by fixed N and one complex. A key feature of this combinatoric problem (and others with a process formulation) is a natural total ordering of the elements involved which is prescribed by 'time'. We note the guiding motif of a partial ordering relation to treat enumeration problems via the Möbius function in the elegent generalisation of the inclusion-exclusion principle developed by Rota [20]. The reader is referred to the survey by R.J. Wilson [25].

4. DISTRIBUTION OF COMPLETION TIME

Let $P(T_m=k)$ be the probability that the event $T_m = k$ occurs. Let $(N,m,\rho)_k$ = the number of (N,m,ρ) sequences for which $T_m = k$. Then trivially

$$P(T_m=k) = (N,m,\rho)_k \frac{\rho!m!(N-m-\rho)!}{N!} \tag{6}$$

Theorem 1. *For* $\rho = 1$

$$P(T_m=k) = \frac{1}{N}\left[1 - \frac{\binom{k-1}{m}}{\binom{N-1}{m}}\right], \quad 0 < k < N. \tag{7}$$

$$P(T_m=0) = \frac{1}{m+1}$$

Proof. $P(T_m=0)$ = the probability that the g follows all m a's $= \frac{1}{m+1}$. For $k > 0$ the event $T_m = k$ can be constructed as follows. Take any of the $\binom{N-1}{m}$ sequences of the m a's and $N - m - 1$ δ's. Insert the g k places before the last of the a's. This can be done in all but the $\binom{k-1}{m}$ sequences in which all the a's are among the first $k - 1$ places. That is

$$(N,m,1)_k = \binom{N-1}{m} - \binom{k-1}{m}$$ and (6) yields (7) □

Note, in particular, that for $0 < k \leqslant m$ equation (7) reduces to

$$P(T_m=k) = \frac{1}{N}. \tag{8}$$

This may be taken as stating that if $m + 1$ integers are chosen by sampling without replacement from $\{1,2,...,N\}$ the probability that a randomly chosen one of them is picked out k places earlier than the last of them is $\frac{1}{N}$ when $0 < k \leqslant m$.

This is a particularly striking result, independent of the precise values of k and m. We also observe that $1 - \frac{\binom{k-1}{m}}{\binom{N-1}{m}}$ is the probability that the m a's and $N - m - 1$ δ's are ordered so that it is possible to insert the g k places before the last of the a's. Hence (7) may be interpreted as saying that for $k > 0$, conditional on the choice of the a's from $N - 1$ ordered elements being such that it is possible to insert g to give $T_m = k$, the probability that $T_m = k$ is $\frac{1}{N}$. This generalises the observation (8).

Corollary.

$$\sum_{s=0}^{j} \binom{s+r}{r} = \binom{j+r+1}{r+1} \qquad r \geq 0, \quad j \geq 0 \tag{9}$$

Proof. The more usual algebraic enumeration, based on the possible positions of g in an $(N,m,1)$ sequence in which $T_m = k$ provides

$$P(T_m=k) = \sum_{\ell=\max(1,m+1-k)}^{N-k} \binom{\ell+k-2}{m-1} \frac{m!(N-m-1)!}{N!} \qquad 0 < k < N \tag{10}$$

Comparing (7) and (10) in the special case $m = k$ yields the standard identity (9) (see, for example, Feller [7], II 12.8), which is usually derived ex nihilo by induction. □

Theorem 2. *For general* ρ

$$P(T_m=k) = \left\{\left[\sum_{s=0}^{\rho-1} (-1)^s \binom{N-k}{\rho-1-s}\binom{N-\rho+s}{m+s}\right] + (-1)^\rho \binom{k-1}{m+\rho-1}\right\} \frac{m!\rho!(N-m-\rho)!}{N!} \qquad 0 < k \leq N-\rho \tag{11}$$

$$P(T_m=0) = \frac{\rho}{m+\rho} \tag{12}$$

Proof. From (6) we must prove that

$$(N,m,\rho)_k = \left[\sum_{s=0}^{\rho-1} (-1)^s \binom{N-k}{\rho-1-s}\binom{N-\rho+s}{m+s}\right] + (-1)^\rho \binom{k-1}{m+\rho-1},$$

a result which follows easily from the identity

$$(N,m,\rho)_k = \binom{N-k}{\rho-1}\binom{N-\rho}{m} - (N,m+1,\rho-1)_k \tag{13}$$

together with

$$(N,m+\rho-1,1)_k = \binom{N-1}{m+\rho-1} - \binom{k-1}{m+\rho-1}$$

obtained earlier for Theorem 1.

(13) can be proved as follows:

Consider an $(N-1,m,\rho-1)$ sequence in which all the g's are in the first $N - k$ places, $k > 0$. If the number of places between the last g and the last a is $\leq k - 1$ another a can be placed k places after the last g. This creates one of the $(N,m+1,\rho-1)_k$ sequences. On the other hand if the number of places between the last g and the last a is $\geq k$ another g can be placed k places before the last a and create one of the $(N,m,\rho)_k$ sequences. Every one of the $(N,m+1,\rho-1)_k$ and $(N,m,\rho)_k$ can be constructed in this manner, therefore, $(N,m+1,\rho-1)_k + (N,m,\rho)_k$ = the number of $(N-1,m,\rho-1)$ sequences in which all of the g's are in the first $N - k$ places

$$= \binom{N-k}{\rho-1}\binom{N-\rho}{m}.$$

This proves (13) and ∴ (11).

Finally, for $T_m = 0$, we require that the last g or a in the (N,m,ρ) sequence is a g. This gives (12) trivially. □

<u>Corollary</u>. For $j \geq 0$, $r \geq 0$, $\ell \geq n$

$$\sum_{s=0}^{j} \binom{s+r}{r}\binom{s+\ell}{n} = \sum_{s=0}^{n} (-1)^{s+n}\binom{j+\ell+1}{s}\binom{j+r+n+1-s}{j} \qquad (14)$$

$$= \left[\sum_{s=0}^{r} (-1)^{s}\binom{j+r+1}{r-s}\binom{s+j+\ell+1}{s+n+1}\right] - (-1)^{r}\binom{\ell}{r+n+1} \qquad (15)$$

<u>Proof</u>. $(N,m,\rho)_k$ can also be evaluated as follows. The last of the g's can be in any position from $\max\{\rho,\rho+m-k\}$ to $N - k$. Let this position be ℓ. The lower limit, $\max\{\rho,\rho+m-k\}$, is derived by noting that ρ g's must occur in the first ℓ positions and ρ g's plus m a's must occur in the first $\ell + k$ positions. Now distribute the $\rho - 1$ remaining g's among the first $\ell - 1$ positions, place one a in position $\ell + k$ and distribute the remaining $m - 1$ a's among the first $\ell + k - \rho - 1$ places not occupied by g's.

For fixed ℓ the number of ways that this operation can be performed is $\binom{\ell+k-\rho-1}{m-1}\binom{\ell-1}{\rho-1}$. Hence

$$(N,m,\rho)_k = \sum_{\ell=\max(\rho,\rho+m-k)}^{N-k} \binom{\ell+k-\rho-1}{m-1}\binom{\ell-1}{\rho-1} \qquad (16)$$

Now equate (16) to the previous version of $(N,m,\rho)_k$.

When $k \leq m$ substitutions $\rho = n + 1$, $m = r + 1$, $k = r + n - \ell + 1$, $N = j + r + n + 2$ yields (14). When $k \geq m$ the substitutions $\rho = r + 1$, $k = \ell + 1$, $m = n + 1$, $N = j + r + \ell + 2$ gives (15). In both the above cases, $k \leq m$ and $k \geq m$, the conditions $j \geq 0$, $r \geq 0$, $\ell \geq n$ can be easily derived by physical considerations. □

The motivation for using (11) in place of (16) is that the former involves less computation. In the application of Hauer and Templeton [9] and the authors [10], typically $N > 1000\rho$. From a more abstract standpoint, the right hand side of (14) can be derived from the left by the same procedure (summation by parts) as for the corresponding sides of (9). (For details of this technique see Jordan [11] §34). Thus equation (14) may be regarded as a natural extension of (9) to a sum of products of binomial coefficients. The identities (14), (15) generalise a number of well known relations amongst binomial coefficients, for example

$$\sum_{s\geq 0} (-1)^s \binom{a}{s}\binom{n-s}{r} = \binom{n-a}{n-r} \qquad (17)$$

(see Feller [7] II 12.15)

An alternative evaluation for $P(T_m=k)$ for $k \geq m$, starting from (16) is

$$\begin{aligned} P(T_m=k) &= \sum_{\ell=\rho}^{N-k} \binom{\ell+k-\rho-1}{m-1}\binom{\ell-1}{\rho-1} m!\rho!(N-\rho-m)!/N! \\ &= \sum_{s=0}^{N-k-\rho} \binom{s+k-1}{m-1}\binom{s+\rho-1}{\rho-1} m!\rho!(N-\rho-m)!/N! \\ &= \sum_{s=0}^{m-1} (-1)^{s+m-1}\binom{N-\rho}{s}\binom{N+m-k-1-s}{N-k-\rho} m!\rho!(N-\rho-m)!/N!\,, \end{aligned} \qquad (18)$$

by virtue of (14).

This formula offers a computational simplification to (16) that is preferable to (11) in the event that $m \leq \rho$. This alternative to (11) corresponds to a variant of (14) produced by interchanging the roles of the two binomial coefficients in the left hand side of (14) when a summation by parts procedure is adopted to obtain the right hand side. The two variants may be derived from the same formula if we generalise (14) to

$$\sum_{s=a}^{b} \binom{s+n_1}{n_2}\binom{s+n_3}{n_4} = \sum_{s=0}^{n_2} (-1)^{s+n_2}\left[\binom{b+n_1+1}{s}\binom{b+n_2+n_3+1-s}{n_2+n_4+1-s} - \binom{a+n_1}{s}\binom{a+n_2+n_3-s}{n_2+n_4+1-s}\right], \quad b \geq a, \qquad (19)$$

which is a more convenient standard version of this identity. It may also be established by induction from $b = a - 1$ as basis. The variant identity now arises on an interchange of the pairs $(n_1,n_2),(n_3,n_4)$ on the right hand side. Equation (16) may now be converted into either (11) or (18) in one step via (19).

A further line of development is invited by the theorem. Suppose we split the sum in (17) as $\Sigma_0^{m+\rho-1} - \Sigma_m^{m+\rho-1}$ and evaluate the first part by an appeal to (14). We obtain

$$P(T_m=k) = (-1)^{m-1}\left[\binom{m+\rho-k-1}{\rho+m-1} - \sum_{r=0}^{\rho-1}(-1)^{r+m}\binom{N-\rho}{r+m}\binom{N-k-r-1}{N-k-\rho}\right] m!\rho!(N-\rho-m)!/N!$$

$$= \left[\sum_{r=0}^{\rho-1}(-1)^{r}\binom{N-\rho}{r+m}\binom{N-k-r-1}{N-k-\rho} - (-1)^{\rho-1}\binom{k-1}{m+\rho-1}\right] m!\rho!(N-\rho-m)!/N! \quad ,$$

on an application of the relation

$$\binom{-a}{\ell} = (-1)^{\ell}\binom{a+\ell-1}{\ell}$$

(Feller [7], II 12.4).

A comparison with (11) finally yields the prototype identity

$$\sum_{s=0}^{\rho-1}(-1)^{s+\rho-1}\binom{N-k}{s}\binom{N-1-s}{N-\rho-m} = \sum_{s=0}^{\rho-1}(-1)^{s}\binom{N-\rho}{s+m}\binom{N-k-1-s}{N-k-\rho} \quad , \quad (20)$$

which again does not appear to have antecedents in the literature.

We remark in conclusion that the above development has been kept deliberately at the level of pencil jottings to emphasise the role of the underlying process interpretation in provoking the inductive discovery of identities such as (19) and (20). The process is, of course, irrelevant to a compact derivation of such identities by, say, the principle of mathematical induction. The point is that these interesting identities arise so naturally in the present discussion.

ACKNOWLEDGEMENT

The authors thank Dr. Jane Pitman for her many valuable comments, especially the present proof to Theorem 2.

REFERENCES

[1] E.S. Andersen, On sums of symmetrically dependent random variables. *Skand. Aktuaretid.* 36 (1953), 123-138.

[2] E.S. Andersen, On the fluctuation of sums of independent random variables. *Math. Skand.* 1 (1953), 263-285.

[3] E.S. Andersen, On the fluctuation of sums of independent random variables II. *Math. Skand.* 2 (1954), 195-223.

[4] A.F. Antippa and N.K. Toan, Topological solution for systems of simultaneous linear equations. *J. Math. Phys.* 20 (1979), 2375-2379.

[5] A.F. Antippa and N.K. Toan, Topological solution of ordinary and partial finite difference equations. *J. Math. Phys.* 21 (1980), 2475-2480.

[6] P. Erdös and J. Spencer, *Probabilistic Methods in Combinatorics.* Academic Press, New York (1974).

[7] W. Feller, *An introduction to probability theory and its applications, Vol. I, 3rd edition.* Wiley and Sons, New York (1968).

[8] H. Furstenberg, *Recurrence relations in ergodic theory and combinatorial number theory.* Princeton U.P., Princeton (1981).

[9] E. Hauer and J.G.C. Templeton, Queueing in Lanes. *Trans. Sci.*, 6 (1972), 247-259.

[10] W. Henderson, R.W. Kennington and C.E.M. Pearce, A second look at a problem of queueing in lanes. To appear in Transportation Science.

[11] C. Jordan, *Calculus of finite differences, 2nd edition.* Chelsea Publ. Co., New York (1947).

[12] S. Karlin, *A first course in stochastic processes.* Academic Press, New York (1966).

[13] V.F. Kolchin and V.P. Chistyakov, Combinatorial problems in probability theory. *Results of Science and Technology, Ser: Prob. Theory, Math. Stats, Theor. Cybernetics (All Union Inst. of Scientific and Tech. Info., Moscow)*, 11 (1974), 5-54 (in Russian).

[14] V.F. Kolchin, B.A. Sevast'yanov and V.P. Chistyakov, *Random Allocations.* V.H. Winston (Wiley), Washington (1978).

[15] D.J. Newman and L. Shepp, The double dixie cup problem. *Amer. Math. Monthly*, 67 (1960), 58-61).

[16] C.E.M. Pearce and M.S. Keane, On normal numbers. *J. Aust. Math. Soc. (A)*, 32 (1982), 79-87.

[17] F. Pollaczek, Fonctions caractéristiques de certaines répartitions definies au moyen de la notion d'ordre. Application à la théorie des attentes. *Compt. Rend. Acad. Sci. Paris*, 234 (1952), 2234-2236.

[18] G. Polya, Eine Wahrscheinlichkeitsaufgabe zur Kundenwerbung. *Zeit. für angewandte Math. und Mech.*, 10 (1930), 96-97.

[19] S.C. Port, An elementary probability approach to fluctuation theory. *J. Math. Anal. and Applic.*, 6 (1963), 109-151.

[20] G.-C. Rota, On the foundations of combinatorial theory I. Theory of Möbius functions. *Z. Wahrscheinlichkeitstheorie*, 2 (1964), 340-368.

[21] J.Th. Runnenburg, On the use of the method of collective marks in queueing theory. *Proceedings of the symposium on congestion theory*, Ed. W.L.Smith and W.E. Wilkinson, Univ. of N. Carolina, Chapel Hill (1965), 399-438.

[22] W. Schmidt, On normal numbers. *Pacific J. Math.*, 10 (1960), 661-672.

[23] F. Spitzer, A combinatorial lemma and its application to probability theory. *Transac. Amer. Math. Soc.*, 82 (1956), 323-339.

[24] L. Takács, *Combinatorial methods in the theory of stochastic processes.* Wiley and Sons, New York (1967).

[25] R.J. Wilson, *Combinatorial Mathematics and its Applications.* Ed. D.J.A. Walsh, Academic Press, New York (1971), 315-333.

FAMILIES ENUMERATED BY THE SCHRÖDER-ETHERINGTON SEQUENCE AND A RENEWAL ARRAY IT GENERATES

S.G. KETTLE

Introduction *'The problems considered here are essentially algebraic; but it is convenient to begin with a picturesque formulation.'* (I.M.H. Etherington, 1940).

Consider the collection $\bar{T}$ of planar subsets determined by the rules

o) $\bullet \in \bar{T}$, and for each $k \geq 0$,

k) if $u_1, u_2, \ldots, u_k \in \bar{T}$ then 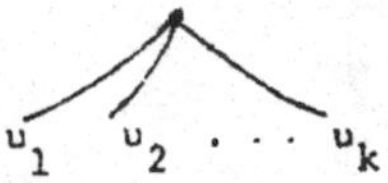 $\in \bar{T}$

If we denote 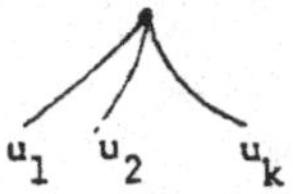 by $\omega(u_1, u_2, \ldots, u_k)$ we have for example

$\omega(.) =$  , $\omega(\cdot, \cdot) =$  ,

$\omega(\omega(\cdot,\cdot), \omega(\cdot), \cdot) =$  , $\omega(\omega(\cdot), \cdot, \omega(\cdot,\cdot)) =$  .

We call such subsets *trees*, their vertices *nodes*, their topmost vertices *roots*, and $u_1, \ldots, u_k$ the *subtrees* of the tree $\omega(u_1, \ldots u_k)$. The *degree* of a node is the number of nodes below it to which it is joined by an edge. A node of degree zero is an *end-node*. In Good's terminology [5] our trees are unlabelled ordered rooted trees.

This paper is concerned with a particular sequence of problems of the form

For each $m \geq 1$, how many trees of some kind are there with m end-nodes?

§1.1 reviews some known answers to problems of this kind. Note that there are infinitely many trees as defined above with n end-nodes for any $n \geq 1$ (consider e.g. • ,  ,  , ...), but only finitely many of these have no nodes of degree one. We call such trees *deleted*. Etherington ([3],1940) observed that a sequence $(e_1,e_2,e_3,\ldots)$ described by Schröder ([11], 1870) is the sequence of answers to the questions -

for each $m \geq 1$, how many deleted trees are there with m end-nodes ?

He also described two other questions with this sequence (which we refer to as the *Schröder-Etherington sequence* or S - E sequence) of answers, and suggested (see the opening quotation) that the underlying problem is 'essentially algebraic'. The S-E sequence is number 1170 in Sloane's handbook [12] where the fourth term is incorrect - it is 45 . The last term quoted by Etherington is also incorrect - it is 103049. §1.2 describes three more problems answered by the S-E sequence.

The main theorem of §3.1 (p.12) makes explicit the algebraic essence of these problem sequences by describing a universal algebra with one generator and a 'null-length morphism' such that the questions -

for each $m \geq 1$, how many elements have null-length m ?

are answered by the S - E sequence. As a prelude to this theorem we show in §2 how to endow a family of Davenport-Shinzel sequences with this structure and carry it to the family of deleted trees via the 'ancestral tree map' (p.8, 9). §3.2 describes two more families which can be endowed with this structure ('deleted' left continuous walks and 'deleted' multi-compartmented folder stacks) and are therefore enumerated by the S - E sequence. In fact every Catalan family has a subset which can be endowed with this structure ([6]).

§4 is devoted to the computation of the S-E sequence and the renewal assay which it generates row-by-row. This is the array of answers to the questions -

for each $m,n \geq 0$, how many forests of n deleted trees are with a total of m end-nodes ?

(A forest is a sequence of trees.) This array (which we refer to as the *Schröder-Etherington array*) turns out (see §4.3) to be generated column-by-column by the sequence $(1,1,2,2,2^2,2^3,\ldots)$ from which we deduce summation formulae for its elements and in particular for the elements of the S-E sequence. §4.3 closes with a survey of formulae for these elements, some old and some new, with special emphasis upon Watterson's formula [15] linking the S-E sequence with Legendre polynomials. §4.4 provides a combinatorial explanation of the association between the S-E sequence and the sequence $(1,1,2,2^2,2^3,\ldots)$ by describing a correspondence between certain discrete random walks (whose steps can take values $-1,1,2,3,\ldots$) and certain coloured discrete random walks whose steps $-1,0,1,2,3,\ldots$ are coloured by $1,1,2,2^2,2^3,\ldots$ colours respectively.

§1. The context of this paper

§1.1 How many trees of some kind are there with m end-nodes?

Unordered rooted trees If we ignore the order in which the subtrees appear below the root we obtain a family of (equivalence classes of) trees which we refer to as unordered rooted trees. The deleted unordered rooted trees with 1,2 and 3 end-nodes are

and those with 4 end-nodes are

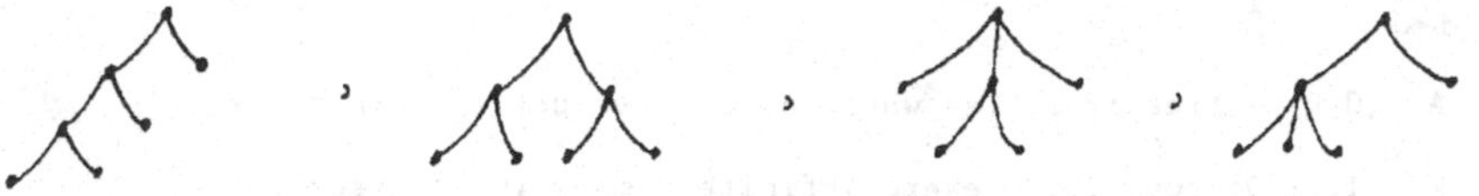

Cayley closed the famous paper [2] in which he discussed the number of unordered rooted trees with a given number of nodes by observing that the sequence of answers $(b_1,b_2,b_3,b_4,\ldots) = (1,1,2,4,\ldots)$ to the questions -

> for each $m \geq 1$, how many deleted unordered rooted trees are there with m end-nodes?

is determined recursively from the equation

$$1 + b_1x + 2b_2x^2 + 2b_3x^3 + \ldots = (1-x)^{-b_1}(1-x^2)^{-b_2}(1-x^3)^{-b_3} \ldots .$$

Graph-theoretic trees A graph-theoretic tree is a connected graph with no cycles. (In our terms it is an unordered unrooted tree). Moon ([8], p.20,21) attributes to V.T. Sós Beinicke the fact that the questions -

> for each pair $\ell \geq 0$, $m \geq 2$, how many *labelled* graph-theoretic trees are there with ℓ internal nodes and m end-nodes?

have the array of answers $M(\ell,m)\, S(\ell+m-2,\ell)$, where M as usual denotes the multinominal coefficient $M(\ell,m,n,\ldots) = \dfrac{(\ell+m+n+\ldots)!}{\ell!m!\,n!\ \ldots}$ and S denotes the array of Stirling numbers of the second kind. Moon also discusses the distribution of the number of end-nodes in a random tree.

Trees in the sense of the introduction

Henceforth by a tree we mean an unlabelled rooted ordered tree as described in the introduction.

A tree is *described by* $\underset{\sim}{f} = (f_0,f_1,f_2,\ldots)$ if it has f_i nodes of degree i, $i \geq 0$. A simple inductive argument shows that a sequence $\underset{\sim}{f}$

of non-negative integers with finite sum is the description of some tree iff $\sum_{i\geq 0} (i-1)f_i = -1$.

A $(0,k)$ - tree is a tree whose nodes have degree zero or k, for some $k \geq 1$. Of course there are infinitely many $(0,1)$-trees with one end-node (and none with more than one end-node!). If $k \geq 2$ the number of $(0,k)$-trees with ℓ nodes of degree k and therefore $(k-1)\ell + 1 = m$ end-nodes is well known to be

$$c_k(\ell) = \frac{1}{\ell+m} M(\ell,m), \quad \ell \geq 0.$$

The sequences $(c_k(0),\ c_k(1),\ c_k(2),\ldots)$, $k \geq 0$ are known as generalised Catalan sequences. $k = 2$ corresponds to the Catalan sequence itself.

§1.2 Some questions answered by the S-E sequence

Etherington [3] shows that the questions -

for $m \geq 1$, how many { deleted trees / dissections of a polygon into $(k+1)$-gons, $k=2,3,4$. / non-associative combinations }

are there if { the trees have m end-nodes? / the polygon has $m+1$ sides? / the combination is of m similar objects? }

are each answered by the same sequence (namely the S-E sequence) by describing bijections between these families which preserve the characteristic of interest. Later we shall see that three more sequences of questions, namely -

for $m \geq 1$ how many { Davenport-Schinzel sequences / deleted left continuous walks satisfying a certain constraint / deleted multi-compartmented folder stacks }

are there having just { m distinct symbols? / m downward steps? / m compartments? }

are also answered by the S-E sequence. The first of these families was considered by Mullin and Stanton [9] who rediscovered the S-E sequence. The association of the other two families with the S-E sequence is we believe new.

§2. The set of Davenport-Schinzel strings endowed with the structure of a universal algebra

We define (modifying the definition in [9] a little) a family D of equivalence classes of 'good' symbol strings which we refer to as DS-strings. A *good string of symbols* satisfies the conditions

1) adjacent symbols are distinct:
2) there is no subsequence of the sequence of symbols formed by the string of the form a, b, a, b ;
3) maximality: no element from the symbol set of the string can be added to the end of the string without violating 1) or 2).

Note that the first and last symbols of a good string necessarily coincide. Two strings u and v are *equivalent* $(u \sim v)$ if one is a relabelling of of the other. A DS-string is an equivalence class of good strings. We shall not distinguish between a string and its equivalence class and shall denote the single element DS-string by $\square$.

We denote by X^k the k-fold Cartesian product of a set X with itself and adopt the convention that X^0 *is the singleton* $\{*\}$. We leave the reader to verify that the map $\omega\colon \bigcup_{k\geqslant 0, k\neq 1} D^i \to D$ defined by $\omega(*) = \square$, and for $k \geqslant 2$ and $(u_1,\ldots,u_k) \in X^k$,

$$\omega(u_1,u_2,\ldots,u_k) = u_1 u_2' \ldots u_k' \,\square$$

where $u_i' \sim u_i$ for $2 \leqslant i \leqslant k$, the symbol sets of u_1, u_2',..., and u_k' are disjoint, and □ is the first symbol of u_1, is well-defined and does indeed have range D.

Now let ω_k denote the restriction of ω to X^k, $k \geq 0$, $k \neq 1$. We can regard ω as endowing D with the structure $\mathcal{D} = (D;\omega) = (D;\ \omega_0,\omega_2,\omega_3,\ldots)$ of a universal algebra with operations of arity $(0,2,3,\ldots)$ respectively (see §3.1 for more details concerning such algebras). The map ω generates a nested sequence $D_0 \subset D_1 \subset D_2 \ldots$ of subsets of D defined by $D_0 = \{\omega(*)\}$ and for each $i \geq 0$

$$D_{i+1} = D_i \cup \{\omega(x);\ x \in \cup\ \{D_i^k;\ 0 \leq k \leq i+1,\ k \neq 1\}\}$$

We call D_i the *i'th generation* (under ω) and define the *rank* $\|x\|$ of $x \in D$ to be the number of the first generation to which x belongs (if any) and to be ∞ otherwise. We refer to $(\bigcup_{i\geq 0} D_i;\omega)$ as the *subalgebra of* $\mathcal{D}$ *generated by* ω.

The *length* $|x|$ [*null-length* $|x|_0$] of a string is the number of [distinct] symbols it contains.

Proposition 1 The algebraic structure $\mathcal{D} = (D;\omega)$ with which the set of DS strings is endowed above satisfies the conditions

(A) $\omega : \bigcup_{k\geq 0,\,k\neq 1} D^k \to D$ is a bijection;

(B) $\bigcup_{i\geq 0} D_i = D$;

(C) The length functions satisfy $|\omega(*)| = 1 = |\omega(*)|_0$, and for $k \geq 2$ and $(u_1,\ldots u_k) \in X^k$,

$$|\omega(u_1,\ldots,u_k)| = 1 + \sum |u_i|$$

$$|\omega(u_1,\ldots u_k)|_0 = \sum |u_i|_0\ ;$$

Proof (C): this is immediate from the definitions of ω, $|\ |$ and $|\ |_0$.

(A), (B): if $w \in D$ and $|w| = 1$, then $w = \square$ and $\omega^{-1}(\square) = \{*\}$. If $|w| > 1$, the following algorithm generates a preimage of w under ω. Let u_1 be the portion of w up to and including the penultimate occurrence of the first symbol ($\square$ say) of w. Thus $w = u_1 v \square$ for some non-empty string v. Let u_2 be the portion of v up to and including the final occurrence of the first symbol of v. Thus $w = u_1 u_2 v \square$ for some non-empty string v. If v is empty, stop; if not, define u_3 as we defined u_2 above. Proceeding thus, $w = u_1 u_2 \ldots u_k \square$ for some $k \geq 2$. Furthermore the symbol sets of $u_1, \ldots, u_k$ are disjoint and each is a DS string so

$$w = \omega(u_1, \ldots, u_k).$$

On the other hand if $w = \omega(u_1, \ldots, u_k)$ and we assume (for convenience) that the symbol sets of $u_1, \ldots, u_k$ are disjoint we leave the reader to check that the first factor of w generated by the above algorithm is u_1 from which an inductive argument establishes that $\omega : \bigcup_{k \geq 2} D^i \to D$ is 1 - 1. (A) follows from this easily and (B) is proved by observing that each factor of w has smaller length than w and arguing by induction.

The ancestral tree morphism

The factorisation procedure of the above proof is conveniently represented by a labelled tree (see figure 1). Label a root node with the DS-string w to be factorised. If w has no factors, stop; if not for some $k \geq 2$ $w = \omega(u_1, \ldots, u_k)$. Join the node labelled w to nodes labelled $u_1, \ldots, u_k$ in order from left to right below their node. Repeat the process at each of these nodes, factorising each factor u_i if possible. Erase labels from the deleted tree obtained when no further factorisation is possible. The result is the *ancestral tree* τw of the string w.

The ancestral tree map $w \to \tau w$ is invertible. For given a deleted tree w label its end-nodes by single element strings (chosen with distinct labels for convenience). Now suppose that the root nodes of all subtrees of a node of w have been labelled by DS-strings $u_1, \ldots, u_k$ say. Label the node itself by $\omega(u_1, \ldots, u_k)$. The label of the root of w ultimately obtained by this procedure is $\tau^{-1}w$.

The map τ satisfies $\tau(\Box) = \cdot$ and

$[\tau(u_1 u_2, \ldots, u_k \Box) =$

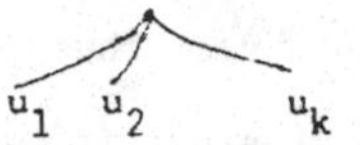

. If we endow the set $\bar{T}$ of trees with the obvious structure $\bar{T} = (\bar{T}, \omega)$ hinted at in the introduction, namely $\omega(*) = \cdot$ and for $k \geq 1$,

$\omega(u_1, \ldots, u_k) =$

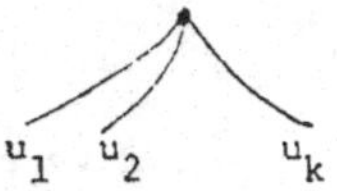

then the set T of deleted trees is the underlying set of the subalgebra $T = (T, \omega)$ of $\bar{T}$ generated by $\omega_0, \omega_2, \ldots$. Note that we have not distinguished the map $\omega: \bigcup_{k \geq 0} \bar{T}^k \to \bar{T}$ from its restriction to $\bigcup_{k \geq 0, k \neq 1} T^k$.

(1) The map τ is thus a *morphism* (p.10) from $\mathcal{D}$ to $\mathcal{T}$ and since it is invertible an *isomorphism*. Note that if we define the [*null-*]*length* of a tree to be its number of [end-] nodes τ *preserves both length and null-length*. (Figure 2 lists strings and trees of null-length between one and four in a fashion which respects the correspondence $w \to \tau w$.)

§3 Deleted Etherington algebras: archetypal families enumerated by the Schöder-Etherington sequence

§3.1 Our aim is to describe the algebraic archetype of Etherington's sequence of problems. We begin by summarising some basic notions of universal algebra (see for example Chapter 1 of [7]). Some of our notation is non-standard since we exploit the fact that the algebras we consider have no more than one operation of a given arity.

Let $\underset{\sim}{\bar{n}} = (0,1,2,\ldots)$ and let $\underset{\sim}{a} = (a_0,a_1,a_2,\ldots)$ be any subsequence of $\underset{\sim}{\bar{n}}$ with $a_0 = 0$. A *universal algebra of type* $\underset{\sim}{a}$ (more briefly an $\underset{\sim}{a}$*-algebra*) is a set X together with a sequence of operations $\omega_0,\omega_1,\omega_2,\ldots$ of arity $a_0, a_1, a_2,\ldots$ respectively. Let $\omega = \bigcup_{k\geq 0} \omega_k : \bigcup_{k\geq 0} X^{a_k} \to X$. We write $X = (X;\ \omega_0,\omega_1,\ldots)$ or $X = (X;\omega)$.

If $X = (X;\omega)$ and $Y = (Y,\eta)$ are $\underset{\sim}{a}$-algebras a *morphism* $\theta: X \to Y$ satisfies $\theta\omega(*) = \eta(*)$ and for each $k \geq 1$ and $(x_1,\ldots,x_k) \in X^k$,

$$\theta\omega(x_1,\ldots,x_k) = \eta(\theta x_1,\ldots,\theta x_k).$$

If X is an $\underset{\sim}{a}$-algebra we define a sequence of *generations* $X_0,X_1,X_2,\ldots$ and a *rank function* $\| \ \| : X \to \mathbb{N} \cup \{0,\infty\}$ by analogy with their definition when $X = \mathcal{D}$ (p.7). If $\bigcup_{i\geq 0} X_i = X$ we say that X is a *one generator* $\underset{\sim}{a}$*-algebra*. We call $\omega(*)$ the *generator* (since X is generated by the non-nullary operations from $X_0 = \{\omega(*)\}$). Note that if X is a one generator $\underset{\sim}{a}$-algebra every element has finite rank.

If we are willing to conceive of the set T_∞ of rooted ordered trees with countably many nodes (maintaining the restriction that each node has finite degree) then the algebra T_∞ obtained by endowing T_∞ with structure in the obvious way is not a one generator $\underset{\sim}{\bar{n}}$-algebra but contains a subalgebra $\bar{T}$ which is .

Proposition 2 Let $X = (X,\omega)$ be a one generator $\underset{\sim}{a}$-algebra and $Y = (Y,\eta)$ be an $\underset{\sim}{a}$-algebra. If there is a morphism $\theta: X \to Y$ it is unique and determined generation by generation.

Proof If $x \in X$ and $\|x\| = 0$ then $x = \omega(*)$ and $\theta x = \eta(*)$. Now let $\|x\| > 0$ and assume inductively that θy is determined if $\|y\| < \|x\|$. Since $x = \omega(x_1,\ldots,x_k)$ with $\|x_i\| < \|x\|$ for each i, $\theta x = \eta(\theta x_1,\ldots,\theta x_k)$ is also determined. Thus θ is determined on all elements of finite rank, i.e. on X. □

Suppose X is a one generator $\underset{\sim}{a}$ - algebra. It follows that ω maps $\bigcup_{k\geq 0} X^{a_k}$ onto X; if ω is also 1 - 1 and therefore a bijection we say that X is a *free* one generator $\underset{\sim}{a}$ - algebra. It is easy to check that free one generator $\underset{\sim}{a}$ - algebras have the following universal property.

Proposition 3 There is a unique morphism between a free one generator $\underset{\sim}{a}$ - algebra and any $\underset{\sim}{a}$-algebra.

Corollary The unique morphism between two free one generator $\underset{\sim}{a}$-algebras is an isomorphism.

Length and null-length

Observe that $\mathbb{N}$ can be endowed with structures $N = (\mathbb{N};\omega)$ and $N_0 = (\mathbb{N};\eta)$ each of which are one generator $\bar{\underset{\sim}{n}}$-algebras by defining $\omega(*) = 1 = \eta(*)$ and for $k \geq 1$ and $(n_1,\ldots,n_k) \in \mathbb{N}^k$, $\omega(n_1,\ldots,n_k) = 1 + \Sigma n_i$, $\eta(n_1,\ldots,n_k) = \Sigma n_i$

If X is a one generator $\underset{\sim}{a}$-algebra for which the morphism $|\ |: X \to N$ exists (as it will if for example X is free) we say that X *has a length morphism* and call $|x|$ the *length of* x. The analogous morphism $|\ |_0: X \to N_0$ is called the *null-length morphism*.

Note that if Y has the length morphism $| \,|_y$ and $\theta : X \to Y$ then $| \,|_y \circ \theta : X \to N$ is the length morphism of X. Thus *morphisms of one generator algebras preserve length.* The same is true of null-length.

Example From Proposition 1 it follows that the algebra $\mathcal{D}$ of DS-strings is a free one generator (0, 2, 3,...)-algebra whose [null-] length morphisms coincides with the [null-] length functions previously defined. Since the algebra T of deleted trees is isomorphic to $\mathcal{D}$ by the ancestral tree isomorphism which preserves [null-] length, the preceding sentence applies to T also. □

Let $\underline{n} = (0, 2, 3,\ldots)$. We call a free one generator [$\underline{n}$ -] $\bar{\underline{n}}$-algebra a [*deleted*] *Etherington algebra* or *E-algebra* for short, Thus $\bar{T}$ is an E-algebra and T and $\mathcal{D}$ are deleted E-algebras.

We denote the cardinality of the set S by #S.

We gloss over the further requirement made in [6] that the map ω be *effective*, i.e that there is an algorithm which computes ω(x) for any x in the domain of ω. We gloss over this point becaus e it is not relevant to our theme. However note that the maps ω of each of our examples *are* effective.

Theorem 1 Let X be a deleted E-algebra and $| \,|_0 : X \to N_0$.

Then $\#\{x \in X;\ |x|_0 = m\} = e_m, \qquad m \geq 1$.

Proof Since T is also a deleted E-algebra there is an isomorphism $\theta: T \to X$. This isomorphism preserves nullary length. Since $\#\{x \in T;\ |x|_0 = m\} =$ by definition (p.2) the result follows. □

§3.2 Two more E-algebras: left continuous walks and multi-compartmented folder stacks

Note that the subalgebra of an E-algebra generated by the operations $\omega_0, \omega_2, \omega_3,\ldots$ is a deleted E-algebra. We now describe two E-algebras

Left continuous walks

A *walk* z is a (possibly empty) sequence $(\delta z_1,\ldots,\delta z_m)$ of numbers called *steps*. The walk with no steps is called the *empty walk*. If the steps are integers $\delta z \geqslant -1$ the walk is *left-continuous* or LC for short.

LC walks crop up naturally in the 'real world' (see e.g. [16] p.494, 495). One context in which they arise is that of a dam with discrete content evolving in discrete time. In each time interval one unit is released from the dam and 0 or 1 or 2,... units flow into the dam. The example motivates the terminology which follows,

The *level* z_i of a walk after i steps is the sum of those steps. We set $z_0 = 0$. A walk z with m steps is *strictly bounded below [above] by its final [initial] level* if $0 \leqslant i < m \Rightarrow z_i > z_m$ $[0 < i \leqslant m \Rightarrow z_i < z_0$ The *dual* ρz of a walk z is its reversal $(\delta z_m, \delta z_{m-1},\ldots,\delta z_1)$. Notice that the two properties just introduced are dual to one another, and that the empty walk is the only walk enjoying both properties. Let $\underset{\sim}{W}$ $[\underset{\sim}{W}(k)]$ denote the set of LC walks strictly bounded below by their final level [of k]. Let $\widetilde{W}$ and $\widetilde{W}(k)$ denote the duals of these sets. Observe that $\underset{\sim}{W}(k)$ and $\widetilde{W}(k)$ are empty if $k > 0$ and $\underset{\sim}{W}(0) = \widetilde{W}(0) = \{\text{empty walk}\}$.

<u>$\underset{\sim}{W}(-1)$ viewed as an E-algebra</u>

Raney [11] defines a set of words $\bar{G}$ on the alphabet $A = \{0,1,2,\ldots\}$ as the set of symbol strings determined by the rules $0 \in \bar{G}$ and for each $k \geqslant 1$ if $(u_1,\ldots,u_k) \in \bar{G}^k$ then the concatenation $ku_1\ldots u_k \in \bar{G}$.

To a symbol string $a = a_1a_2\ldots a_m$ on the alphabet A corresponds the LC walk $\theta a = (a_1-1,\ldots,a_m-1)$. It is plausible that if we endow $\bar{G}$ with the obvious operations the resulting structure $\bar{G}$ is an E-algebra. A convenient way of checking this is to show that the structure carried to the image of $\bar{G}$ by θ is an E-algebra.

We denote by $u \cdot v$ the concatenation $(\delta u_1,\ldots,\delta u_m,\delta v_1,\ldots,\delta v_n)$ of two sequences $u = (\delta u_1,\ldots,\delta u_m)$, $v = (\delta v_1,\ldots,\delta v_n)$. Let $\bar{\mathcal{W}} = (\bar{W};\omega)$ denote the one generator $\underset{\sim}{\bar{n}}$-algebra defined by the operations $\omega(*) = (-1)$ and for $k \geq 1$ and sequences $u_1,\ldots,u_k$, $\omega(u_1,\ldots,u_k) = (k-1)\cdot u_1\cdot u_2\cdots\cdot u_k$.

Proposition 4 (A) $\bar{W} = \bar{G} = \underset{\sim}{W}(-1)$

(B) $\bar{\mathcal{W}} = (\bar{W},\omega)$ is an E-algebra;

(C) the [null-] length of a walk is its number of [downward] steps.

Proof $\bar{W}$ is defined so that $\bar{W} = \theta\bar{G}$. To check the remaining assertions we need the following simple lemma whose proof we leave to the reader.

Lemma For each $k \geq 1$ the concatenation map from $\underset{\sim}{W}(-1)^k$ to $\underset{\sim}{W}(-k)$ (i.e. the map : $(u_1,u_2,\ldots,u_k) \mapsto u_1\cdot u_2\cdot\ldots\cdot u_k$) is a bijection.

Now to show that $\bar{W} \subset \underset{\sim}{W}(-1)$ use induction on the rank of $w \in \bar{W}$, and to show that $\underset{\sim}{W}(-1) \subset \bar{W}$ use induction on the step number of $w \in W(-1)$. The lemma now implies that $\omega: \bigcup_{k\geq 0} \underset{\sim}{W}(-1)^k \to \underset{\sim}{W}(-1)$ is $1-1$ so establishing (B). (C) is straightforward. □

Corollary The map $\theta : \bar{G} \to \bar{W}$ defined on p.13 is an isomorphism $\theta : \bar{\mathcal{G}} \to \bar{\mathcal{W}}$. Thus $\bar{\mathcal{G}}$ is an E-algebra.

The subalgebra of deleted walks in $\underset{\sim}{W}(-1)$: a deleted E-algebra

$w \in \bar{W}$ is in the subalgebra of $\mathcal{W}$ generated by $\omega_0,\omega_2,\omega_3,\ldots$ iff w has no zero steps. We call an LC walk with no zero steps a *deleted LC walk*. If we denote this subalgebra by $\mathcal{W} = (W;\omega)$ (identifying ω with its restriction to $\bigcup_{k\geq 0,\, k\neq 1} W^k$) then W is the set of deleted LC walks in $\underset{\sim}{W}(-1)$.

Stacks of multi-compartmented folders

These stacks are the brainchild of the notorious statistician - cum - bureaucrat Henry Finucan [4]. We shall represent a k-leaf manila folder by its cross-section 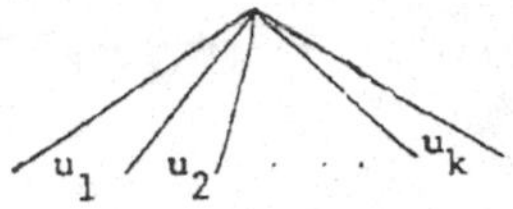

with spine uppermost. Unlike Finucan we permit a single leaf folder .
The collection $\bar{F}$ of multi-compartmented folder stacks is generated by the operation

$\omega(*)$ = the empty folder stack,

$\omega(u) = \;|u$ [i.e. put a single leaf to the left of the stack] ,

and for each $k \geq 2$ and stacks $u_1,\ldots,u_k$

$\omega(u_1,\ldots,u_k) =$ 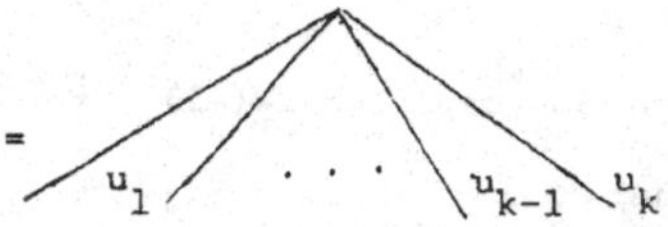

Thus for example denoting the empty stack by 0,

$\omega(0) = \;|$, $\omega(0,0) =$  , $\omega(0,\omega(0),\,\omega(0,0)) =$  .

We leave the reader to convince himself that $\bar{F} = (\bar{F};\omega)$ is an E-algebra and that the [null -] length of a stack is the number of folders [compartments plus one] in the stack.

A stack $w \in \bar{F}$ is in the subalgebra of $\bar{F}$ generated by $\omega_0, \omega_2, \omega_3, \ldots$ iff w contains no single leaf folders. We refer to such stacks as *deleted* - they coincide with the set of stacks as defined by Finucan - and denote the corresponding deleted E-algebra by $F = (F;\omega)$

A correspondence between stacks and left continuous walks in $\underset{\sim}{W}(-1)$

There is a simple correspondence between a stack and a walk in $\underset{\sim}{W}(-1)$ which generalises Finucan's correspondence between stacks of single compartment folders and walks with steps ± 1 bounded below by their final level of zero ([4], p.3].

Imagine a beetle boring its way from left to right through the stack and capable of discerning how many leaves are in the folder through whose leaf it is boring and whether or not this leaf is the first belonging to that folder which it has encountered.

Each time the beetle bores through the first [a subsequent] leaf of a folder with k leaves write $k - 1$ [(-1)] is the step sequence of the corresponding walk. Add a final downward step when the beetle emerges from the stack.

Figure 2 lists stacks and walks in a fashion respecting this correspondence. We leave the reader to check that it is in fact the *isomorphism* from $\bar{F}$ to $\bar{W}$.

§4.1 On lists of elements from deleted E-algebras

If $\chi = (X;\omega)$ is a free one generator $\underset{\sim}{a}$-algebra lists of elements of X crop up naturally via the map $\omega^{-1} : X \to \bigcup_{k\geq 0} X^{a_k}$

$$: \omega(x_1,\ldots,x_k) \mapsto (x_1,\ldots,x_k) .$$

For example, deleting the root node and the edges incident with it from a tree $\omega(x_1,\ldots,x_k)$ produces a list of k-trees $x_1,\ldots,x_k$ usually referred to as a *forest* of k trees. Likewise deleting the first step from the walk $\omega(x_1,\ldots,x_k) = (k-1) \cdot x_1 \cdot \ldots \cdot x_k$ in $\underset{\sim}{W}(-1)$ we obtain the concatenation $x = x_1 \cdot \ldots \cdot x_k \in \underset{\sim}{W}(-k)$ which is a convenient representation (by the lemma of p.14) of a list of k walks. Notice that $\omega^{-1}(\omega(*)) = *$ which we refer to as the *empty list*.

We now define the algebraic archetype of the array of forests of n trees with m end-nodes. Let $\chi = (X;W)$ be a deleted E-algebra. Extend the domain of $|\ |_0$ from X to $\bigcup_{k\geq 0} X^k$ by defining $|*|_0 = 0$ and $|(x_1,\ldots,x_k)|_0 = \Sigma|x_i|_0$ for each $k \geq 1$.

For $x \in \bigcup_{k\geq 0} X^k$ we call $|x|_0$ the *null-length of the list x.*
Let

$$X^n_m = \{x \in X^n;\ |x|_0 = m\}, \qquad m,n \geq 0.$$

Observe that X^n_0 and X^0_n are empty unless $n = 0$ in which case $X^o_0 = \{*\}$.

§4.2 The Schröder-Etherington array

As usual $\#S$ denotes the cardinality of the set S. Let $e(x) = e_1x + e_2x^2 + e_3x^3 + \ldots$ denote the generating function of the S-E sequence, and let $e(x,y) = \sum_{m\ n\geq 0} \#X^n_m x^m y^{-n}$ denote the generating function of
↑
NB!

the array $\#X^n_m$. For any series $a(x) = a_0 + a_1x + a_2x^2 + \ldots$ we adopt the convention that $a(x)^n = a^n_0 + a^n_1x + a^n_2x^2 + \ldots, n \geq 0$. Here n on the right hand side is an *index* not an exponent. The potential ambiguity does not arise in what follows.

Proposition 5 $e(x,y) = 1 + y^{-1}e(x) + y^{-2}e(x)^2 + \ldots$

or equivalently, $\# X^n_m = e^n_m$.

Proof. Observe that for each $m \geq 0$, $X^{m+1}_n = \bigcup_{i+j=n} X^m_i \times X^1_j$

so taking cardinalities,

$$\# X^{m+1}_n = \sum_{i+j=n} \# X^m_i\ e_j$$

and an inductive argument completes the proof. □

Since the Schröder-Etherington sequence determines the above array, we refer to it as the *Schröder-Etherington array.*

Renewal arrays

Let $b(x) = b_1x + b_2x^2 + \dots$. The array c_m^n, $m,n \geqslant 0$ is *generated row-by-row* by the sequence $\underset{\sim}{b} = (b_1,b_2,\dots)$ if

$$\sum c_m^n x^m y^{-n} = 1 + y^{-1}b(x) + y^{-2}b(x)^2 + \dots ,$$

or equivalently if $c_m^n = b_m^n$, $m,n \geqslant 0$.

It is a remarkable fact that most arrays generated row-by-row are also generated column-by-column (in a fashion we shall now describe). The following discussion is essentially a summary of Rogers' remarks ([12], §2) on renewal arrays. To obtain a renewal array in Rogers' sense from one as defined below, interchange rows and columns and delete the first row and column, reindexing those remaining by 0, 1, 2, ...

Let $a(y) = \sum_{i\in Z} a_iy^i$ and $b(y) = \sum_{i\in Z} b_iy^i$ be two formal Laurent's series. We define their *truncated product* $a(y) * b(y)$ to be the truncation of the usual product $a(y)\,b(y)$ to negative powers of y. The *i'th truncated power* $a^{(i)}(y)$ of $a(y)$ is defined inductively by $a^{(0)}(y) = 1$, $a^{(i+1)}(y) = a^{(i)}(y) * a(y)$. We adopt the convention that $a(y)^{(n)} = \dots a_{-1}^{(n)}y^{-1} + a_0^{(n)} + a_1^{(n)}y + \dots$.

Let $a(y) = a_{-1}y^{-1} + a_0 + a_1y + ,\dots$.

The array c_m^n. $n,m \geqslant 0$ is *generated column-by-column* by the sequence $\underset{\sim}{a} = (a_{-1},a_0,a_1,\dots)$ if

$$\sum c_m^n x^m y^{-n} = 1 + x a^{(1)}(y) + x^2 a^{(2)}(y) + \dots$$

or equivalently $c_m^n = a_{-n}^{(m)}$, $m,n \geqslant 0$.

Theorem (Rogers)

(A) The sequences $(b_1,b_2,b_3,\dots)$ and $(a_{-1},a_0,a_1,\dots)$ generate the same array (c_m^n say) row-by-row and column-by-column respectively iff the generating functions $b(x) = b_1x + b_2x^2+\dots$ and $\tilde{a}(x) = a_{-1} + a_0x + a_1x^2+ \dots$

satisfy the equation

$$b(x) = x\tilde{a}(b(x)) \ldots \quad (1)$$

The terms of the array c are given by

$$c^n_m = b^n_m = \frac{n}{m} a^m_{-n} .$$

(B) For any sequence a the equation (1) has a unique solution $\underset{\sim}{b}$ $(= \lambda(\underset{\sim}{a})$ say) given by $b_m = \frac{1}{m} a^m_{-1}$, $m \geq 1$.

On the other hand if $b_1 \neq 0$ $\lambda^{-1}(\underset{\sim}{b})$ is unique and determined from the equation (1) by recursion. If $b_1 = 0$, $\lambda^{-1}(\underset{\sim}{b})$ is empty unless $\underset{\sim}{b}$ vanishes in which case $\lambda^{-1}(\underset{\sim}{b}) = \{\underset{\sim}{a};\ a_{-1} = 0\}$ and the array c vanishes.

An array which is both row and column generated as above is a *renewal array*. Rogers (op cit) describes many contexts in which these arrays arise (and in particular a sequence of arrays generated row-by-row by the sequences $\underset{\sim}{c}_0 = (1,1,0,0,0,\ldots)$, $\lambda(\underset{\sim}{c}_0)$, $\lambda^2(\underset{\sim}{c}_0),\ldots$; $\lambda^k(\underset{\sim}{c}_0)$ turns out to be the k'th generalised Catalan sequence! (See [11], §5 and p of this paper.))

The formula $b^n_m = \frac{n}{m} a^m_{-n}$ is a simple form of the Lagrange inversion formula. Note that since by definition $c^n_m = a^{(m)}_{-n}$ we have $a^{(m)}_{-n} = \frac{n}{m} a^m_{-n}$; i.e. Lagrange inversion provides a formula relating the coefficients of the m'th truncated power to those of the m'th power of the series $\sum_{i \geq -1} a_i y^i$. For direct combinatorial proofs of the Lagrange inversion formula based on the same idea see [11] or [16]. For a more recent proof based on Joyal's theory of species of structures and a description of the deeper algebraic significance of the Lagrange inversion formula see [8].

On the terms of the S-E array

The generating function $e(x)$ of the S-E sequence satisfies the functional equation $y = x + y^2 + y^3 + \ldots$, an observation which Etherington attributes to Schröder [12]. To see this let $\chi = (X,\omega)$ be a deleted E-algebra. Now $|\omega(*)|_0 = 1$ and for $n \geq 2$ $|\omega(u_1,\dots u_m)|_0 = \Sigma|u_i|_0$ so $\omega: X_0^o \to X_1^1$ and for $m \geq 2$, $\omega: \bigcup_{n\geq 2} X_m^n \to X_m^1$. These maps are bijections so taking cardinalities $e_1 = \# X_1^1 = \# X_0^o = 1$ and for $m \geq 2$, $e_m = \# X_m^1 = \# \bigcup_{n\geq 2} X_m^n = \Sigma_{n\geq 2} e_m^n$ from which it follows that $e(x)$ satisfies Schröder's functional equation.

Schröder's equation can be written $y = x + y^2(1-y)^{-1}$ which is equivalent to (multiply by (1-y) and manipulate the result) $y = x(1-y)(1-2y)^{-1} = x\,\tilde{d}(y)$ where $\tilde{d}(y) = 1 + y(1 + 2y + (2y)^2 + \dots)$. Thus $e(x) = x\tilde{d}(e(x))$ so from Rogers' Theorem the sequences $(e_1,e_2,\dots)$ and $(1,1,2,2^2,\dots) = (d_{-1},d_0,d_1,\dots)$ say generate the S-E array row-by-row and column-by-column respectively.

Hence

$$e_m^n = \frac{n}{m} d_{-n}^m \quad \text{where} \quad d(y) = y^{-1} + (1-2y)^{-1}$$

To compute the coefficients d_{-n}^m let $c(y) = \frac{1}{1-2y}$. Observe that $d_{-n}^m = 0$ unless $n \leq m$ so assuming this

$$\begin{aligned} d_{-n}^m &= \text{coefficient of } y^{-n} \text{ in } (y^{-1} + c(y))^m \\ &= \sum_{n\leq i\leq m} \binom{m}{i} c_{-n+i}^{m-i} \\ &= \sum_{j+k=m-n} \binom{m}{j} c_k^j . \end{aligned}$$

Now $c_k^j = \binom{-j}{k}(-2)^k = [{j \atop k}]2^k$ where $[{j \atop k}] = (-1)^k \binom{-j}{k}$.

Thus replacing j and k by i and j,

$$e_m^n = \frac{n}{m} \sum_{i+j=m-n} \binom{m}{i}[{i \atop j}]2^j, \quad n \geq 0,\ m > 0$$

and in particular taking $n = 1$ and replacing m by $m+1$,

$$e_{m+1} = \frac{1}{m+1} \sum_{i+j=m} \binom{m+1}{i}[{i \atop j}]2^j, \quad m \geq 0 \quad \dots\dots\dots \quad (1)$$

the summation in each formula extending over non-negative pairs i, j.
The S-E array is tabulated in Fig 3 for $0 \leq m, n \leq 8$.

A survey of formulae for the terms of the S-E array and the S-E sequence

A string of symbols on the alphabet $\{0,1,2,\ldots\}$ is *described by* $\underset{\sim}{f} = (f_0, f_1, f_2, \ldots)$ if it contains f_i occurrences of the symbol i, $i \geq 0$. Let $\mu(\underset{\sim}{f}) = \Sigma(i-1)f_i$. Raney defines a *list* of the words which we mentioned on p.13 as a (possibly empty) concatenation of words. He shows that the number of lists of words with description $\underset{\sim}{f}$ is one if $\underset{\sim}{f} = \underset{\sim}{0}$, and when $\underset{\sim}{f} \neq \underset{\sim}{0}$ is zero if $\mu(\underset{\sim}{f}) \geq 0$ and $\frac{-\mu(f)}{f_0+f_1+f_2+\ldots} M(f_0, f_1, f_2, \ldots)$ if $\mu(\underset{\sim}{f}) < 0$. The map θ previously defined (p.13) establishes a bijection between the set of lists of words with description $\underset{\sim}{f}$ and that of lists of walks in $\bar{W} = \underset{\sim}{W}(-1)$ containing a total of f_i steps of height $i - 1$, $i \geq 0$. Since a walk in $\underset{\sim}{W}(-1)$ described by $\underset{\sim}{g}$ has final level $-1 = \mu(\underset{\sim}{g})$, the list above is of $-\mu(\underset{\sim}{f})$ walks. Hence the number of lists of n deleted walks in $\bar{W}$ with a total of m downward steps is

$$\# W_m^n = e_m^n = \sum \frac{n}{m+f_2+f_3+\ldots} M(m, 0, f_2, f_3, \ldots), \quad m,n \geq 0, \; m + n > 0,$$

the summation extending over all sequences $(f_2, f_3, \ldots)$ of non-negative integers of finite sum satisfying $-m + f_2 + 2f_3 + 3f_4 + \ldots = -n$

i.e. $$f_2 - 2f_3 + 3f_4 + \ldots = m - n .$$

We now quote six more formulae for the terms of the S-E sequence together with the discoverers of each. We comment very briefly on all but Watterson's formulae linking these terms with the Legendre polynomials. These we justify in detail.

For $m \geq 0$ e_{m+1} is given by the expressions

$$\frac{1}{m(m+1)} \sum_{1 \leq i \leq m} i\, M(i,i,m-i) \quad . \quad . \quad . \quad . \quad . \quad . \quad (2)$$ [the author]

$$\frac{1}{4} \int_{-1}^{3} P_m(x)\,dx \quad . \quad . \quad . \quad . \quad . \quad . \quad (3)$$ [Watterson]

(where $P_m(x)$ denotes the m'th Legendre polynomial).

For $m \geq 1$, e_{m+1} is given by the expressions

$$\frac{1}{4} (-1)^m \sum_{0 \leq i \leq \frac{1}{2}(m+1)} \binom{\frac{1}{2}}{m+1-i} \binom{m+1-i}{i} 6^{m+1-2i} \quad . \quad . \quad . \quad (4)$$ [Schröder]

$$\frac{1}{4\pi} \int_{-2\sqrt{2}}^{2\sqrt{2}} (x+3)^{m-1} (8-x^2)^{\frac{1}{2}} dx \quad . \quad . \quad . \quad . \quad . \quad (5)$$

$$3^{m-1} \sum_{0 \leq i \leq \frac{1}{2}(m-1)} \binom{m-1}{2i} \left(\frac{\sqrt{2}}{3}\right)^{2i} c_i \quad . \quad . \quad . \quad . \quad . \quad (6)$$

[Mullin and Stanton] (for (5) and (6))

(where c_i denotes the i'th Catalan number $\frac{1}{i+1}\binom{2i}{i}$).

$$\frac{1}{4(2m+1)} (P_{m+1}(3) - P_{m-1}(3)) \quad . \quad . \quad . \quad . \quad . \quad . \quad . \quad (7)$$ [Watterson].

The expression (2) is obtained by using a combinatorial argument and the idea of cycling a walk in the same way as Raney and Wendel cycle words and walks respectively. (4) is obtained by solving Schröder's equation to obtain $e(x) = \frac{1}{4}(1+x - (1-6x+x^2)^{\frac{1}{2}})$ and expanding the square root binomial series. (5) and (6) are obtained by observing that $e(x) - 1 - x = y$ say satisfies the differential equation $(1 - 6x + x^2)y' - (x - 3)y = 0$ which can be solved using a method of Laplace. Assuming instead a series solution $y = \sum_{n \geq 0} a_n x^n$ Mullin and Stanton deduce that the numbers a_n generated by the difference equation

$$(n+1)a_{n+1} - (6_{n-3})a_n + (n-2)a_{n-1} = 0$$

from the initial conditions $a_0 = 1/4$, $a_1 = 3/4$ coincide with the numbers e_n for $n \geq 2$.

Watterson's formulae (3) and (7)

The following discussion is entirely due to the Monash statistician G.N. Watterson [15]. The facts stated about the Legendre polynomials can be found for example in [1].

The generating function of the sequence $P_0(x)$, $P_1(x)$,... of Legendre polynomials is

$$R(x,y)^{-1} = (1-2xy + y^2)^{-\frac{1}{2}} = \sum_{n\geq 0} P_n(x)y^n .$$

Notice that $4\, e(y) = 1 + y - R(3,y)$. Since $\frac{\partial R}{\partial x} = -yR^{-1}$,

$$\int_{-1}^{x} -yR^{-1}dx = R(x,y) - R(-1,y) = \sum_{n\geq 0} \int_{-1}^{x} - P_n(x)y^{n+1}dx$$

so

$$4\, e(y) = -R(3,y) + 1 + y = \sum_{n\geq 0} \int_{-1}^{3} P_n(x)y^{n+1}dx .$$

Equating coefficients, we obtain the expression (3). For $n \geq 2$ $\int_{-1}^{1} P_n(x)dx = 0$ and so for $n \geq 1$

$$e_{n+1} = \int_{1}^{3} P_n(x)dx$$

$$= \frac{1}{4(2n+1)} [P_{n+1}(x) - P_{n-1}(x)]_1^3 \quad \text{(a standard integral formula)}$$

which yields (7) since $P_n(1) = 1$, $n \geq 0$.

§4.4 Renewal arrays and weighted constrained left-continuous walks

The main purpose of this section is to point out that every renewal array has a natural combinatorial interpretation (so elaborating on Rogers' remarks of [12], p.340) as a prelude to providing a combinatorial explanation for the fact that E's array is generated column-by-column by $(1,1,2,2^2,\ldots)$.

Weighted walks

Let $\underset{\sim}{a} = (a_{-1},a_0,a_1,\ldots)$. We say that $a(i) = a_i$ is the *($\underset{\sim}{a}$-)weight* of i, that $\prod_{1\leq i\leq n} a(\delta z_i)$ is the *weight $a(z)$ of the LC walk z* $= (\delta z_1,\ldots,\delta z_m)$, and that $a(S) = \sum_{z\in S} a(z)$ is the *weight of the set of LC walks S*. We assign a weight of 1 to the empty walk.

If $a_i \geq 0$ and $\sum_{i\geq -1} a_i = 1$ the weight of a step can be regarded as its probability. For a discussion of weighted LC walks of this kind see [16]. If the a's are non-negative integers weights are associated with colours as we shall see.

Let d_m^n be the total weight of all LC walks whose graphs have endpoint (m,n). (The *graph* of a walk with levels $z_0 = 0, z_1,\ldots,z_n$ is the piecewise linear subset of $\mathbb{R}^2$ obtained by joining successive vertices (i,z_i) by straight lines.) Clearly

$$\sum_{m\geq 0}\sum_{n\in Z} d_m^n x^m y^n = 1 + xa(y) + x^2a(y)^2 + \ldots .$$

Now let $\tilde{c}_m^n$ denote the total weight of all walks to (m,n) strictly bounded *above* by their *initial* level (p. 13). We leave the reader to check that

$$\sum_{m\geq 0}\sum_{n\in Z} \tilde{c}_m^n x^m y^n = 1 + xa^{(1)}(y) + x^2a^{(2)}(y) + \ldots .$$

Theorem 2 Let $c_m^n = \tilde{c}_m^{-n}$ above, $m, n \geq 0$. Then the array c is a renewal array generated column-by-column by the weight sequence.

<u>Corollary</u> $\tilde{c}^{-n}_{m} = \frac{n}{m} a^{m}_{-n}$, $\qquad m, n \geq 0,\ m \geq 0.$

Note that every renewal array has the above combinatorial interpretation.

A correspondence between deleted LC walks and LC walks coloured by $(1,1,2,2^2,\ldots)$

If the weight sequence consists of non-negative integers we interpret the weight $\underset{\sim}{a}(\delta z)$ of a step δz as the number of colours available to colour steps of height δz. The $\underset{\sim}{a}$-weight of a walk is thus the number of $\underset{\sim}{a}$-coloured walks with the same step sequence as the walk.

Etherington's array therefore enumerates two families of walks: e^{n}_{m} is the number of deleted LC walks with m downward steps strictly bounded above by their initial level and with final level -n and the number of $\underset{\sim}{d}$-coloured LC walks to (m,-n) strictly bounded above by their initial level.

To see that these two sets have the same cardinality observe that a walk in the first is either empty when $m = 0$ or can be expressed in the form $(-1, u_1,\ldots, -1, u_m)$ when $m > 0$ and where each u is a (possibly empty) sequence of positive steps. There are as many deleted walks of the form $(-1,u)$ with final level $s \geq 0$ as there are ordered partitions of $s + 1$ into $1, 2,\ldots, s + 1$ positive parts and there are 2^s of these. For each $s \geq 0$ choose a bijection ϕ_s between the above set of walks and the set of 2^s coloured steps of height s available to a $\underset{\sim}{d}$-coloured walk. Let $\phi_{-1}(-1) = -1$, $\phi = \bigcup_{i \geq -1} \phi_i$ and the map

$$\begin{aligned}\phi &: \text{empty walk} \to \text{empty } (\underset{\sim}{d}\text{-coloured) walk}\\ &: (-1, u_1,\ldots,-1,u_m) \to (\phi(-1,u_1),\ldots, \phi(-1,u_m)),\ m \geq 1\end{aligned}$$

is a bijection between the two sets enumerated by e^{m}_{n}, $m,n \geq 0$.

References

[1] M. Abramowitz and I.A. Stegun,
Handbook of Mathematical Functions, U.S. Government Printing Office, 1968.

[2] A. Cayley, On the theory of the analytical forms called trees, Philosophical Magazine, 13 (1857) 172-176.

[3] I.H.M. Etherington, some problems of non-associative combinations (1), Edinburgh Mathematical Notes 32 (1940) I - IV.

[4] H.M. Finucan, Some decompositions of generalised Catalan Numbers, preprint to appear in the Proceedings of the Ninth Australian Conference of Combinatorial Mathematics.

[5] I.J. Good, The generalisation of Lagrange's expansion and the enumeration of trees, Proc. Camb. Phil. Soc., 61(1965) 499-517.

[6] S.G. Kettle, Every Catalan family is an Etherington family and vice versa, in preparation.

[7] H. Lausch and W. Nöbauer, Algebra of Polynomials, North-Holland Publishing Co., London/New York, 1973.

[8] G. Labelle, Une Nourelle démonstration combinatoire der formules d'inversion de Lagrange, Advances in Maths. 42(1981) 217-247.

[9] J.W. Moon, Counting labelled trees, Canadian Mathematical Monographs, 1970.

[10] R.C. Mullin and R.G. Stanton, A map-theoretic approach to Davenport-Schinzel sequences, Pacific Journal of Mathematics (1) 40(1972) 167-172.

[11] G.N. Raney, Functional composition patterns and power series reversion, Trans. Am. Math. Soc., 94(1960), 441-451.

[12] D.G. Rogers, Pascal triangles, Catalan numbers and renewal arrays, Discrete Mathematics 22 (1978) 301-310.

[13] E. Schröder, Vier combinatorische Probleme, Zeitschrift für Mathematik und Physik, 15 (1870), 371-376.

[14] N.J.A. Sloane, A handbook of integer sequences, Academic Press, New York, 1973.

[15] G.N. Watterson, private communication.

[16] J.G. Wendel, Left-continuous random walk and the Lagrange expansion, Am. Math. Monthly, 82 (1975) 494-499.

Figure 1 The factorisation of a DS string

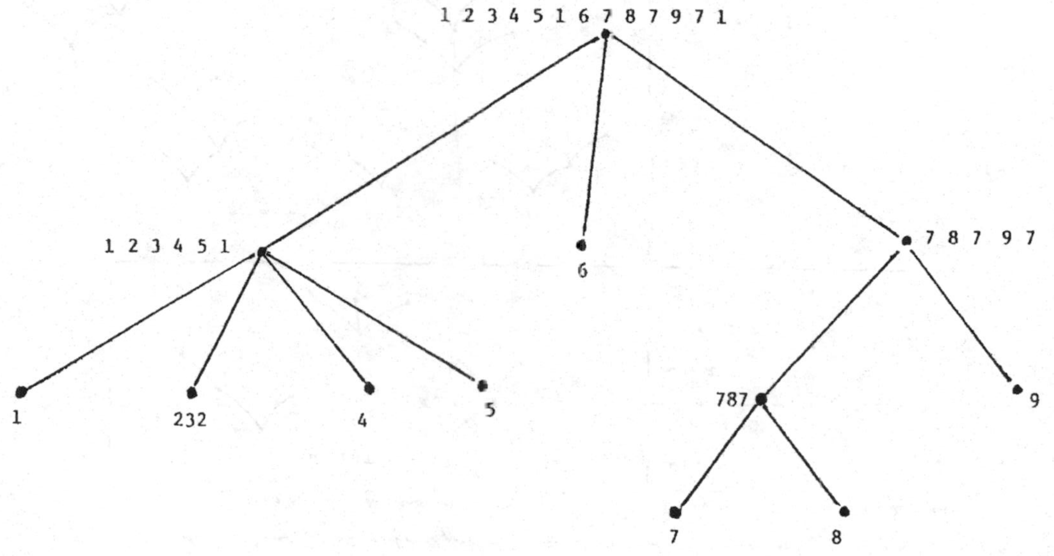

Figure 2 (a) Deleted trees and Davenport - Schinzel strings listed by null - length.

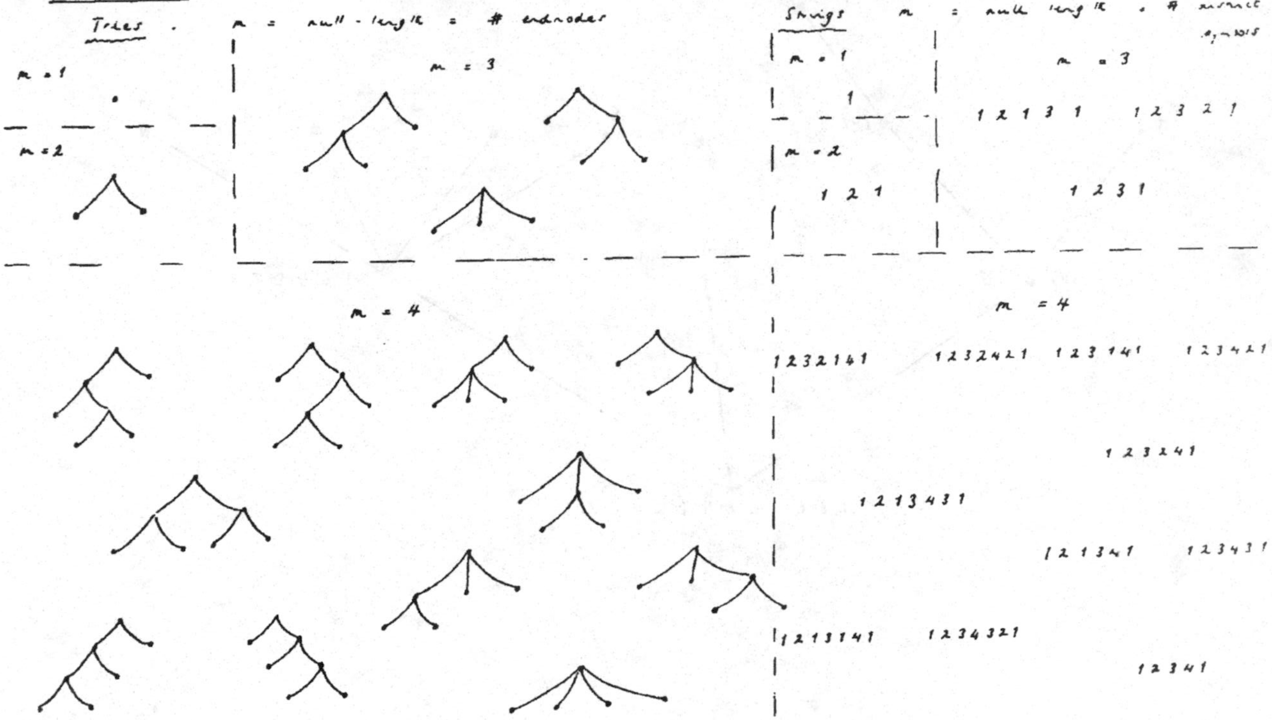

Figure 2(5) Deleted walks and folded stacks listed by null length

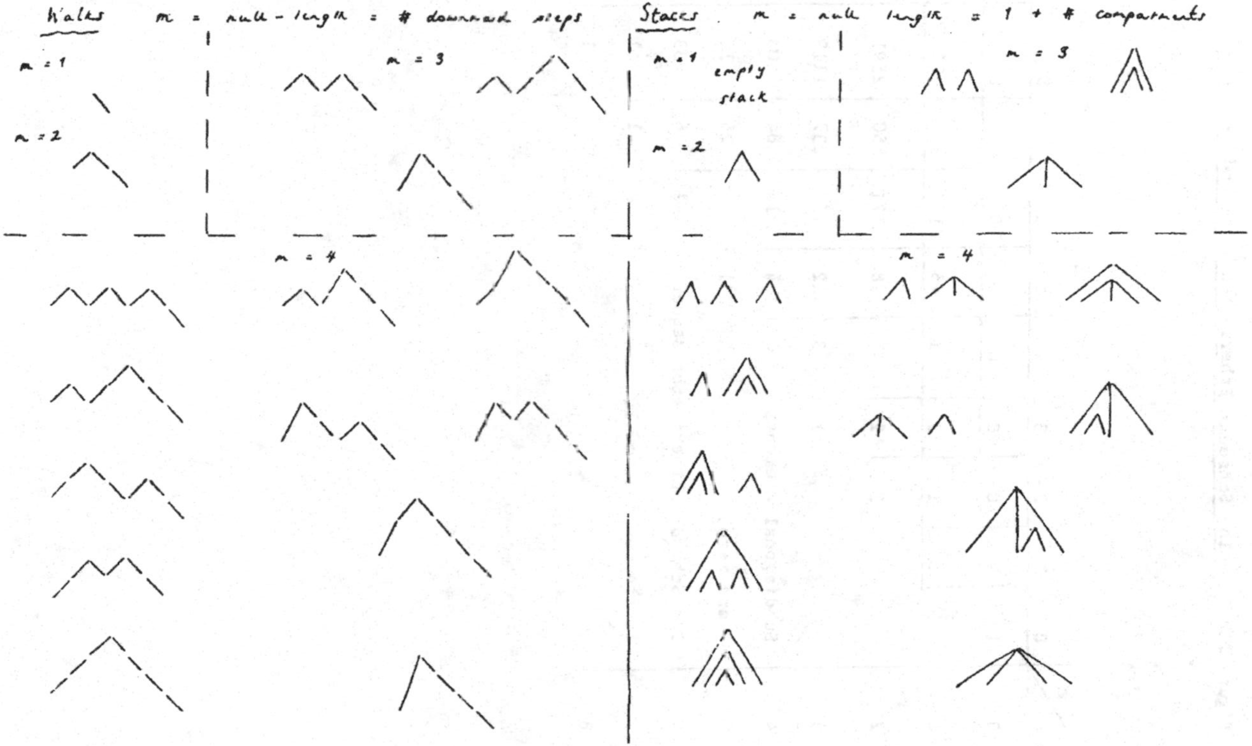

Figure 3 The Schröder Etherington array and sequence

n \ m	0	1	2	3	4	5	6	7	8
0	1	0	0	0	0	0	0	0	0
1		1	1	3	11	45	197	903	4279
2			1	2	7	28	121	550	2591
3				1	3	12	52	237	1119
4					1	4	18	84	403
5						1	5	25	12
6							1	6	33
7								1	1
8									1

Subdiagonal elements are zero.

The Schröder Etherington sequence is the row $n=1$, $m \geq 1$.

CLASSIFYING AND ENUMERATING SOME FREELY GENERATED FAMILIES OF OBJECTS

S.G. KETTLE

The essence of this paper lies in the appendix where we classify families of objects ranging from Douglas Rogers' bushes to Henry Finucan's folder stacks according to the way in which these families are generated. Objects in each family have certain features and our generation procedure is selected so that the question 'how many objects are there with a prescribed number of features of each kind?' can be rephrased 'how many objects are generated by a prescribed number of partial operations of each kind?'

We deduce the answer to this question for arbitrary freely generated universal algebras from an array of Raney's. We find that a wide range of families of objects can be viewed as such algebras or subsets thereof obtained by appropriately restricting or colouring the generation procedure. We investigate relationships between the procedures generating these families and deduce relationships between the arrays answering the question above for each procedure.

Our message is that a classification by generation procedure is a useful tool in arranging the combinatorial information which has been amassed concerning such families.

0. INTRODUCTION

The prime inspiration for this paper comes from a sequence of papers by Douglas Rogers ([7] - [12] : some coauthored by L.W. Shapiro) which discuss a broad range of families of objects - from trees, foliated trees and bushes at the botanical end of the spectrum to certain families of discrete random walks at (I suppose) the statistical end thereof. Objects in these families have certain distinctive features and Rogers' main concern is to count those objects in a family having a specified number of features of each kind - say a of the first, b of the second, c of the third and so on. The resulting array is usually determined either by describing a 'feature preserving' correspondence between the family in question and another for which the array has been determined, or by reducing the enumerational problem to the solution of others which have been dealt with, or by determining a recurrence relation for the coefficients of a (one-dimensional) array, converting it into a fixed-point equation for the generating function of those coefficients and solving this using the Lagrange Inversion Formula. The problem has variations : for example we may determine the array corresponding to some natural subfamily of a family and seek to relate this array to that associated with the family itself, or we may permit certain features of an object to be coloured and count coloured objects having a prescribed number of features of each kind.

Our message is that many of the families considered by Rogers can be viewed as *freely generated from a collection of generators by (partial) operations* $\alpha, \beta, \gamma, \ldots$ *say in such a way that an object having* a *features of one kind,* b *of another,... is generated by* a *applications of* α, b *of* $\beta, \ldots$. Jo-Anne Growney has already shown in her thesis [4] that an object in a Catalan family of 'length' n can be viewed as generated from the unique object of length 0 by the application of a binary operations.

Advantages of our viewpoint. The relative ease with which the enumerational problems are solved is explained by the fact that the families are *freely* generated. An object in such a family can be reviewed as a structure imposed on an ordered set which has no non-trivial automorphisms, and so we do not need to invoke Burnside's lemma to enumerate such a family.

We can characterise generation procedures in an abstract fashion - i.e. without reference to a particular family - and then discuss the class of families generated by a certain procedure. Which existing families are in it? Can we construct others? (Note that all families in the same class will be associated with the same array.) We can investigate the relationships between the generation procedures themselves. For example two such procedures may be *equivalent* in the sense that any family generated by one is generated by the other and vice versa. Again to any procedure corresponds a *sub-procedure* obtained by selecting a subset of the generating set and another of the set of partial operations, and a *coloured procedure* obtained by permitting some of the generators and partial operations to be coloured. We may *specialise* a procedure by restricting it in some way - perhaps by restricting the domain of some of its partial operations, or excluding some (cf the notion of sub-procedure).

For example we find (see Theorem 7) that the procedure generating a certain family described by Growney is equivalent to that generating a free one generator universal algebra having one operation of degree i for each $i \geq 1$, and that many of the families considered by Rogers in the papers [7] - [12] can be viewed as appropriately coloured specialisations of such an algebra.

Outline of contents. §1 poses our analogue of Rogers' thematic problem, namely - 'how many elements in a freely generated set of arise by the application of a prescribed number of partial operations of each kind?' - and explains precisely what this means. When looked at from our viewpoint a result of Raney's solves this problem provided that all partial operations are operations in the usual sense (see the second Corollary to Theorem 1), and this is the content of §2. §3 is devoted to the detailed consideration of 4 procedures which generate many of the families considered by Rogers in the papers [7] - [12]. We focus on the relationships between these procedures obtained via specialisation (eg. Theorem 3) or colouring (Theorem 2), and on the relationships between these and other procedures (Theorems 1, 7). From the relationships between the procedures flow relationships between their associated arrays which we describe. The appendix lists a selection of the many families generated by each of these 4 procedures together with some remarks which include reference to recent papers on these families. §3.4 is a case study in which we reconsider some of the relationships between a hierarchy of families of relations discussed by Rogers in [11] in terms of the relationships between their generating procedures.

A reader intent upon an overview should now consult the introduction to sections 2, 3.1 - 3.4, and *survey the lists of the appendix*, where he may find his favourite family mentioned.

1. FAMILIES FREELY GENERATED BY PARTIAL OPERATIONS, OR MILD GENERALISATIONS OF FREELY GENERATED UNIVERSAL ALGEBRAS

Suppose that X_0 is a set of objects called *generators* and that P is a set of *procedures* which generate objects from certain finite sequences (which we call *lists*) of objects. By successively applying the procedures we obtain a sequence of generations Y_i and their accumulations X_i defined by

$$Y_0 = X_0, \quad Y_{i+1} = \{\omega x \; ; \; \omega \in P, \text{ and } x \text{ is a list of objects drawn from } X_i \text{ and in the domain of } \omega\}$$

$$X_{i+1} = X_i \cup Y_{i+1} .$$

We call $X = \bigcup_{i \geq 0} X_i$ the *set generated by the procedures* P *from the generating set* X_0. Denoting the domain and range of a map f by dom f and im f respectively, X is *freely* generated if

1) for each $\omega \in P$, im $\omega \cap X_0 = \phi$

2) for each $y \in X \setminus X_0$ there is a unique (list, procedure) pair (x,ω) such that $y = \omega x$.

If X and Y are generated from X_0 and Y_0 by P and Q respectively a *morphism* θ between X and Y (with respect to these generation procedures) is a map

$$\theta : \begin{cases} X \to Y \\ P \to Q \end{cases} \quad \text{such that for each } \omega \in P$$

and list $x = (x_1,\ldots,x_k) \in$ dom ω, $\theta x = (\theta x_1,\ldots,\theta x_k) \in$ dom $\theta\omega$ and $\theta.x = \theta\omega.\theta x$.

We may view a procedure ω as the union $\bigcup_{k \geq 1} \omega_k$ of a sequence of *partial operations* where ω_k is the restriction of ω to the set X^k of lists of length

k drawn from X. If $\mathrm{dom}\,\omega_k = X^k$, ω_k is an operation in the usual sense. If each $\omega \in P$ decomposes into sequences of operations then X is the universal algebra generated from X_0 by these operations.

The ancestral tree map. An element may be generated in several different ways and each of these is conveniently represented by a labelled rooted ordered tree (in the terminology of [3]). If X is freely generated then each $x \in X$ arises in just one way and the tree τx representing this way is called the *ancestral tree* of x. Let $\|x\|$ denote the index of the first generation to which x belongs, a number which we call the *rank* of x. The ancestral tree map τ is defined by induction on $\|x\|$ as follows.

If $\|x\| = 0$, $\tau x = \overset{x}{\bullet}$

If $\|x\| > 0$ then for some unique (list, procedure) pair $((x_1,\ldots x_k), \omega)$ say, $x = \omega(x_1,\ldots,x_k)$ and

$\tau x =$

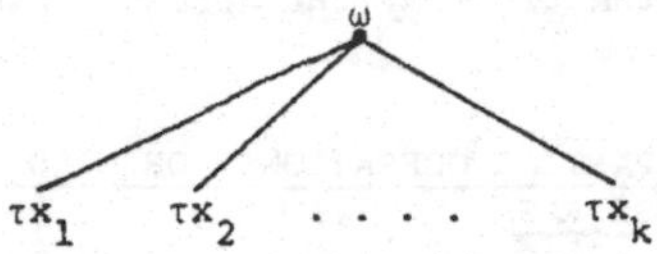

The first of the above pictures represents a tree whose single root node is labelled x; the second a tree whose root node, labelled ω, has the list of subtrees $\tau x_1,\ldots,\tau x_k$ appearing in order from left to right below it. In this case we say that the node labelled ω has *degree* k.

Example 1. The set of trees T whose nodes of degree k may be labelled $\alpha_k, \beta_k, \ldots,\ k \geq 0$ is freely generated from $\{\alpha_0, \beta_0, \gamma_0, \ldots\} = T_0$ by the operations $\alpha_k, \beta_k, \ldots,\ k \geq 1$ defined by eg.,

$\alpha_k : (x_1,\ldots,x_k) \to$

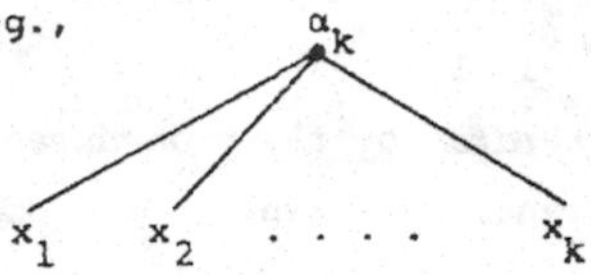

$,\ k \geq 1.$

Note that if X is any algebra generated by the operations $\alpha_k, \beta_k, \ldots,\ k \geq 1$ from the generating set $\{\alpha_0, \beta_0, \ldots\} = X_0$ then the inverse of the ancestral tree map is a morphism from T to X.

Suppose that x is a tree, labelled by the elements of some set L, which has f_λ labels λ and f_k nodes of degree k, $\lambda \in L$ and $k \geq 0$. The *ancestral description* of x is the formal sum $\sum_{\lambda \in L} f_\lambda \lambda$. The *degree-description* of x is the sequence $\underset{\sim}{f} = (f_0, f_1, f_2, \ldots) = \sum_{k \geq 0} f_k \underset{\sim}{\delta_k}$ where $\underset{\sim}{\delta_k} = (\delta_{k0}, \delta_{k1}, \ldots)$ and δ_{ij} denotes the kronecker δ. Suppose that X is *freely* generated by the partial operations $\alpha_k, \beta_k, \ldots, k \geq 1$, from some generating set X_0. The *ancestral description* and *degree-description* of $x \in X$ are those of its ancestral tree.

For example, $x = \alpha_3(\alpha_1(\alpha_0)), \beta_0, \alpha_2(\alpha_0, \alpha_1(\beta_0)))$ has ancestral description $2\alpha_0 + 2\beta_0 + 2\alpha_1 + \alpha_2 + \alpha_3$, degree description $4\delta_0 + \delta_1 + \delta_2 + \delta_3$, and the ancestral tree shown in Figure 1.

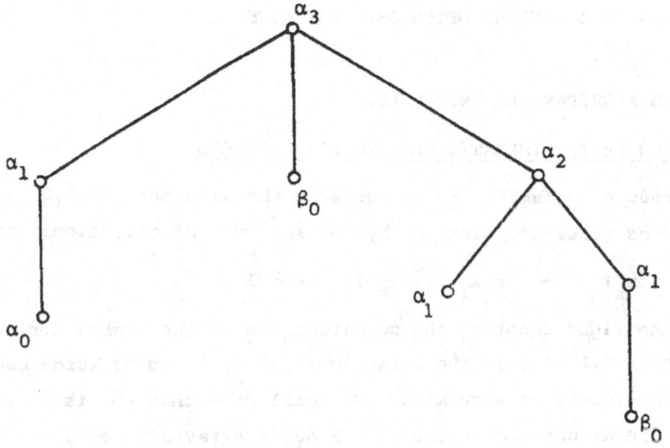

F I G U R E 1

The thematic problem. Supposing still that X is freely generated, this paper revolves around the following thematic problem -

count the number of elements in X having a presribed ancestral description.

The solution to this problem is an array of numbers. We shall investigate two marginal arrays derived from it such as that counting elements with a prescribed degree description and the sequence counting elements generated by a fixed number of partial operations. We may also weight the elements of X as follows. Let $a : X_0 \cup P \to \mathbb{R}$ be an arbitrary real function called the weight function (or the colouring function if $\mathrm{im}\, a \subset \{0,1,2,\ldots\}$). Extend the domain of a to X by defining (inductively)

$$a : \alpha(x_1,\ldots,x_k) \to a(\alpha) \prod_{1 \leq i \leq k} a(x_i).$$

We say that X has been *weighted by* a. Since elements with the same ancestral description have the same weight, the array above also solves the problem -

determine the total weight of elements in X having a prescribed ancestral description.

For example if a is a colouring function we can interpret $\alpha(x)$ as the number of colours available to colour α, $\alpha \in X_0 \cup P$, and so for $x \in X$ $a(x)$ is the number of ways of colouring x.

CONVENTIONS: (1) HENCEFORTH ALL GENERATED SETS ARE FREELY GENERATED (unless we explicitly contradict this) AND WE OFTEN OMIT THE ADVERB.

(2) ALL ALGEBRAS ARE UNIVERSAL.

RANEY'S SOLUTION OF THE THEMATIC PROBLEM FOR ALGEBRAS

Raney [6] introduces a family $\underline{G}$ of words on the alphabet $\{0,1,2,\ldots\}$ which can be viewed as generated from 0 by the sequence of operations defined by

$$\omega_k : (x_1,\ldots,x_k) \rightarrow k\, x_1 \ldots x_k, \qquad k \geq 1,$$

the expression on the right denoting the concatenation of the symbol strings k, $x_1,\ldots x_k$. He determines (Theorem 2.2 of [6]) the array enumerating lists of words having a prescribed number of symbols of each kind. We shall show that $\underline{G}$ is freely generated by these operations and deduce that Raney's array affords a solution to our thematic problem in the case of an algebra generated from a single element by at most one operation of each degree, and indicate how it can be modified to solve generalisations of this problem (see Theorem 1 and its corollaries).

Words and walks form free algebras. To a symbol string $a = a_1a_2\ldots a_k$ on the above alphabet corresponds the sequence $\theta a = (a_1 - 1, a_2 - 1,\ldots,a_k - 1)$ of integers ≥ -1. We shall view a sequence of numbers as the step sequence of a walk and show that $\theta\underline{G}$ is freely generated from (-1) by the operations carried to $\theta\underline{G}$ by θ, namely

$$\omega_k : (x_1,\ldots,x_k) \rightarrow (k-1) \wedge x_1 \wedge \ldots \wedge x_k, \quad k \geq 1,$$

where $x \wedge y = (x_1,\ldots, x_m, y_1, \ldots, y_n)$ denotes the concatenation of the sequences x and y.

Walks - some terminology. Walks crop up throughout this paper so we now inflict upon the reader the language we shall use to describe them.

A *list* is a finite (possibly empty) sequence of objects. We denote the empty list by $*$. A *walk* is a list of numbers called *steps*. If the steps are drawn from some subsequence $s(i)$, $i \geq 0$, of $(-1, 0, 1, 2, \ldots)$ the walk is a $(\sum_{i\geq 0} \delta_{s(i)+1})$ - walk. The set W of $(\sum_{i \geq 0} \delta_i)$ - walks are more commonly known as left-continuous walks. They have been the subject of considerable study because they crop up naturally in 'the real world'. One context in which they arise is that of a dam with discrete content evolving in discrete time. In each time interval 1 unit flows out of the dam and 0 or 1 or 2 ... units flow into the dam. This example motivates the terminology which follows.

The *level* z_i of a walk z after i steps is the sum of those steps. We set $z_0 = 0$. A walk z with m steps is *lower* [*upper*] *constrained* if $0 \leq i < m \Rightarrow z_i > z_m$ $[0 < i \leq m \Rightarrow z_i < z_0]$. The *dual* ρz of a walk is the walk obtained by reversing the step sequence of z. Notice that the two properties just introduced are dual to one another. Let $\underline{W}$ $[\underline{W}(k)]$ denote the set of lower constrained $(\sum_{i \geq 0} \underset{\sim}{\delta_i})$ - walks [which have final level k], and $\overline{W}$ $[\overline{W}(k)]$ the duals of these sets. Observe that $\underline{W}(k)$ and $\overline{W}(-k)$ are empty if $k > 0$ and $\underline{W}(0) = \overline{W}(0) = \{\text{empty walk}\}$. The *step-description* of a walk having f_k steps $k-1$, $k \geq 0$, is $\sum_{k \geq 0} f_k \underset{\sim}{d_k}$.

The *symbol-description* of a word having f_k symbols k, $k \geq 0$ is $\sum_{k \geq 0} f_k \underset{\sim}{\delta_k}$.

<u>Proposition 1.</u> *(A)* $\theta\underline{G} = \underline{W}(-1)$

(B) $\underline{W}(-1)$ *and* $\underline{G}$ *are freely generated from* (-1) *and* 0 *respectively by the procedures described above, and* $\theta : \underline{G} \to \underline{W}(-1)$ *is the isomorphism between them with respect to these procedures.*

(C) The degree-description and step - [symbol]- description of a walk [word] coincide.

<u>Remarks on the proof.</u> To check that $\theta\underline{G} \subset \underline{W}(-1)$ use induction on the rank of a word, and to check the reverse inclusion use induction on the step number of a walk in $\underline{W}(-1)$, together with the observation that if $y \in \underline{W}(-1)$ and $y \neq (-1)$ then for some $k \geq 1$ $y = (k-1) \wedge x$ where $x \in \underline{W}(-1)^k$, and the

<u>Concatenation Lemma.</u> *The concatenation map from*

$$\underline{W}(-1)^k \to \underline{W}(-k) \quad \textit{defined by}$$

$$(x_1, x_2, \ldots, x_k) \to x_1 \wedge x_2 \wedge \ldots \wedge x_k, \quad k \geq 1,$$

is a bijection.

(B) is a consequence of the concatenation lemma and the fact that the procedure generating $\underline{W}(-1)$ was carried to $\underline{W}(-1)$ by θ, and (C) is easy to check.

<u>Remarks.</u> We say for example that $\underline{G}$ is an $((0)\delta_0 + \sum_{k \geq 1} \omega_k \underset{\sim}{\delta_k})$ - algebra, meaning that $\underline{G}$ is freely generated from 0 by the sequence of operations ω_k, $k \geq 1$. The ancestral tree map establishes an isomorphism between $\underline{G}$ and the set of unlabelled rooted ordered trees with respect to the obvious generation procedure (cf example 1). We call such trees $(\sum_{i \geq 0} \underset{\sim}{\delta_i})$ - trees since they have nodes of degree 0,1,2,... .

$\overline{G}$ is often regarded (see for example [13]) as a collection of words in forward Polish notation. If $\overline{G}$ denotes the corresponding collection of words in reverse

Polish notation (eg. 002, 0102,...) we have that

$$\begin{array}{ccc} \underline{G} & \xrightarrow{\quad\theta\quad} & \underline{W}\ (-1) \\ \rho \updownarrow & & \updownarrow -\rho \\ \overline{G} & \xrightarrow{\quad-\theta\quad} & \overline{W}\ (+1) \end{array}$$

where ρ denotes the map reversing a symbol string or sequence. In particular all of these sets are $(\sum_{k \geq 0} \omega_k \underset{\sim}{\delta}_k)$ - algebras for an appropriate generator ω_0.

Raney's array. We denote the array of multinomial coefficients by

$$M[f_0, f_1, f_2, \ldots] = \frac{(f_0 + f_1 + f_2 + \ldots)!}{f_0!\ f_1!\ f_2!\ \ldots}$$

and the corresponding multi-variable generating function (GF) by

$$M(x_0, x_1, x_2, \ldots) = \sum M[f_0, f_1, f_2, \ldots]\ x_0^{f_0}\ x_1^{f_1}\ x_2^{f_2} \ldots$$

$$= \sum M[\underset{\sim}{f}]\ \underset{\sim}{x}^{\underset{\sim}{f}} \text{ say,}$$

the summation extending over all sequences of non-negative integers of finite sum. We denote by $X(d)$ the subset consisting of elements of X having the description d, so that eg. $X(7\omega_0 + \omega_1 + 3\omega_3)$ is the subset of X generated from 7 instances of ω_0 by 1 application of the partial operation ω_1 and 3 of ω_3. Throughout this paper $\#S$ denotes the cardinality of the set S.

Theorem 1 (Raney). *Suppose* X *is an* $(\sum_{k \geq 0} \omega_k \underset{\sim}{\delta}_k)$*-alg.*

$$\#X(f_0\omega_0 + f_1\omega_1 + \ldots) = \begin{cases} \dfrac{1}{f_0 + f_1 + \ldots} M[f_0, f_1, \ldots] & \text{if } \sum_{i \geq 0} (i-1) f_i = -1 \\ 0 & \text{otherwise} \end{cases}$$

$$= R[f_0, f_1, \ldots] \text{ say.}$$

The GF $R(x)$ *of Raney's array is determined by*

$$R(\underset{\sim}{x}) = x_0 + x_1 R + x_2 R^2 + \ldots\ .$$

Remarks on the proof. Raney deduces his more general form of this theorem (Theorem 2.2 of [6]) from a simple combinatorial result (Theorem 2.1 of [6]) which is in some sense equivalent to the Lagrange Inversion Formula and which has reappeared frequently in the literature. The equation determining the GF of Raney's array is just a reexpression of the fact that a Raney word whose first symbol is k is expressible as a concatenation $ku_1 \ldots u_k$ for some uniquely determined list of k words. The GF is readily computed by successive approximations from the equation.

Corollary 1. *Suppose* X *is an algebra as above weighted by the function* $a : \omega_k \to a_k$, $k \geq 0$. *The weight of* $X(f_0\omega_0 + f_1\omega_1 + \ldots)$ is $R[f_0, f_1, \ldots]\ a_0^{f_0}\ a_1^{f_1} \ldots$, *and the* GF *of this array is* $R(a_0x_0, a_1x_1, a_2x_2, \ldots)$ *which is determined by the equation.*

$$R(a_0x_0,\ a_1x_1,\ a_2x_2,\ldots) = a_0 + a_1x_1R + a_2x_2R^2 + \ldots .$$

The following corollary solves the thematic problem in the case of an algebra having an arbitrary number of generators and operations of each degree.

Corollary 2. *Suppose* X *is an algebra generated from* $\{\alpha_0, \beta_0, \ldots\}$ *by operations* $\alpha_k, \beta_k, \ldots,$ *of degree* k, $k \geq 1$. *Then*

$$\#X\Big(\sum_{k\geq 0} (a_k\alpha_k + b_k\beta_k + \ldots)\ \underset{\sim}{\delta_k}\Big) = \prod_{k\geq 0} M[a_k, b_k, \ldots]\ R\Big[\sum_{k\geq 0} (a_k + b_k + \ldots)\ \underset{\sim}{\delta_k}\Big]$$

for any non-negative integers for which $\sum_{k>0} (a_k + b_k + \ldots)$ *is finite.*

3. FOUR GENERATION PROCEDURES - THEIR RELATIONSHIPS WITH OTHER PROCEDURES AND ONE ANOTHER, AND THEIR ASSOCIATED ARRAYS.

Introduction. We focus upon procedures generating 'Catalan algebras', 'Schröder generated sets', 'Rogers algebras' and 'Rogers generated sets'. §3.1 - §3.3 emphasise the interplay between these procedures whilst §3.4 and the appendix emphasise the range of families generated by each of these procedures.

The interplay between these procedures has bizarre consequences. For example,

$$\text{every}\begin{cases}\text{Catalan algebra}\\ \text{Rogers algebra}\\ \text{Schröder algebra}\end{cases}\quad \text{contains a subset which is a}\quad \begin{cases}\text{Rogers algebra}\\ \text{Schröder generated set}\\ \text{Catalan algebra}\end{cases}$$

these assertions following from Theorems 8, 6 and 4 respectively. It follows that a set generated by any of these procedures contains an infinite nested sequence of subsets generated by each of the others! In particular from Growney's observation the every Catalan family can be viewed as a Catalan algebra we deduce that every Catalan family contains such infinite nested sequences.

Each of the 4 procedures listed above determines an array solving the thematic problem for that procedure, these being the Catalan sequence $c[n]$, the Schröder array $s[i,j_0k]$, the Rogers array $r[i,j]$ and its partner $q[i,j]$. We shall emphasise the interplay between these arrays and others resulting from the interplay between the corresponding generation procedures. For example, $c[n] = r[0,n] = s[0,n,0] = s[0,0,n]$, $n \geq 0$ and the GF of Rogers array is related to that of Raney's array by

$$r(x,y) = R(1,x,y,0,0,\ldots) .$$

3.1. ON THE INTERPLAY BETWEEN THE CATALAN GENERATION PROCEDURES AND THE SCHRÖDER GENERATION PROCEDURE

The sequence of Catalan algebras. A *Catalan family* is a set X with an associated length function $\| : X \quad \{0,1,2,\ldots\}$ such that $\#\{x \in X;\ |x| = n\} =$

$$\#\{x \in X;\ |x| = n\} = \frac{1}{(n+1)+n} M[n+1, n],\ n \geq 0,$$

$$= c[n], \text{ the n'th Catalan number.}$$

Growney has shown that every Catalan family is freely generated from its sole element of length 0 (ω_0 say) by an appropriately chosen binary operation (ω_2 say) in such a way that

$$X((n+1)\omega_0 + n\,\omega_2) = \{x \in X;\ |x| = n\},\ n \geq 0.$$

This motivates the terminology describing the generalisation which follows.

The *t'th Catalan algebra* is an $(\omega_0\underset{\sim}{\delta}_0 + \omega_t\underset{\sim}{\delta}_t)$ - alg, $t \geq 1$. The thematic problem is solved by a subarray of Raney's array :

$$\# X(f_0 \underset{\sim}{0} + f_t \underset{\sim}{t}) = R[f_0 \underset{\sim}{0} + f_t \underset{\sim}{t}]$$

$$= \begin{cases} \dfrac{1}{f_0 + f_t} M[f_0, f_t] & \text{if } -f_0 + (t-1)f_t = -1 \\ 0 & \text{otherwise} \end{cases} .$$

The generating function $R(x_0, 0, \ldots, x_t, 0, \ldots)$ of this subarray is determined by

$$R = x_0 + x_t R^t .$$

The *t'th Catalan sequence* is

$$c_t[n] = R[((t-1)n+1)\underset{\sim}{\delta_0} + n\underset{\sim}{\delta_t}] = \frac{1}{tf_t + 1} M[(t-1)f_t + 1, n], \; n \geq 0.$$

Our especial interest is in the *Catalan algebra* and *Catalan sequence* corresponding to the case $t = 2$.

By 'decomposing' the binary operation generating a Catalan algebra into a unary operation and partial binary operation via the equations below we can obtain more information about the way in which an element in a Catalan algebra is generated. In most Catalan families this information is of combinatorial interest.

<u>Theorem 2.</u> *"Catalan algebras are equivalent to Catalan generated sets".*

(A) Suppose X *is freely generated from* 0 *by the binary operation* '$\cdot$' *which generates* X *freely from* 0.

(B) *If* $x \in X$ *has ancestral description* $(j+1)0 + i\theta + j(*)$ *with respect to the second generating procedure then* x *has ancestral description* $(i+j+1)0 + (i+j)(\cdot)$ w.r.t. *the first.*

<u>Remarks.</u> We leave the reader to check this. Note that the equations obtained from the above by left-right interchange :

$$0 \cdot x = \theta x \quad , \quad x \in X$$
$$x \cdot y = x * y, \quad (x,y) \in (X\backslash\{0\}) \times X$$

also yield partial operations generating X freely from 0. We say that X is *left (right) Catalan generated by* θ *and* $*$ according as the domain of $*$ is $(X\backslash\{0\}) \times X$ or $X \times (X\backslash\{0\})$.

We define an array

$$c[i,j] = \# X((j+1)0 + i\theta + j(*)), \quad i,j \geq 0$$

for X as above.

Example 2 "($\underset{\sim}{\delta}_0$ + $\underset{\sim}{\delta}_2$)-stacks". Finucan's stacks of two-leaf (or single compartment) folders are freely generated from the empty stack by the unary operation

θ: y →  y

and the partial binary operation

*: (x,y) → 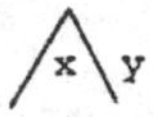

, where x is non-empty.

For example, denoting the empty stack by 0 , $\theta 0$ =  , $(\theta 0)*0$ =  and Figure 2 shows a stack and its corresponding ancestrial tree.

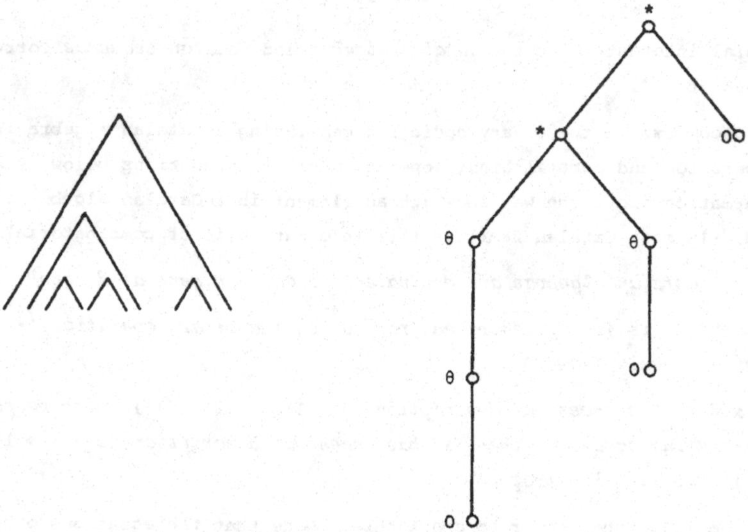

FIGURE 2

[These pictures can be thought of as cross-sections of stacks of two-leaf manila folders. For more details see [2].] Observe that a stack with ancestral description $(j + 1)0 + i\theta + j(*)$ has i empty folders and j non-empty ones. Thus $c[i,j]$ is the number of such stacks and this has been determined by Finucan ([2], §2.2, formula (13) - Finucan's array $c(i + j,i,+)$ coincides with our array $c[i,j]$).

On the array $c[i,j]$. From the definition of a Catalan generated set it follows that the GF $c(x,y)$ of the array c is determined by

$$c = 1 + xc + yc(c - 1)$$

and from part (B) of the preceding theorem we deduce that

$$c(x,x) = c(x)$$

where $c(x)$ is the GF of the Catalan sequence.

If we set $e(x,y) = c(x,y) - 1$ it follows from the first equation above that e is determined by

$$e = 1 + (x + y)e + xy\, e^2.$$

This should be compared with the GF $m(x,y)$ of the binomial coefficients $M(i,j)$ which is determined by

$$m = 1 + (x + y)m.$$

In particular the array e is *symmetric* which implies that

$$c[i + 1,j] = c[j + 1,i], \; i,j \geqslant 0,$$

a fact with some curious combinatorial consequences (consider e.g. folder stacks). The explicit formulae are

$$c[i,j] = \frac{1}{i + j + 1}\begin{bmatrix} i \\ j \end{bmatrix}\begin{bmatrix} j + 2 \\ i \end{bmatrix}, \; e[i,j] = \frac{1}{i + j + 1} M[i + 1,j]M[j + 1,i], \; i,j \geqslant 0,$$

where $\begin{bmatrix} i \\ j \end{bmatrix} = (-1)^i\binom{-j}{i}$. Some of the many sets enumerated by $c[i,j]$ are described in the appendix, List 1, and some recent references to the array are mentioned in the remarks on this list. According to Rogers this array was known to Kirkman.

Schröder generated sets. These are so named because they have a natural partition enumerated by the Schröder sequence (Sequence 1163 of [14]). A *right Schröder generated set* X is freely generated from 0 by a unary operation θ and a pair of partial binary operations $\overset{+}{\cdot}, \overset{-}{\cdot}$ each with domain $X \times (X\backslash\{0\})$. The *Schröder array*

$$s[i,j,k] = \#X((j + k + 1)0 + i\theta + j(\overset{+}{\cdot}) + k(\overset{-}{\cdot}))$$

has GF $s(x,y^+,y^-)$ determined by

$$s = 1 + xs + y^+s(s - 1) + y^-s(s - 1)$$

and the marginal arrays with GF's

$$s(x,y) = s(x,y,y), \quad s(x) = s(x,x,x).$$

$s(x)$ turns out to the the Schröder sequence. Note that

$$s(x,y^+,y^-) = s(x,y^-,y^+),$$

a symmetry which has unexpected combinatorial consequences.

Some Schröder generated sets are described in List 2 of the appendix. We now describe in detail another Schröder generated family - that of $((-1)\delta_0 + \sum_{i\geq 2} \delta_i)$-walks in $\underline{W}(-1)$ or 'deleted' walks in $\underline{W}(-1)$. This family is an analogue of example (B) of the list.

Example 3. Let X denote this set. Note that if $y \in X$ and $y \neq (-1)$, y has a non-empty initial sequence of upward steps or an initial 'jump'. The set X is right Schröder generated by the partial operations $\theta: x \to (+1,-1) \wedge x$, $x \in X$ and for $y = (j_1,\ldots,j_k,-1) \wedge y'$, $j_i > 0$ and $k \geq 1$,

$$x \overset{+}{\cdot} y = (j_1,\ldots,j_k,+1,-1) \wedge x \wedge y'$$

$$x \overset{-}{\cdot} y = (j_1,\ldots,j_{k-1},j_k + 1,-1) \wedge x \wedge y'.$$

The *upward motion* of a walk is the sum of its upward steps. A deleted walk has ancestral description $(j + k + 1)(-1) + i\theta + j(\overset{+}{\cdot}) + k(\overset{-}{\cdot})$ if it has i upward steps followed by a step -1, j remaining upward steps, and k is its upward motion less its number of upward steps. Note that such a walk has upward motion $i + j + k$ and therefore $i + j + k + 1$ downward steps.

This set of deleted walks is just an example of an $(w_0 \delta_0 + \sum_{i\geq 2} w_i \delta_i)$-algebra. Since all such algebras are isomorphic each can be viewed as a Schröder generated set.

Schröder generated sets are two-coloured Catalan generated sets. The ancestral tree of a Schröder generated element is a $(\delta_0 + \delta_1 + \delta_2)$-tree (i.e. a tree having nodes of degree 0, 1 and 2) whose nodes of degree 2 are marked $\overset{+}{\cdot}$ or $\overset{-}{\cdot}$. Suppose that X is right Schröder generated. We may define a set Y of equivalence classes ($[x]$ say) obtained by identifying elements whose ancestral trees are the same if we *ignore markers*.

Theorem 3. (A) Suppose that X is right Schröder generated by θ and $\overset{+}{\cdot}$, $\overset{-}{\cdot}$ from 0. The set Y obtained from X by ignoring markers is right Catalan generated from 0 by θ and the binary partial operation '$\cdot$' defined by

$$[x]\cdot[y] = [x \overset{+}{\cdot} y] (= [x \overset{-}{\cdot} y]).$$

Conversely by introducing markers in the obvious way on the ancestral trees of Catalan generated elements we obtain a Schröder generated set of trees.

(B) If $x \in X$ has ancestral description $(j+k+1)0 + i\theta + j(\dot{+}) + k(\dot{-})$ then $[x]$ has ancestral description $(j+k+1)0 + i\theta + (j+k)(\cdot)$.

<u>Corollary</u>. $s[i,j,k] = M[j,k]c[i,j+k]$,
and hence $s[i,\ell] = 2^{\ell}c[i,\ell]$
or equivalently $s(x,y) = c(x,2y)$ $(= s(x,y,y))$.

Often one can mark certain features of elements of a Catalan generated set rather than their ancestral trees to obtain a Schröder generated set. For example we shall show (see Proposition 3 of §3.4) that the set of connective relations can be Catalan generated so that a relation generated by j binary partial operations has j edges. Marking these edges + or − we obtain a Schröder generated set. Rogers ([7], formula (13)) gives a summation for the number of connective relations with n vertices and edges coloured by k different colours. Taking $k = 2$, his formula is equivalent to $s(x) = c(x,2x)$, a consequence of the last formula above.

<u>Theorem 4</u>. Suppose that X is Schröder generated from 0 by θ and $\dot{+}$, $\dot{-}$. The subsets X^+, X^- of X generated from 0 by θ and $\dot{+}$, θ and $\dot{-}$ respectively are Catalan generated by these pairs of partial operations.

<u>Corollary</u>. $s(x,0,y) = s(x,y,0) = c(x,y)$.

The reader might consider the subsets of the family of deleted walks in $\underline{W}(-1)$ arising in this way. The case study (§3.4) mentions another example.

3.2. <u>ON THE INTERPLAY BETWEEN THE ROGERS GENERATION PROCEDURES AND THE SCHRÖDER GENERATION PROCEDURE</u>.

In a sequence of papers ([11], [12] (with Shapiro) and especially [8]) Rogers has studied the interplay between the Schröder sequence - which is the sequence with GF $s(x)$ already introduced - and the sequence $r(x)$ given by $r(x) + 1 = 2s(x)$. We shall also consider the doubled Schröder sequence $q(x) = 2s(x)$ as of independent interest and view the interplay between these sequences as a consequence of the interplay between three corresponding generation procedures - those of a Rogers algebra $(r(x))$, a Rogers generated set $(q(x))$ and a Schröder generated set $(s(x))$.

Rogers [11] refers to the pair of sequences $r(x), s(x)$ as the Schröder numbers. We call $r(x)$ Rogers sequence because it is associated with a Rogers algebra (see below) in the same way as the Catalan sequence is associated with a Catalan algebra - and because the alliteration is appealing.

The t'th Rogers algebra defined below is so named because the family of subdiagonal walks on "the lattice L_{t-1}" which return to the diagonal discussed in [8] is an example thereof, and a range of examples of the Rogers algebra $(t = 2)$

are introduced and related in [12]. No doubt examples of these algebras have been discussed before.

The sequence of Rogers algebras. The *t'th Rogers algebra* is an $(\omega_0\underset{\sim}{\delta_0} + \omega_1\underset{\sim}{\delta_1} + \omega_t\underset{\sim}{\delta_t})$-algebra, $t \geqslant 2$. The thematic problem is solved by the subarray of Raney's array

$$\#X(f_o\omega_o + f_1\omega_1 + f_t\omega_t) = R[f_0\underset{\sim}{\delta_0} + f_1\underset{\sim}{\delta_1} + f_t\underset{\sim}{\delta_t}]$$

$$= \begin{cases} \dfrac{1}{f_0 + f_1 + f_t} M[f_0,f_1,f_t], & \text{if } -f_0 + (t-1)f_t = -1, \\ 0 & \text{otherwise.} \end{cases}$$

The GF of this array is determined by

$$R = x_0 + x_1R + x_tR^t.$$

The *t'th Rogers array* is

$$r_t[i,j] = \frac{1}{i + tj + 1} M[(t-1)j + 1,i,j], \quad i,j \geqslant 0$$

and the *t'th Rogers sequence* is

$$r_t(x) = r_t(x,x).$$

Our especial interest is in the *Rogers algebra* and *Rogers array* corresponding to the case $t = 2$. Examples of these algebras are described in List 3 of the appendix, and discussed in the remarks on the lists.

Symmetrising a Rogers algebra. Suppose Y is freely generated from 0 by the unary operations θ and binary operation '$\cdot$', so Y is a Rogers algebra. By repeatedly factorising the right hand factor (if any) of an element in Y it can be expressed in just one of the forms

$0,\ x\cdot 0,\ x\cdot(y\cdot 0),\ x\cdot(y\cdot(z\cdot 0))\ ,\ldots$ $(\hat{Y})$

$\theta x\ ,\quad x\cdot\theta y\quad ,\ x\cdot(y\cdot\theta z),\ldots$ $(\check{Y})$

where $x,y,z,\ldots \in Y$. We say that an element whose right factorisation is in the first (second) sequence of forms above is *tied (untied)*, and denote the sets of tied and untied elements by $\hat{Y}$ and $\check{Y}$ respectively.

We can symmetrise Y by introducing an extra untied element $\check{0}$ to correspond with 0 in the second sequence above. We set $\check{X} = \check{Y} \cup \{\check{0}\}$, $\hat{X} = \hat{Y}$ and $X = Y \cup \{\check{0}\}$. We refer to the obvious correspondence $v\colon \hat{X} \to \check{X}$ as the *tie deletion map* and its inverse $\wedge$ as the *tie insertion map*. If we extend the domain of '$\cdot$' to $Y \times X$ by defining $x \cdot \check{0} = \theta x$ then we have

Theorem 5. *The symmetrised set* X *is freely generated from* 0 *and* $\check{0}$ *by the partial operation* '$\cdot$' *with domain* $Y \times X$ *defined above.*

An element in Y *having ancestral description* $h0 + i\theta + j(\cdot)$ *has ancestral description* $h0 + i\check{0} + (i + j)(\cdot)$ *when viewed as generated from* 0 *and* $\check{0}$.

Denote the arrays $\# X(i0 + j\check{0} + (i + j - 1)(\cdot))$ and $\# \bigcup_{i+j=k+1} X(i0 + j\check{0} + k(\cdot))$ by $q[i,j]$ and $q[k]$ respectively.

Corollary. *The* GF's *of the above array and sequence are related to Rogers' array and sequence by*

$$q(x,y) = y + xr(y,x)$$
$$q(x) = r(x) + 1 \; (= x^{-1}q(x,x)).$$

We say that a set X generated as above is *right Rogers generated.*

Tied elements are Schröder generated. This discussion of the relation between the set $\hat{Y}$ of tied elements and the Rogers algebra $\hat{Y}$ itself grew out of a concrete discussion of the same relationship discussed in [11], where $Y(\hat{Y})$ is the set of (tied) left Schroder relations S_ℓ. (We discuss this relationship in Proposition 4 of §3.4.) Our notation, terminology and the numbers $r[i,j]$ quoted below are borrowed (with thanks) from [11].

If for example $q[i,j]$ is an array associated with some two-parameter partition of Y then $\hat{q}$ ($\check{q}$) are the arrays associated with the induced partitions of $\hat{Y}$ ($\check{Y}$).

Proposition 2. "*Interplay between* $r[i,j]$ *and* $\hat{r}[i,j]$, $\check{r}[i,j]$." *Suppose* Y *is a Rogers algebra generated from* 0 *by* θ *and* '$\cdot$'.

(A) *The tie deletion map is a bijection between*
$\hat{Y}(j + 1)0 + i\theta + j(\cdot))$ and $\check{Y}(j0 + (i + 1)\theta + (j - 1)(\cdot))$,
$i \geq 0$, $j \geq 1$.

(B) *The* GF $r(x,y)$ *is determined by the equation*

$$r = 1 + xr + yr^2 \quad \ldots \quad (B1)$$

and determines $\hat{r}$, $\check{r}$ *via the equations*

$$\hat{r}(x,y) = 1 + y\hat{r}r$$
$$\check{r}(x,y) = xr + y\check{r}r \quad \ldots \quad (B2)$$

(Of course, $r(x,y) = \hat{r}(x,y) + \check{r}(x,y)$. *)*

(C) $\hat{r}[0,0] = 1,\ \check{r}[0,0] = 0$ *and*

$$\hat{r}[i,j] = \hat{r}[i+1, j-1],\ i \geq 0,\ j \geq 1.$$

$\hat{r}[0] = 1,\ \check{r}[0] = 0$ *and*

$$\hat{r}[n] = \check{r}[n] = \frac{1}{2} r[n],\ n \geq 1.$$

(D) *For* $i + j = n \geq 1$,

$$\frac{\hat{r}[i,j]}{r[i,j]} = \frac{i(i+1)}{n(n+1)} \quad \text{and} \quad \frac{\check{r}[i,j]}{r[i,j]} = \frac{i(j+n+1)}{n(n+1)}.$$

Remarks on the proof. (A) follows from the definition of tie deletion and (C) is a corollary of (A). (B1) follows from the fact that

$$r(x,y) = R(1, x, y, 0, 0, \ldots)$$

and (B2) from the way in which the binary operation '·' generates the symmetrisation of Y.

(D) is a consequence of the formula immediately after (12) of [11] : Rogers' array $\hat{r}$ and our array $\tilde{r}$ are related by $\tilde{r}[i+j+1,j] = \hat{r}[i,j]$.

The subset of tied elements is Schröder generated. With Y as above an element $z \in \hat{Y}\backslash\{0\}$ either has a tied left-hand factor, in which case it is expressible in one of the two forms

$$\hat{x}\cdot 0,\quad \hat{x}\cdot y,\quad \hat{x} \in \hat{Y},\quad \hat{y} \in \hat{Y}\backslash\{0\},$$

or it does not, which case it is expressible in one of the sequence of forms

$$\theta x_1\cdot\hat{y},\ [x_1\cdot\theta x_2]\cdot\hat{y},\ x_1\cdot(x_2\cdot\theta x_3)]\cdot\hat{y},\ \ldots,\ x_1, x_2, \ldots \in Y,\ \hat{y} \in \hat{Y}.$$

If we denote the first pair of forms above by

$\eta\hat{x},\ \hat{x} \dotplus \hat{y}$ and the sequence of forms above by

$$0 \cdot [x_1\cdot\hat{y}],\ [x_1\cdot 0] \mathbin{\bar{\cdot}} [x_2\cdot\hat{y}],\ [x_1\cdot(x_2\cdot 0)] \mathbin{\bar{\cdot}} [x_3\cdot\hat{y}],\ \ldots$$

then

Theorem 6. *(A) The unary operation* η *and binary partial operations* $\dotplus, \mathbin{\bar{\cdot}}$ *each with domain* $\hat{Y} \times (\hat{Y}\ \{0\})$ *defined above generate* $\hat{Y}$ *freely from* 0, *or in other words* $\hat{Y}$ *is right Schroder generated by these partial operations.*

(B) A tied element having ancestral description $(j+k+1)0 + i\theta + j(\dotplus) + k(\bar{\cdot})$ *has ancestral description* $(i+j+1)0 + k\eta + (i+j)(\cdot)$ *when viewed as generated by the operations generating* Y.

<u>Corollary</u>. $s(x,x,y) = \hat{r}(x,y)$

$$s(x) = \hat{r}(x).$$

From the Corollary to Theorem 4, part (C) of Proposition 2 and the last formula above we deduce that $q(x) = r(x) + 1 = 2\, s(x)$. This link between the sequences $r(x)$ and $s(x)$ is just one of a number explored in [8], §3.

<u>Example 4. 'Duplicating' the Schroder generated set of deleted trees.</u> The set $\check{X}$ of $(\underset{\sim}{\delta_0} + \sum_{i \geq 2} \underset{\sim}{\delta_i})$ - trees or deleted trees is Schröder generated since it is isomorphic to the set of deleted walks described in Example 3. We 'duplicate' $\check{X}$ by defining $\hat{X} = \{\overset{\bullet}{\underset{x}{|}}\ ;\ x \in \check{X}\}$, $X = \hat{X} \cup \check{X}$, and call $\overset{\bullet}{\underset{x}{|}}$ a *planted* deleted tree. The set Y formed by excluding the tree with a single root node from X is a Rogers algebra whose symmetrisation is $\hat{X}$ and whose set of tied elements is $\hat{X}$. The subsets of Y, X and $\hat{X}$ enumerated by $r[i,j]$, $g[i,j]$ and $\hat{r}[i,j]$ are described (respectively) in example (C) of Lists 3, 4 and 2 of the appendix. □

3.3 RENEWAL ARRAYS AND ALGEBRAS EQUIVALENT TO THE FOUR GENERATION PROCEDURES UNDER RIGHT FACTORISATION

Rogers has shown that with each of the sequences $c(x)$, $r(x)$, $s(x)$ ($b(x)$ say) is associated another ($\alpha b(x)$ say) which is the 'a-sequence' of the renewal array whose 'b-sequence' is $b(x)$. The a - and b- sequences of a *non-zero* renewal array each determine the array and one another, the explicit relation between them being $b(x) = a(xb(x))$... (1). The sequences are

$$\alpha c(x) = 1 + x + x^2 + \ldots = \frac{1}{1-x} \quad \text{([9], p.308; note that Rogers' } c_0(x) = \frac{1}{1-x}\text{)}$$

$$\alpha\, r(x) = 1 + 2x + 2x^2 + \ldots = 1 + \frac{2x}{1-x} \quad \text{([8], formula (29))}$$

$$\alpha\, s(x) = 1 + x + 2x^2 + 2^2x^3 + \ldots = 1 + \frac{x}{1-2x} \quad \text{([8], formula (32))}$$

and we shall show that

$$\alpha q(x) = 2 + x + x^2 + x^3 + \ldots = 2 + \frac{x}{1-x} \quad \text{(Theorem 7).}$$

Rogers typically determines $\alpha b(x)$ from $b(x)$ by putting the functional equation determining $b(x)$ into the form (1). For example, Rogers sequence $r(x)$ is determined by $r = 1 + xr + xr^2$ (see p.16) so $r = 1 + \frac{2xr}{1-xr}$ which yields αr as above. He also provides 'combinatorial' explanations for these pairs of sequences. For example the second formula after formula (4) of [12], p.296 asserts that

$r[m] = \#\{(\sum_{i\geq 0} \underset{\sim}{\delta_i})$ - trees whose 'eldest' edges are two-coloured and which have m edges$\}$, $m \geq 0$,

where the 'eldest' edge of a node is that joining it to its first subtree. By shifting the colour from the eldest edge to its parent node we deduce that

$r[m] = \#\{(\sum_{i\geq 0} \underset{\sim}{\delta_i})$ - trees whose nodes of degree 0, 1, 2, 3,... are coloured by 1, 2, 2, 2, ... colours respectively and which have m edges$\}$, $m \geq 0$.

Our aim is to link these explanations of the above sequence pairs via the following two results. Say that an $(\sum_{i\geq 0} \omega_i \underset{\sim}{\delta_i})$ - algebra is *weighted by* $a(x)$ if its *weight* function a has GF $\sum_{i\geq 0} a(\omega_i)x^i = a(x)$, and that an element in such algebra whose degree-description is $\sum_{i\geq 0} f_i \underset{\sim}{\delta_i}$ has *total degree* $\sum_{i\geq 0} if_i$.

<u>Theorem 7.</u> *The following are equivalent* -

(A) $a(x)$ *and* $b(x)$ *are* a- *and* b-*sequences of the same renewal array.*

(B) $b(x) = a(xb(x))$.

(C) $b(x) = R(a_0, a_1x, a_2x^2, \ldots)$ *where* R *is Raney's array.*

(D) b[m] *is the total weight of these elements in an* $(\sum_{i\geq 0} \omega_i \delta_i)$ - alg *weighted by* a(x) *which have total degree* m, $m \geq 0$.

Remarks on the proof. We refer to [9], § 2, for the equivalence of (A) and (B). The equivalence of (B), (C) and (D) follows from the first Corollary to Theorem 1.

Rogers (private communication) arrived at the equivalence of (A) and (D) above by considering a family of walks with weighted diagonal steps which turns out to be an $(\sum \omega_i \underset{\sim}{\delta_i})$ - alg.

Theorem 8. *(A) Catalan algebras, Rogers algebras, Schröder generated sets and Rogers generated sets can be viewed as algebras freely generated from* a_0 *generators by* a_i *options of degree* i, $i \geq 1$ *where* $a(x) = \alpha c(x)$, $\alpha r(x)$, $\alpha s(x)$, *and* $\alpha q(x)$ *respectively, these being the sequences defined on* p.20 . *The converse is also true.*

(B) In each case an element generated by a total of m *operations has total degree* m *when viewed as generated by the operations of the corresponding algebra.*

Remarks on the proof. Consider for example a Rogers algebra Y. If we denote each of the two possible sequences of forms for the right factorisation of an element shown on p.16 by

$$\hat{\omega}_1(x),\ \hat{\omega}_2(x,y),\ \hat{\omega}_3(x,y,z),\ \ldots\quad (\hat{Y})$$

and $$0,\ \check{\omega}_1(x),\ \check{\omega}_2(x,y),\ \check{\omega}_3(x,y,z),\ \ldots\quad (\check{Y})$$

respectively, we define sequences of operations $\hat{\omega}_1, \hat{\omega}_2, \hat{\omega}_3, \ldots$ and $\check{\omega}_1, \check{\omega}_2, \check{\omega}_3, \ldots$, which generate Y freely from 0. Much the same argument applies in each of the other cases - in the case of a right Rogers generated set consider the possible left factorisations of an element.

Remarks. In particular every Catalan algebra can be viewed as a $(\sum_{i\geq 0} \omega_i \underset{\sim}{\delta_i})$ - alg and so contains a subset generated by ω_1 and ω_2 from ω_0 which is a Rogers algebra. In short - every Catalan algebra contains a Rogers algebra.

As a Corollary of Theorems 7 and 8 we have eg.

that $\quad r(x) = R(1, 2x, 2x^2, 2x^3, \ldots)$

whilst $q(x) = r(x) + 1 = R(2, x, x^2, x^3, \ldots)$!

Explicit summation formulae for the terms of the sequence b(x) in each of our cases may be deduced from the Lagrange Inversion formula which solves $b(x) = a(xb(x))$ given $a(x)$.

3.4 A CASE STUDY - SOME FAMILIES OF RELATIONS ON ORDERED SETS

[11] is the most comprehensive of a number of papers in which Rogers discusses the hierarchy of 4 families of reflexive symmetric relations which can be imposed on ordered sets shown in Figure 3. S, S_ℓ, S_r and C are the families of superconnective relations, left and right Schroder relations and connective relations respectively. A reflexive symmetric relation x on the ordered set $[1, 2, \ldots, n]$ can be represented by a graph - join vertices i and j by an edge if $i\ x\ y$ and $i \neq j$ - and each of these families contains natural subfamilies of relations whose graphs have nice properties such as being connected, or 'tied', or ... . We shall not trouble to repeat the definitions which are concisely summarised in the introduction of [11].

Given some (sub) family X of relations Rogers' basic concern is to enumerate the sets $X(n)$ $[X(n,m)]$ of relations in X having n vertices [and m edges]. For example he describes a correspondence between $S_\ell(n,m)$ and $S_r(n,m)$, $n \geq 1$, $n \geq 0$ and determines the numbers $\#S_\ell(n,m)$, $\#S_\ell(n)$, $\#C(n,m)$ and $\#C(n)$ obtaining (in our notation) $r[n-1-m,m]$, $r[n-1]$, $c[n-m,m]$, and $c[n]$, $n \geq 1$, $m \geq 0$, respectively.

We shall survey Rogers' results in a different light by focussing on the way in which these families are generated (see Figure 4). For example we shall view his correspondence between S_ℓ and S_r as an isomorphism between two Rogers' algebras, and show that $\bar{C} = C \cup \{\text{empty relation}\}$ forms a Catalan algebra. We shall also investigate the hierarchy of sets obtained from the Rogers' algebra S_ℓ via Theorems 6 and 4, and deduce the enumeration of some families of relations (see Figure 5).

We do not pretend to have summarised the wealth of information on relations in these papers. For example we have ignored the toughest nut - superconnective relations - entirely. However we think that our viewpoint is a step toward *cataloguing that information.*

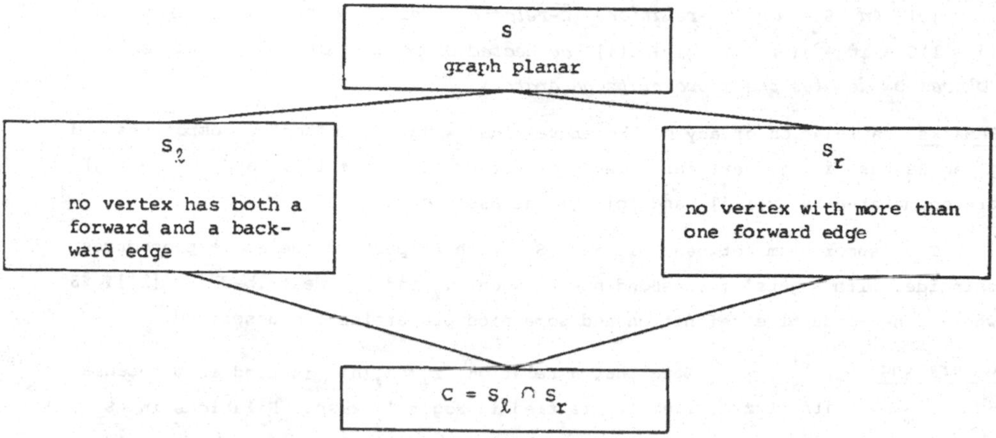

F I G U R E 3

<u>Generating S_ℓ, S_r and C.</u> It suffices to describe the inverse to the generation procedure, i.e. the procedure for factorising a non-generator. Note especially that S_ℓ and S_r consist of relations on non-empty sets but that we have included the empty relation in $\overline{C}$.

For any family of relations X let X(n) denote the set of X-relations on an n-element ordered set (usually [1,...,n]. For any $z \in X(n)$, $n \geq 1$, let $j = \min\{i \geq 1;\ 1\ z\ i\}$ and if $j \geq 2$ let $x = z\,|\,[2,\ldots,j]$ where $z|S$ denotes the restriction of the relation z to S.

We factorise a non-empty relation $z \in \overline{C}(n)$ as follows.

Let $y = z\,|\,[j+1,\ldots,n]$, so y is empty if $j = n$. If $j = 1$, $z = \theta y$.

If $j > 1$, $z = x \cdot y$.

We factorise a relation $z \in S_\ell(n)$ or $S_r(n)$, $n \geq 2$ as follows.

If $j = 1$, let $y = z\,|[2,\ldots,n]$ and $z = \theta y$.

If $j > 1$, and if $z \in S_\ell$, let $y = z\,|[1, j+1,\ldots,n]$ and $z = x \cdot y$.

If $j > 1$, and if $z \in S_r$, let $y = z\,|[j,\ldots,n]$ and $z = x \cdot y$.

<u>Proposition 3</u>. (A) S_ℓ *and* S_r *form Rogers algebras generated from* $S_\ell(1) = S_r(1)$ *by the operations* θ *and* '$\cdot$' *defined above.* $\overline{C}$ *is left Catalan generated* Y *from* $\overline{C}(0)$ *by the partial operations* θ *and* '$\cdot$' *defined above.*

(B) *An* S_ℓ- *or* S_r*-relation* [$\bar{C}$*-relation*] *having ancestral description* $(j+1)0 + i\theta + j(\cdot)$ *has* $i+1[i]$ *connected components and* $j[j]$ *edges, (where* 0 *denotes the appropriate generator).*

Remarks. A relation of any of the above kinds with i connected components and j edges has $i+j$ vertices. Thus the sets $S_\ell(i+j+1,j)$ and $\bar{C}(i+j,j)$ are enumerated by $r[i,j]$ and $c[i,j]$ as asserted.

The isomorphism between S_ℓ and S_r with respect to the above procedures coincides with Rogers' correspondence between S_ℓ and S_r described in [11], §3 where a non-inductive definition and some nice properties are described.

Generating $\hat{S}_\ell$, $\hat{S}^+_\ell$, $\hat{S}^-_\ell$. Note that a relation $z_\ell \in S_\ell(n)$ is tied in our sense iff $1\,z\,n$, i.e. z is tied in Rogers' sense. Relations in S_r which are tied in our sense turn out to be 'chained' in Rogers' sense. If for some $1 \leq i \leq n$, $j = \min\{k > i;\ i\,z\,k\}$ exists we say that vertex i has a *bottom forward edge,* namely that joining vertices i and j.

We factorise a relation $z \in \hat{S}_\ell(n)$, $n \geq 2$ as follows.

Let $j = \min\{k \geq 1;\ 1\,z\,k\}$, $k = \max\{i;\ 2\,z\,i\}$ (so $2 \leq k \leq j$), and $x = z|[2,\dots,k]$.

If $k = j = n$, $z = \eta x$.

If $k < n$, let $y = z|[1, k+1,\dots,n]$ and

if $k = j$, let $z = x \overset{+}{\cdot} y$

if $k < j$, let $z = x \overset{-}{\cdot} y$.

We advise the reader to interpret the above graphically!

Proposition 4 (A) $\hat{S}_\ell$, *is right Schröder generated from* $\hat{S}_\ell(1)$ *by the partial operations defined above.*

(B) *An* $\hat{S}_\ell$*-relation having ancestral description* $(j+k+1)0 + i\eta + j(\overset{+}{\cdot}) + k(\overset{-}{\cdot})$ *has* i *bottom forward edges,* j *remaining edges and* $k+1$ *connected components.*

Remarks. The Catalan generated subsets $\hat{S}_\ell^+$, $\hat{S}_\ell^-$ (see Theorem 4) obtained from $\hat{S}_\ell$ are described in Figure 5.

The set S_ℓ of left Schröder relations is a Rogers algebra

The set S_r of right Schröder relations is a Rogers algebra

The set $\overline{C} = S_\ell \cap S_r \cup \{\text{empty relation}\}$ of connective relations is Catalan generated

F I G U R E   4

The set S_ℓ of left Schröder relations from a Rogers algebra

see   *Theorem 6*

The set $\hat{S}_\ell$ of tied left Schroder relations is Catalan generated

see Theorem 4

The set $\hat{S}_\ell^{+}$ of connected left Schröder relations is Catalan generated

The set $\hat{S}_\ell^{-}$ of tied connective relations *is* Catalan generated

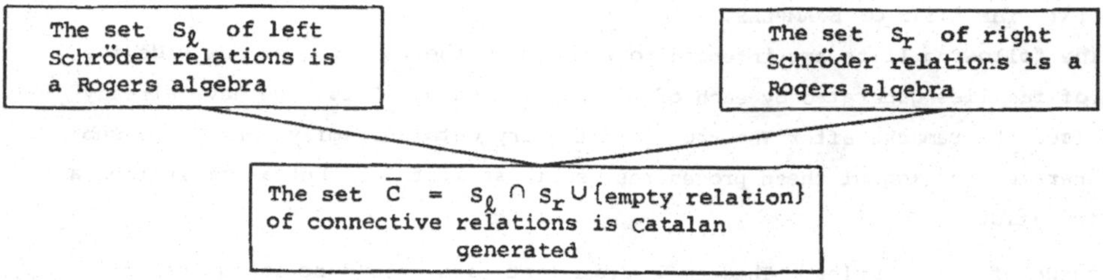

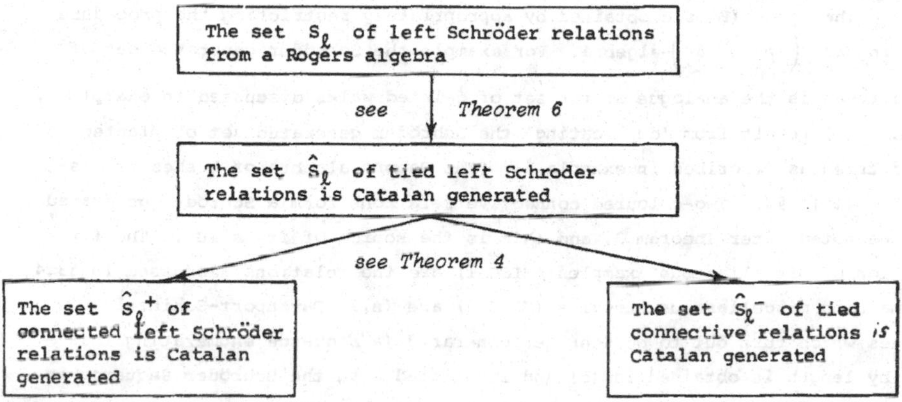

F I G U R E   5

APPENDIX: THE LISTS OF EXAMPLES.

The following lists are intended to suggest to the reader just how wide the range of families generated by each of the procedures of §3 is. We have already noted (see the remarks after Theorem 7) that every Catalan family contains a subset generated by each of these procedures, so these lists are but a sample from a big population.

The source of the examples. Theorem 7 shows that each procedure is equivalent (via right (or left) factorisation) to a certain algebra. The items labelled (A) in each list are examples of such algebras. $\left(\sum_{i\geq 0} \underset{\sim}{\delta}_i\right)$-trees are defined after Proposition 1, and a $\left(\sum_{i\geq 0} \underset{\sim}{\delta}_i\right)$-folder stack consists of folders with one or more leaves. The items (B) are obtained by appropriately restricting the procedure generating an $\left(\sum_{i\geq 0} \omega_i \underset{\sim}{\delta}_i\right)$-algebra. For example the Schröder generated set of deleted trees is the analogue of the set of deleted walks discussed in example 3. The items (C) result from 'duplicating' the Schröder generated set of planted deleted trees as described in example 4. The Rogers algebra of bushes is discussed in [10], §2. Two-coloured connective relations form a Schröder generated set as we noted after Theorem 3, and this is the source of items (D). The first collection of miscellaneous examples (MISC 1) are the relations discussed in §3.4. The remaining miscellaneous examples (MISC 2) are (n,3) Davenport-Schinzel sequences which turn out to be Schröder generated (a sequence enumerating this family by length is obtained in [5] and identified with the Schröder sequence in [12], Ex.4), and the Rogers algebras of foliated trees and output restricted deque permutations which are two of a number of Rogers algebras discussed in [10] and [12].

The enumeration of the examples. The subsets described in each of the 4 lists which follow are enumerated by the arrays $c[i,j]$, $s[i,j,k]$, $r[i,j]$ and $q[i,j]$ respectively. The arrays q and s are essentially derived from Rogers array r and the array c respectively. We now refer the reader to papers which associate some of the subsets listed with the appropriate array.

The association of connective relations, $\left(\sum_{i\geq 0} \underset{\sim}{\delta}_i\right)$-trees and 2-leaf folder stacks with the array c can be found (respectively) in [11], (formula (1) and especially §3), [12], p.297 (where some consequences of the symmetry of the array $e[i,j]$ are mentioned) and [2]. The association of the Rogers algebras of the bushes, foliated trees, deque permutations, 'trees with marked eldest edges' (cf our example A), left and right Schröder relations with the array r can be found in [2], [2], [6], [6], [4] and [4] respectively.

The author has not seen mention of the array $s[i,j,k]$. However the association of deleted trees with the Schröder sequence goes back to Etherington

(1940) [1], and its association with two coloured connective relations, tied left Schröder relations, chained right Schröder relations and Davenport Schinzel strings can be found in [4], [4], [4] and [6] respectively. The author has not seen explicit mention of the array q but Rogers often gives hints of the symmetrisation of a Rogers algebra. For example he describes a correspondence between foliated trees and trees with marked endnodes (excluding roots); omitting this qualification we obtain the Rogers generated set of example (A).

<u>List 1. Some examples of subsets $X((j + 1)0 + i\theta + j(\cdot))$ of Catalan generated sets X.</u>

(A1) $\left(\sum_{i\geq 0} \underset{\sim}{\delta_i}\right)$-walks in $\underline{W}(-1)$ having $\begin{cases} i & \text{steps } \geq 0 \\ j+1 & \text{steps } -1. \end{cases}$

(A2) $\left(\sum_{i\geq 0} \underset{\sim}{\delta_i}\right)$-trees having $\begin{cases} i & \text{nodes of degree } \geq 1 \\ j+1 & \text{end-nodes.} \end{cases}$

(A3) $\left(\sum_{i\geq 0} \underset{\sim}{\delta_i}\right)$-folder stacks having $\begin{cases} i & \text{folders} \\ i+j & \text{leaves.} \end{cases}$

(B1) $(\underset{\sim}{\delta_0} + \underset{\sim}{\delta_2})$-walks in $\underline{W}(-1)$ having $\begin{cases} i \text{ steps } +1 \text{ followed by a step } -1 \\ j \text{ steps } +1 \text{ followed by a step } +1, \\ \text{i.e. } i \text{ local maxima and } i+j \text{ steps} \\ +1. \end{cases}$

(B2) $(\underset{\sim}{\delta_0} + \underset{\sim}{\delta_2})$-trees having $\begin{cases} i \text{ nodes of degree 2 with a trivial} \\ \text{left subtree} \\ j \text{ nodes of degree 2 with a non-trivial} \\ \text{left subtree.} \end{cases}$

(B3) $(\underset{\sim}{\delta_0} + \underset{\sim}{\delta_2})$-folder stacks having $\begin{cases} i \text{ empty folders} \\ j \text{ non-empty folders.} \end{cases}$

(D) Connective relations having $\begin{cases} i \text{ connected components} \\ j \text{ edges.} \end{cases}$

(MISC 1) Connective relations, tied connective relations and connected left Schröder relations are also Catalan generated sets.

<u>List 2. Some examples of subsets $X((j+k+1)0 + i\theta + j(\overset{+}{\cdot}) + k(\overset{-}{\cdot}))$ of Schröder generated sets X.</u>

(A)	$\left(\sum_{i\geq 0} \underset{\sim}{\delta_i}\right)$-trees whose nodes of degree $k+1$ are marked by sequence of k +'s and -'s, $k \geq 0$, having	i internal nodes j markers + k markers -.
(B)	$(\underset{\sim}{\delta_0} + \sum_{i\geq 2} \underset{\sim}{\delta_i})$-trees (i.e. deleted trees) having	i internal nodes with a trivial first subtree j remaining internal nodes k edges which are not first or second edges.
(C)	Planted deleted trees having	" " " " " "
(D)	Connective relations whose edges are marked + or - having	i connected components j markers + k markers -.

(MISC 1) Tied left and chained right Schröder relations
(MISC 2) 'deleted' walks (cf (B) above), (n,3) Davenport-Schinzel sequences
are also Schröder generated sets.

<u>List 3. Some examples of subsets $X((j+1)0 + i\theta + j(\cdot))$ of Rogers algebras X.</u>

(A)	$\left(\sum_{i\geq 0} \underset{\sim}{\delta_i}\right)$-trees whose internal nodes are marked ∧ or ∨ having	i nodes marked ∧ j + 1 remaining nodes.
(B)	$(\underset{\sim}{\delta_0} + \underset{\sim}{\delta_1} + \underset{\sim}{\delta_2})$-trees having	i nodes of degree 1 j nodes of degree 2.
(C)	'bushes', (i.e. deleted trees which may be planted having at least one edge) having	i + j + 1 endnodes j + 1 remaining nodes.

(MISC 1) Left and right Schröder relations are also Rogers algebras.

(MISC 2) Foliated trees, permutations which can be obtained from an output restricted deque are also Rogers algebras.

List 4. Some examples of subsets $X(i0 + j\check{0} + (i + j - 1)(\cdot))$ of Rogers generated sets.

(A)	$\left(\sum_{i\geq 0} \delta_j\right)$-trees whose endnodes may be marked $\vee$ having	$\begin{cases} i & \text{unmarked endnodes} \\ j & \text{endnodes marked } \vee. \end{cases}$
(C)	Deleted trees which may be planted having	$\begin{cases} i + j & \text{endnodes} \\ i & \text{remaining nodes.} \end{cases}$

REFERENCES

[1] I.H.M. Etherington, Some problems of non-associative combinations, (1), *Edinburgh Mathematical Notes* 32 (1940) I-IV.

[2] H.M. Finucan, Some decompositions of generalised Catalan numbers, *Proc. 9th Australian Conf. on Combinatorial Math., 1981* (Springer-Verlag, Lecture Notes in Mathematics 952 (1982)) 275-293.

[3] I.J. Good, The generalisation of Lagrange's expansion and the enumeration of trees, *Proc. Camb. Phil. Soc.*, 61 (1965) 499-517.

[4] J.A. Growney (née Simpson), Finitely generated free groupoids, *Ph.D. thesis, University of Oklahoma*, 1970. An abstract of this thesis appeared in *Dissertation Abstracts International* 31(B) (1970) 3543.

[5] R.C. Mullin and R.G. Stanton, A map-theoretic approach to Davenport - Schinzel sequences, *Pacific Journal of Mathematics* (1) 40 (1972) 167-172.

[6] G.N. Raney, Functional composition patterns and power series reversion, *Trans. Am. Math. Soc.*, 94 (1960) 441-451.

[7] D.G. Rogers, The enumeration of a family ladder graphs. Part I : Connective Relations, *Quart. J. Math. (Oxford)* (2), 28 (1977) 421-431.

[8] D.G. Rogers. A Schröder Triangle : three combinatorial problems, *Proc. 5th Australian Conf. on Combinatorial Math., 1977* (Springer-Verlag, Lecture Notes in Mathematics 622 (1977)) 175-196.

[9] D.G. Rogers, Pascal triangles, Catalan numbers and renewal arrays, *Discrete Math.* 22 (1978) 301-311.

[10] D.G. Rogers and L.W. Shapiro, Some correspondences involving the Schröder numbers, *Proc. 5th Australian Conf. on Combinatorial Math., 1977* (Springer-Verlag, Lecture Notes in Mathematics, 686 (1978)) 240-247.

[11] D.G. Rogers, The enumeration of a family of ladder graphs by edges. Part II : Schröder and superconnective relations, *Quart. J. Math., (Oxford)* (2) 31 (1980) 491-506.

[12] D.G. Rogers and L.W. Shapiro, Deques, trees and lattice paths, *Proc. 8th Australian Conf. on Combinatorial Math., 1980* (Springer-Verlag, Lecture Notes in Mathematics 884 (1981)) 293-303.

[13] A.D. Sands, Notes on generalised Catalan numbers, *Discrete Math.* 21 (1978) 219-221.

[14] N.J.A. Sloane, *A handbook of integer sequences*, (Academic Press, New York, (1973)).

COMPOSITE GRAPHS WITH EDGE STABILITY INDEX ONE

K.L. McAVANEY

We show that for any connected composite graph C *the following statements are equivalent:*

1. C *is pair edge transitive,*
2. C *has edge stability index one or* $C=C_4$,
3. $C=G^n$ $(n > 1)$ *where* G *is a connected prime pair transitive and pair edge transitive graph.*

Analogous but more complicated results are found for disconnected composite graphs.

1. INTRODUCTION

First some definitions. Graph R is *semistable* at vertex v if the neighbourhood of v is fixed under all automorphisms of R-v [5]. R is *edge semistable* at edge e if the line graph of R is semistable at e [2]. R has *stability index one* if R is semistable at some vertex v and, for all such v, R-v is not semistable [1]. R has *edge stability index one* if the line graph of R has stability index one [2]. R is *pair transitive* if for all vertices u and v in R there is an automorphism of R that interchanges u and v. (This is called Property A in [6].) R is *pair edge transitive* if the line graph of R is pair transitive.

We relate these ideas here to (cartesian) products of graphs [10]. The vertices in the *product* G×H of graphs G and H are all the ordered pairs (g_i,h_j) where $g_1,g_2,\ldots$ and $h_1,h_2,\ldots$ are the vertices of G and H respectively. For brevity we denote (g_i,h_j) by ij. In G×H, ij and $k\ell$ are adjoint if and only if either i=k and h_j is adjacent to $g\ell$ or $j=\ell$ and gi is adjacent to gk. Clearly a product is regular if and only if its factors are. Likewise for correctedness. A *composite* graph is isomorphic to the product of two non-trivial graphs. A graph is *prime* if neither trivial nor composite. The following three lemmas are fundamental.

Lemma 1. [10]. *Every non-trivial graph is a unique product of prime graphs.*

Lemma 2. [10]. *If* G *and* H *are connected relatively prime graphs then the automorphism group of* $G \times H$ *is the set of permutations* $\gamma = (\alpha, \beta)$ *where* α *and* β *are automorphisms of* G *and* H *respectively and for all vertices* ij *in* $G \times H$ $\gamma ij = \alpha i \beta j$.

Lemma 3. [4]. *For connected prime graph* G *the automorphism group of* G^n *is the set of permutations* $\gamma = (\alpha; \beta_1, \ldots, \beta_n)$ *where* α *is a permutation on* $\{1,2,\ldots,n\}$ *and* $\beta_1, \ldots, \beta_n$ *are automorphisms of* G *and, for all vertices* $u = (u_1, \ldots, u_n)$ *in* G^n, $\gamma_u = (\beta_1 u_{\gamma 1}, \ldots, \beta_n u_{\gamma n})$.

Let C be a connected composite graph. In [13], Sims and Holton showed that C is semistable and indeed, with few exceptions, is semistable at every vertex. In [6], they showed that if each factor of C is pair transitive and C is not the 4-cycle C_4 then C has stability index one. It turns out that the converse is true except for $P_3 \times P_2$ and $P_4 \times P_2$ [8,9]. We can extend these results a little with the help of

Lemma 4. *A connected composite graph is pair transitive if and only if each of its factors is pair transitive.*

Proof. Let $C = G \times H$ be the product in question. As noted in [6, Lemma 6] if G and H are each pair transitive then C is.

To examine the converse, suppose C is pair transitive and G and H are non-trivial and relatively prime. Let g1 and g2 be any two vertices in G. There is an automorphism γ of C that interchanges 11 and 21. From Lemma 2, $\gamma = (\alpha, \beta)$ and therefore α interchanges g1 and g2. Thus G is pair transitive.

Now suppose C is pair transitive and $C = G^n$ where G is a prime graph. Let $x = (1, u_2, \ldots, u_n)$ and $y = (2, u_2, \ldots, u_n)$ be two vertices in C. There is an automorphism γ of C that interchanges x and y. From Lemma 3, α is the identity and β_1 interchanges vertices g1 and g2 of G. Thus G is pair transitive. Also, from the first paragraph, G^r is pair transitive for $r=1,2,\ldots,n$.

Finally suppose $C = K \times L$ where K and L are not relatively prime. From Lemma 1 and the argument above, K is pair transitive.

Lemma 4 and the reworks prior to it give us

Theorem 5. *A connected composite graph* C *has stability index one if and only if* $C = P_3 \times P_2$ *or* $P_4 \times P_2$ *or* C *is pair transitive and* $C \neq C_4$.

The main purpose of this paper is to demonstrate the edge analogue of these results. First the easy part:

Theorem 6. *A connected composite pair edge transitive graph* $C \neq C_4$ *has edge stability index one.*

Proof. It was shown in [6, Theorem 4] that the stability index of a pair transitive graph R is one if and only if, for all distinct vertices u and v of R, the neighbourhood of u in R-v does not equal the neighbourhood of v in R-u. Let C = G×H where neither G nor H is trivial. Let e and d be two distinct edges in C. Without loss of generality let e = [11,21]. There are nine possibilities for d: [21,31], [31,41], [12,22], [22,32], [32,42], [21,22], [31,32], [22,23], [32,33]. In each case there is a neighbour h2 of h1 in H such that the edge [11,12] in C is adjacent to e but not to d, unless d = [12,22] and $H = P_2$. As $G \neq P_2$ there is an edge, [11,31] say, adjacent to e but not to d.

To prove the converse of Theorem 6 we use the following ideas and results. For edge e = [u,v] in a graph R, end(e) = {u,v}. R is *stable* at e if end(e) is fixed by all automorphisms of R-e[11,7]. The *section set* of G in the product C = G×H is the set of all the edges [ik,jk] in C[7,13]. Similarly for H. Holton and Sims showed in [7] that all connected composite graphs are stable and moveover at precisely which edges. The following lemma is a special case of this result but is general enough for our needs.

Lemma 7. *For any connected composite graph* C = G×H, *if* G *and* H *contain more than two vertices then* C *is stable at all edges and if* $G = P_2$ *then* C *is stable at all edges* e *in the section set of* G *unless* $H = P_3$ *and the vertices in* end(e) *have degree three.*

There is a likeness between stability and edge semistability. Indeed Grant showed in [3, Theorems 3.1 and 3.2] that with few exceptions any graph R is stable at edge e if and only if R is edge semistable at e. The next two lemmas are immediate corollaries to this result.

Lemma 8. *A connected composite graph* C *is stable at edge* e *if and only if* C *is edge semistable at* e.

Lemma 9. *Let* e *be an edge in a connected composite graph* $C \neq C_4$. C-e *is stable at edge* d *if and only if* C-e *is edge semistable at* d.

2. IRREGULAR PRODUCTS

We dispatch connected irregular composite graphs with

Theorem 10. *A connected composite graph with edge stability index one is regular.*

Proof. Let $C = G \times H$ be an irregular connected composite graph with neither G nor H trivial. We will find two edges e and d of C so that C is edge semistable at e and $C-e$ is edge semistable at d. G or H is irregular; suppose it is G. Let m and n be the minimum degrees in G and H respectively.

Case 1. H has two adjacent vertices of degree n.

Let $g1$ be a vertex in G with maximum degree r. Let $g2$ be a vertex in G with degree m. Let $h1$ and $h2$ be two adjacent vertices in H with degree n. From Lemmas 7 and 8, C is edge semistable at $[11,12]$ unless $C = P_3 \times P_2$. But we may assume $C \neq P_3 \times P_2$ since this graph obviously does not have edge stability index one. It follows from Lemma 9 and the minimality of m and n that $C - [11,12]$ is edge semistable at $[21,22]$.

Case 2. No two vertices of minimum degree in H or in G are adjacent.

Let M be the set of vertices in G of degree m. Let R be the set of neighbours of vertices in M that have the least degree r (greater than m). Let $g1$ be a vertex in M that is adjacent to a vertex $g2$ in R. We define similarly adjacent vertices $h1$ and $h2$ in H with $h1$ having degree n and $h2$ having degree s (greater than n) which is minimum among all the neighbours of vertices of degree n. Without loss of generality we assume the degree of 21 is less than or equal to the degree of 12, that is $r+n \leq m+s$. From Lemmas 7 and 8, C is edge semistable at $[21,22]$. We show by contradiction that $C - [21,22]$ is edge semistable at $[11,21]$. Suppose $C - [21,22]$ is not edge semistable at $[11,21]$. By Lemma 9, there is an automorphism α of $J = C-[21,22]-[11,21]$ for which $\alpha\{11,21\} \neq \{11,21\}$. But only 11 and possibly 21 have the least degree in J, so $\alpha 11 = 11$ and $\alpha 21 = i_1 j_1 \neq 21,11$. Also $i_1 j_1 \neq 22$, because the degree of 22 in J is $r+s-1 > r+n-2 =$ the degree of 21 in J. Also degree $gi_1 < r$ because otherwise $r+n-2 =$ degree of $i_1 j_1$ in $J =$ degree of $i_1 j_1$ in $C \geq r+n$. Similarly degree $hj_1 < s$ because otherwise $r+n-2 =$ degree of $i_1 j_1$ in $J =$ degree of $i_1 j_1$ in $C \geq m+s \geq r+n$. This gives us two useful observations.

Observation 1. $g1 \not\sim gi_1$ and $h1 \not\sim hj_1$.

Observation 2. $i_1 \neq 2 \neq j_1$.

We now show that $\alpha 22 = 22$. Suppose $\alpha 22 = i_2j_2 \neq 22$. Since 22 is distance two from 11 so is i_2j_2. Hence i_2j_2 and 11 are in a common square unless $i_2 = 2$ or $i_2 = 1$ or $j_2 = 1$. But 22 and 11 are not in a common square, so indeed $i_2 = 2$ or $i_2 = 1$ or $j_2 = 1$.

Case A. $i_2 = 2$.

Degree $hj_2 < s$ since degree of i_2j_2 $(= 2j_2)$ in J = degree 22 in J = $r+s-1 < r+s$ = degree of 22 in C. By the minimality of s, hj_2 and $h1$ are not adjacent. But i_2j_2 is distance two from 11, so $j_2 = 1$. Thus $i_2j_2 = 21$. This is impossible since degree 21 in $J = r+n-2 < r+s-1$ = degree 22 in J = degree i_2j_2 in J. Hence $i_2 \neq 2$.

Case B. $j_2 = 1$.

Since $\alpha 11 = 11$ and $i_2 \neq 2$ we may let $i_2 = 3$. Since $12 \sim 11$ and 22 and $11 \not\sim 21$ we may let $\alpha 12 = 41 \sim 11$ and 31. Thus 11, 41, 31 lie in a subgraph of J which is isomorphic to $P_2 \times P_3$ and contains the vertices 11, 12, 41, 42, 31, 32. Now $\alpha^{-1}12 = 13$ or $k1$ $(\neq 21, 31)$. If $\alpha^{-1}12 = 13$ then $\alpha^{-1}42 = 14$ and $\alpha^{-1}32 = 24$. Since $23 \sim 13$ and 24, $\alpha 23 = \gamma 2 \sim 12$ and 32. Also $23 \sim 21$ so $\gamma 2 \sim \alpha 21 = i_1j_1$. Hence $i_1 = \gamma$ or $j_1 = 2$. From Observation 2, $j_1 \neq 2$ so $i_1 = \gamma \neq 2$. Thus $g1 \sim g\gamma = gi_1$ which contradicts Observation 1.

Consequently $\alpha^{-1}12 = k1$ $(\neq 21, 31)$. Therefore $\alpha^{-1}42 = k2$ and $\alpha^{-1}32 = \ell 2$ $(\neq 12) \sim k2$ and 22. Since $\ell 1 \sim k1$ and $\ell 2$, $\alpha \ell 1 = s2 \sim 12$ and 32. Also $\ell 1 \sim 21$ so $s2 \sim \alpha 21 = i_1j_1$. This gives a contradiction as above. Hence $j_2 \neq 1$.

Case C. $i_2 = 1$.

Since $\alpha 11 = 11$ and $11 \not\sim 22$ we may let $j_2 = 3$. Since $12 \sim 11$ and 22, $\alpha 12 = 1k \sim 11$ and 13. Now degree $g_2 >$ degree $g_1 \geq 1$ so there exists $g3 \sim g2$. Thus $22 \sim 32 \sim 31 \sim 21$. Hence $\alpha 32 = \ell 3 \sim 13$ or $\alpha 32 = 14 \sim 13$.

Suppose $\alpha 32 = \ell 3$. Now 22, 32, and 31 do not lie in a common square so neither do $\alpha 22 = 13$, $\alpha 32 = \ell 3$, and $\alpha 31$. Therefore, either $\alpha 31 = 21$ and $\ell = 2$, or $\alpha 31 = \gamma 3$ $(\neq 13) \sim \ell 3$. If $\alpha 31 = 21$ and $\ell = 2$ then $11 \sim 13$ which is impossible since $\alpha^{-1}11 = 11 \not\sim 22 = \alpha^{-1}13$. So $\alpha 31 = \gamma 3$. Thus 13, $\ell 3$, $\gamma 3$ lie in a subgraph of J which is isomorphic to $P_2 \times P_3$ and contains the vertices 13, 1k, $\ell 3$, ℓk, $\gamma 3$, γk. So $\alpha^{-1}\ell k = s2$ ($\neq 22$) ~ 12 and 32, and $\alpha^{-1}\gamma k = s1$. But $s1 \sim 11$ which is impossible since $\alpha s1 = \gamma k \not\sim 11 = \alpha 11$. So $\alpha 32 \neq \ell 3$.

Suppose $\alpha 32 = 14$. Again $\alpha 22$ ($= 13$), $\alpha 32$ ($= 14$), and $\alpha 31$ do not lie in a common square so $\alpha 31 = 1t \sim 14$. Also 32, 31, 21 do not lie in a common square so neither do $\alpha 32 = 14$, $\alpha 31 = 1t$, and $\alpha 21$. Hence $\alpha 21 = 15 \sim 1t$. Thus 13, 14, 1t, 15 lie in a subgraph of J which is isomorphic to $P_2 \times P_4$ and contains the vertices 13, 23, 14, 24, 1t, 2t, 15, 25. Now $\alpha^{-1}25 = 41$ or 2u. If $\alpha^{-1}25 = 41$ then $\alpha^{-1}2t = 51 \sim 41$ and 31, $\alpha^{-1}24 = 52$, $\alpha^{-1}23 = 42$. But $42 \sim 41$, hence $23 \sim 25$ and $13 \sim 15$. So $22 = \alpha^{-1}13 \sim \alpha^{-1}15 = 21$, a contradiction. If $\alpha^{-1}25 = 2u$ then $\alpha^{-1}2t = 3u$, $\alpha^{-1}24 = 3v$ ($\neq 31$) $\sim 3u$ and 32, $\alpha^{-1}23 = 2v$. But $2v \sim 2u$, hence $23 \sim 25$, $13 \sim 15$, $22 \sim 21$, the same contradiction. Hence $i_2 \neq 1$.

So $\alpha 22 = 22$.

We can now show that α also fixes 21. As observed in Case C there exists $g3 \sim g2$. Let $\alpha 31 = i_3 j_3$. Since $21 \sim 31$ we have $i_1 j_1 \sim i_3 j_3$. Also $\alpha 31$ and $\alpha 22 = 22$ have a unique common neighbour $\alpha 32 = i_4 j_4 \neq 21$. Moreover $i_3 j_3$ is the unique common neighbour of $i_1 j_1$ and $i_4 j_4$.

Suppose $i_4 \neq i_1$ and $j_4 \neq j_1$. Since $i_1 j_1$ and $i_4 j_4$ have a common neighbour $i_3 j_3$ we have $i_3 j_3 = i_1 j_4$ or $i_4 j_1$. Thus $gi_1 \sim gi_4$ and $hj_1 \sim hj_4$. Moreover $i_1 j_1$ and $i_4 j_4$ are none of 11, 21, 22 so they have two common neighbours $i_1 j_4$ and $i_4 j_1$. This is a contradiction. So either $i_4 = i_1$ or $j_4 = j_1$.

Suppose $j_4 = j_1$. Then $j_3 = j_1$. Since $i_4 j_4 \sim 22$, i_4 or $j_4 = 2$. By Observation 2, $j_1 \neq 2$ so $i_4 = 2$. Thus $2j_1 \sim i_3 j_1$ and 22 and $i_3 2 \sim i_3 j_1$ and 22. But $i_3 j_1$ ($= i_3 j_3$) and 22 have only one common neighbour so we again have a contradiction.

Suppose $i_4 = i_1$. Then $i_3 = i_1$. Since $i_4 j_4 \sim 22$, i_4 or $j_4 = 2$. By Observation 2, $i_1 \neq 2$ so $j_4 = 2$. Now $j_3 \neq 1$ because otherwise $i_1 1 = i_3 j_3 \sim i_1 j_1$ and hence $h1 \sim hj_1$ which contradicts Observation 1. Therefore $2j_3$ and $i_1 2 \sim i_1 j_3$ and 22. But $i_1 j_3 = i_3 j_3$ and 22 have only one common neighbour. Another contradiction.

Hence α fixes 21 and $C - [21,22]$ is indeed edge semistable at [11,21]. This completes Case 2 and the proof of Theorem 10.

3. REGULAR PRODUCTS

We tighten the net with

Theorem 11. *If* C *is a connected composite graph with edge stability index one then* $C = G^n$ $(n > 1)$ *where* G *is a connected prime pair transitive and pair edge transitive graph.*

Proof. By Theorem 10, C is regular of degree m say. Suppose $C = G \times H$ where G and H are non-trivial relatively prime graphs with H having at least as many vertices as G. Let g1, g2 and h1, h2 be adjacent vertices in G and H respectively. By Lemmas 7 and 8, C is edge semistable at [11,12]. So C - [11,12] is not edge semistable at [11,21]. By Lemma 9, there is an automorphism γ of J = C - [11,12]-[11,21] for which $\gamma\{11,21\} \neq \{11,21\}$. But 11 is the only vertex in J of degree m-2 and 12,21 are the only vertices in J of degree m-1. So γ fixes 11 and interchanges 21 and 12. Thus γ is also an automorphism of C. From Lemma 2, $\alpha g1 = g1 = \alpha g2$ which is impossible. Hence $C = G^n$ for some prime graph G and integer $n > 1$ (Lemma 1).

To show G is pair edge transitive first consider any two adjacent edges e = [g1,g2] and d = [g2,g3] of G. Then $x = (1, u_2, \ldots, u_n)$ is adjacent to $y = (2, u_2, \ldots, u_n)$ which is adjacent to $z = (3, u_2, \ldots, u_n)$ in G^n. As in the previous paragraph there is an automorphism γ of G^n which fixes y and interchanges x and z. From Lemma 3, α is the identity and β_1 is an automorphism of G that fixes g2 and interchanges g1 and g3. Thus β_1 induces an edge automorphism of G that interchanges c and d.

We can now show that G is pair transitive and pair edge transitive. This is obvious if G is P_2 or a cycle so we may assume the degree of G is at least 3. G must have two non-adjacent edges since otherwise, by a simple induction argument on the number of edges, G would be $K_3 \cup sK_1$, which is disconnected or a cycle, or G is $K_{1,r} \cup sK_1$ which is disconnected or irregular or P_2.

Let e = [g1,g2] and d = [g3,g4] be any two non-adjacent edges in G. Let gi and gj be any two vertices in G and consider the edges $e' = [(i,1,u_3,\ldots,u_n),(i,2,u_3,\ldots,u_n)]$ and $d' = [(j,3,u_3,\ldots,u_n),(j,4,u_3,\ldots,u_n)]$ in G^n. We will find an automorphism γ of G^n that interchanges end(e') and end(d'). From Lemma 3, α is the identity, β_1 is an automorphism of G that interchanges gi and gj, and β_2 is an automorphism of G that interchanges end(e) and end(d). Thus G is pair transitive and pair edge transitive.

To simplify notation we perceive e' and d' as edges in $G \times G^{n-1} = G^n$ and denote e' as $[i1,i2]$ and d' as $[j3,j4]$. From Lemmas 7 and 8, G^n is edge semistable at e'. Since $G^n - e'$ is not edge semistable at d' it follows from Lemma 9 that there is an automorphism γ of $J = G^n - e' - d'$ for which $\gamma \mathrm{end}(d') \neq \mathrm{end}(d')$. Since $i1, i2, j3, j4$ are the only vertices in J of degree $m-1$ this means that $\gamma \mathrm{end}(d') = \mathrm{end}(e')$ or $\{x,y \mid x \in \mathrm{end}(d'),\ y = \mathrm{end}(e')\}$. In the former case $\gamma \mathrm{end}(e') = \mathrm{end}(d')$ and hence γ is an automorphism of G^n that interchanges $\mathrm{end}(e')$ and $\mathrm{end}(d')$. In the latter case $\gamma \mathrm{end}(e') = (\mathrm{end}(d') - \{x\}) \cup (\mathrm{end}(e') - \{y\})$. Without loss of generality, let $\gamma \mathrm{end}(d') = \{i1, j3\}$ and $\alpha \mathrm{end}(e') = \{i2, j4\}$. Let $K = J+b+c$ where b joins $i1$ to $j3$ and c joins $i2$ to $j4$. Then $\alpha^{-1}: K \to G^n$ is an isomorphism so K is composite. Let E and $\bar{E}$ denote the section sets of K. Without loss of generality, b is in E.

We now use some of the section set properties listed in [13,7]. By Property 1, there is an edge g in $\bar{E}$ incident with $i1$. If $g = [i1, k1]$ then $k=j$, otherwise, by Property 3, there is vertex $k3$ adjacent to $k1$ and $j3$ and vertex $j1$ adjacent to $k1$ and $j3$ which contradicts Property 5. Similarly if $g = [i1, ik]$ then $k=4$. Thus there are at most two edges in $\bar{E}$ incident with $i1$. Since K is regular, and hence each of its factors is regular, this means the factor F induced by $\bar{E}$ is P_2 or a cycle (Property 4). But, from Lemma 1 and the fact that G is prime, it follows that F is isomorphic to G^r for some $r = 1, 2, \ldots, n-1$. Hence G has degree less than three, a contradiction.

To complete our task we need the following lemma which is an immediate corollary to Theorem 1 Case B in [14].

Lemma 12. *Each edge automorphism of a regular graph* R *is induced by an automorphism of* R.

Theorem 13. *If* $C = G^n$ $(n > 1)$ *where* G *is a connected prime pair transitive and pair edge transitive graph then* C *is pair edge transitive.*

Proof. Consider any two edges $e' = [(u_1, \ldots, u_{r-1}, i, u_{r+1}, \ldots, u_n), (u_1, \ldots, u_{r-1}, j, u_{r+1}, \ldots, u_n)]$ and $d' = [(v_1, \ldots, v_{s-1}, k, v_{s+1}, \ldots, v_n), (v_1, \ldots, v_{s-1}, \ell, v_{s+1}, \ldots, v_n)]$ in G^n. Note that $e = [gi, gj]$ and $d = [gk, g\ell]$ are edges in G. For each $t \neq r, s$ there is an automorphism β_t of G that interchanges gu_t and gv_t. Also there is an edge automorphism δ of G that interchanges e and d. Since G is pair transitive and therefore regular, δ is induced by an automorphism β of G that interchanges end(e) and end(d) (Lemma 12). Let $\beta_r = \beta_s = \beta$ and $\alpha = (rs)$ then the permutation γ of vertices of G^n defined by $\gamma(w_1,\ldots,w_n) = (\beta_1 w_{\alpha 1},\ldots,\beta_n w_{\alpha n})$ is an automorphism of G^n that interchanges end(e') and end(d') (Lemma 3). Thus γ induces an edge automorphism of G^n that interchanges e' and d'. So G^n is pair edge transitive.

Theorem 14. *For any connected composite graph* C *the following three statements are equivalent:*

1. C *is pair edge transitive,*
2. C *has edge stability index one or* $C = C_4$,
3. $C = G^n$ $(n > 1)$ *where* G *is a connected prime pair transitive and pair edge transitive graph.*

Proof. Combine Theorems 6, 11, 13.

4. DISCONNECTED PRODUCTS

Difficulties arise in characterising disconnected composite graphs with edge stability index one because they may contain prime components. Theorem 16 is the analogue to Theorem 6. We use

Lemma 15. [14, Theorem 1]. *If two connected graphs are edge isomorphic and neither is* $K_{1,3}$ *then they are isomorphic.*

Theorem 16. *A composite graph* C *that is pair edge transitive has edge stability index one or* $C = R \cup tK_1$ *where* $t \geq 0$ *and* $R = nK_1$, nC_4, *or* nG *for some prime connected pair edge transitive graph* G, *with* $n \geq 1$, *or* $R = nP_2$ *with* $n \geq 2$ *or* $R = rK_{1,3} \cup sK_3$ *with* r *and* $s \geq 0$ *and* $r+s \geq 1$.

Proof. From Lemma 15, either all non-trivial components of C are isomorphic or $C = rK_{1,3} \cup sK_3 \cup tK_1$. The latter case satisfies the theorem. In the former case let $C = nS \cup tK_1$ for some connected graph $S \neq K_1, K_{1,3}, K_3$. S is pair edge transitive. Hence S has edge stability index one or $S = C_4$ or S is prime (Theorem 14). Therefore C has edge stability index one or it is one of the exceptions listed in the Theorem.

The following lemmas take us towards a converse of Theorem 16. An immediate corollary to Lemma 15 is

Lemma 17. *If each of* R *and* S *is a connected composite graph, or is one with just one edge deleted, and* R *and* S *are edge isomorphic, then they are isomorphic.*

Lemma 18. [7, Theorem 3.1]. *A connected composite graph with just one edge deleted is connected and prime.*

The next lemma follows directly from Lemma 2.1 in [7] and appears as Theorem 4.1.2 in [12].

Lemma 19. *If* e *and* d *are edges of connected composite graphs* R *and* S *respectively and* R-e *is isomorphic to* S-d *then* R *is isomorphic to* S.

Theorem 20. *If* C *is a composite graph with edge stability index one and no prime component then* $C = sG^n \times tK_1$, *where* $s,n,t+1 > 0$ *and* G *is a connected prime pair transitive and pair edge transitive graph, and hence* C *is pair edge transitive.*

Proof. If C is connected the result comes from Theorem 14. Suppose C is disconnected. Each non-trivial component of C is composite and therefore edge semistable (Lemmas 7,8). Let E and D be two distinct non-trivial components of C edge semistable at e and d respectively. C itself is edge semistable at e (lemmas 17, 18). But C-e is not edge semistable at d. Hence E-e and D-d are isomorphic (Lemmas 17,18). Therefore E and D are isomorphic (Lemma 19). Thus all non-trivial components of C are isomorphic. Hence each has edge stability index one. By Theorem 14, C has the required form and is pair edge transitive.

REFERENCES

[1] D.D. Grant, The stability index of graphs, *Combinatorial Mathematics* (Proceedings of the Second Australian Conference), Lecture Notes in Mathematics 403, (Springer-Verlag, Verlin-Heidelberg-New York, 1974), 29-52.

[2] D.D. Grant, *The Stability Index of Graphs*, (M.Sc. Thesis, University of Melbourne, 1974).

[3] D.D. Grant, Stability of line graphs, *J. Austral. Math. Soc.*, 21A (1976), 457-466.

[4] F. Harary and E.M. Palmer, On the automorphism group of a composite graph, *Studia Sci. Math. Hungar.*, 3, (1968), 439-441.

[5] D.A. Holton and D.D. Grant, Regular graphs and stability, *J. Austral. Math. Soc.*, 20A, (1975), 377-384.

[6] D.A. Holton and J.A. Sims, Graphs with stability index one, *J. Austral. Math. Soc.*, 22A, (1976), 212-220.

[7] D.A. Holton and J. Sims, The cartesian product of two graphs is stable, *Theory and Applications of Graphs* (Proceedings, Michigan, 1976), Lecture Notes in Mathematics 642 (Springer-Verlag, Berlin-Heidelberg-New York, 1978), 286-303.

[8] K.L. McAvaney, Some even composite graphs with stability index greater than one, *Combinatorial Mathematics IX* (Proceedings Brisbane, Australia 1981), Lecture Notes in Mathematics, (Springer-Verlag, Berlin-Heidelberg-New York, 1982).

[9] K.L. McAvaney, Composite graphs with stability index one, in preparation.

[10] G. Sabidussi, Graph multiplication, *Math. Z.*, 72, (1960), 446-457.

[11] J. Sheehan, Fixing subgraphs, *J. Comb. Th.*, 128, (1972), 226-244.

[12] J.A. Sims, *Stability of the Cartesian Product of Graphs*, (M.Sc. Thesis, University of Melbourne, 1976).

[13] J. Sims and D.A. Holton, Stability of cartesian products, *J. Comb. Th.*, 25 (Series B), (1978), 258-282.

[14] H. Whitney, Congruent graphs and the connectivity of graphs, *Amer. J. Maths.*, 54, (1932), 150-168.

A NUMBER-THEORETICAL NOTE ON CORNISH'S PAPER

Jane Pitman and Peter Leske

Abstract/Introduction

In his paper, Cornish gave an expression for a_n, the number of ways of placing $n \geq 0$ different balls in r distinct cells so that, for $j = 1,\ldots,r$, the number of balls in the jth cell is congruent to h_j modulo k_j. In this note, which stems from discussion following Cornish's talk, we give an alternative proof of Cornish's theorem and discuss the connection between conditions for a_n to be non-zero and Frobenius's problem in number theory.

1. Alternative proof of Cornish's theorem

We write $e^{2\pi iy} = e(y)$. The proof is based on the fact that if $x, h,$ and k are integers with $k > 0$ then

$$\sum_{s=0}^{k-1} e\left(s(x-h)/k\right) = \begin{cases} k & \text{if } x \equiv h \pmod{k} \\ 0 & \text{otherwise.} \end{cases}$$

From this and the combinatorial interpretation of $n!/(n_1!n_2!\ldots n_r!)$, where $n = n_1+\ldots+n_r$, as in Cornish, we see that, using Cornish's notation,

$$a_n k_1 k_2 \ldots k_r = \sum_{\substack{n_1,\ldots,n_r \geq 0 \\ n_1+\ldots+n_r = n}} \frac{n!}{n_1!\ldots n_r!} \prod_{j=1}^{r} \sum_{0 \leq s_j \leq k_j - 1} e\left(s_j(n_j-h_j)/k_j\right)$$

$$= \sum_{\substack{s_1,\ldots,s_r \geq 0 \\ 0 \leq s_j \leq k_j - 1}} \prod_{j=1}^{r} \omega_j^{-h_j s_j} \sum_{\substack{n_1,\ldots,n_r \geq 0 \\ n_1+\ldots+n_r = n}} \frac{n!}{n_1!\ldots n_r!} \prod_{j=1}^{r} \omega_j^{s_j n_j} .$$

By the multinomial theorem, the inner sum is

$$\left(\sum_{j=1}^{r} \omega_j^{s_j}\right)^n ,$$

which gives us Cornish's expression.

2. Conditions for a_n to be non-zero

Writing the condition $n_j \equiv h_j \pmod{k_j}$ in the form $n_j = h_j + x_j k_j$, we see that a_n is non-zero if and only if the linear Diophantine equation

$$k_1x_1 + k_2x_2 + \ldots + k_rx_r = N$$

is solvable in non-negative integers $x_1,\ldots,x_r$ for $N = n - \sum_{j=1}^{r} h_j$. This equation is solvable in *integers* for all N if and only if $\text{g.c.d}(k_1,\ldots,k_r) = 1$ (see, for example, Leveque [2], §2.3). If $\text{g.c.d.}(k_1,\ldots,k_r) = 1$, then the equation is solvable in *non-negative* integers for all N sufficiently great, and we are led to Frobenius's problem of determining the largest N for which the equation is *not* solvable in non-negative integers. An account of the problem with complete bibliography up to 1977 is given by Selmer [5]. More recent papers on the problem have mostly been concerned with algorithms for finding its solution; see, for example, Rødseth [4], Nijenhuis [3], Greenberg [1] and the papers they cite.

REFERENCES

[1] Greenberg, H., An algorithm for a linear diophantine equation and a problem of Frobenius, Numer. Math. 34 No.4 (1980), 349-352

[2] Leveque, W.J., Fundamentals of Number Theory, Addison - Wesley (1977)

[3] Nijenhuis, A., A minimal path algorithm for the "money-changing problem, Amer. Math. Monthly 86 (1979), 832-835.

[4] Redseth, O.J., On a linear diophantine problem of Frobenius II, J.reine angew. Math. 307 (1979), 431-440.

[5] Selmer, E.S., On the linear diophantine problem of Frobenius, J.reine angew. Math. 29314 (1977), 1-17.

ON THE AUTOMORPHISMS OF ROOTED TREES WITH HEIGHT DISTRIBUTIONS

CHERYL E. PRAEGER AND P. SCHULTZ

Let $UB(\underline{v})$ be the supremum of the orders of automorphism groups of rooted trees with height distribution $\underline{v} = (v_0,\ldots,v_h)$. It is shown that among the set of rooted trees with height distribution $\underline{v}$ and automorphism group of order $UB(\underline{v})$ there is always one with a very simple structure. The problem of finding an efficient algorithm to determine $UB(\underline{v})$ in terms of the parameters $v_0,\ldots,v_h$ is considered and partial results obtained for some classes of height distributions $\underline{v}$.

INTRODUCTION

The *height* of a vertex in a rooted tree is the length of the longest path from a leaf to the vertex. The *exponent* of a rooted tree is the height of the root. In [1], A.W. Hales posed the problem of determining the number of isomorphism classes of rooted trees with a fixed number v_i of vertices of height i for $i=0,\ldots,h$, that is the number $N(\underline{v})$ of rooted trees with *height distribution* $\underline{v} = (v_0,v_1,\ldots,v_h)$. In Hales' paper $N(\underline{v})$ is determined for a few classes of height distributions $\underline{v}$, and in our paper [2] two general methods are described for evaluating $N(\underline{v})$. However both of these methods involve considerable computational difficulty. Because of this we also obtained in [2] upper and lower bounds for $N(\underline{v})$, one set being the following: $K(\underline{v})LB(\underline{v}) \le N(\underline{v}) \le K(\underline{v})UB(\underline{v})$, where $LB(\underline{v})$ and $UB(\underline{v})$ are the minimum and maximum values respectively of the orders of automorphism groups of rooted trees with height distribution $\underline{v}$, and $K(\underline{v})$ is an explicit function of the parameters v_i, (namely

$$K(\underline{v}) = \prod_{0\le i\le h-1} \frac{1}{v_i!} \sum_{0\le n\le v_{i+1}} (-1)^n \binom{v_{i+1}}{n} \Big(\sum_{i+1\le j\le h} v_j - n\Big)^{v_i},$$

see [2] Theorem 2.2 and Lemma 5.1). In this paper we consider the problem of determining the maximum $UB(\underline{v})$ of the orders of automorphism groups of rooted trees with a given height distribution $\underline{v}$. First we show that there are rooted trees with height distribution $\underline{v}$ and automorphism group of order $UB(\underline{v})$ having a particularly simple structure. We obtain an efficient algorithm for finding $UB(\underline{v})$ in terms of the parameters v_i for a certain class of height distributions $\underline{v}$, but such an algorithm for general $\underline{v}$ has not as yet been obtained.

1. NOTATION, DEFINITIONS, AND STATEMENT OF RESULTS

In investigating rooted trees with a given height distribution it is often convenient to work with an alternative definition, due to Hales, of this class of trees, (especially when investigating their automorphisms).

Definition 1.1. A *Hales' distribution of exponent* h, where $h \geq 0$, is a vector $\underline{u} = (u_1,\ldots,u_h)$ where the u_i are non-negative integers and $u_h > 0$ if $h \geq 1$, or the vector $\underline{u} = (0)$ if $h = 0$.

There is a 1-1 correspondence between the set of height distributions $\underline{v}$ (for trees with at least one vertex) and the set of Hales' distributions $\underline{u}$. Given a Hales' distribution $\underline{u}$ of exponent h we define the corresponding height distribution $\underline{v}$ by

$$v_h = 1$$

and, if $h \geq 1$,

$$v_i = \sum_{i+1 \leq j \leq h} u_j \,, \quad \text{for } i=0,\ldots,h-1.$$

Conversely, given $\underline{v}$ of exponent h we define the corresponding $\underline{u}$ by $\underline{u} = (0)$ if $h=0$, and if $h \geq 1$,

$$u_h = v_{h-1}$$

and, if $h \geq 2$,

$$u_i = v_{i-1} - v_i, \quad \text{for } i=1,\ldots,h-1.$$

To explain why we have introduced Hales' distributions we need the concept of a direct sum of rooted trees.

Definition 1.2 The *direct sum* $T_1 \oplus \ldots \oplus T_r)$ of rooted trees $T_1,\ldots,T_r$, $r \geq 1$, is the rooted tree which is the set theoretic union of $T_1,\ldots,T_r$, amalgamating their roots.

Given a Hales' distribution $\underline{u} = (u_1,\ldots,u_h)$ of exponent $h \geq 1$, the direct sum of u_1 chains of length $1,\ldots,u_h$ chains of length h, is a rooted tree with height distribution $\underline{v}$, where $\underline{v}$ corresponds to $\underline{u}$ as above. Moreover Hales [1] showed that every tree with height distribution $\underline{v}$ can be obtained from this tree by a finite sequence of operations, each of which detaches one of the chains of length k say from the root and re-attaches it to any vertex of height at least k. Conversely any tree obtained in this way has height distribution $\underline{v}$. So we shall speak of a tree T having height distribution $\underline{v} = \underline{v}(T)$ and Hales' distribution $\underline{u} = \underline{u}(T)$ and call T a $\underline{v}$-tree or $\underline{u}$-tree; we shall write $UB(\underline{u}) = UB(\underline{v})$ etc., and use these notations interchangeably. Also in our *proofs* we allow Hales' distributions to have last entry zero, (the exponent being the last non-zero entry), to avoid needless special cases. A $\underline{u}$-tree T with $|\mathrm{Aut}\ T| = UB(\underline{u})$, where $\mathrm{Aut}\ T$ is the automorphism group of T, will be called a *maximal* $\underline{u}$-tree.

Our first theorem asserts that maximal $\underline{u}$-trees exist which are very "simple" in shape and we must first make this notion of simplicity precise.

Definition 1.3. (a) A *component* of a tree T is the direct sum of all principal subtrees of T isomorphic to a given one.

(b) A rooted tree T is called an (n,k,U)-*wreath with bud* U, where n and k are positive integers and U is either a single vertex or a rooted tree with at least two principal subtrees, if T is the direct sum of n rooted trees of exponent k, each isomorphic to a rooted tree S say, where S consists of a chain of length $k-r$ extended by U, r being the exponent of U, see Figure 1.

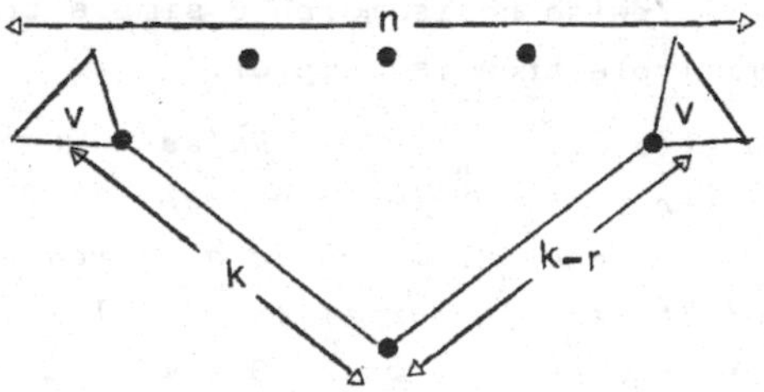

Figure 1.

If U is a single vertex we call T an $(n,k,\cdot)$-*wreath* or a 0-*wreath with parameters* $[n,k]$, see Figure 2.

If U is an $(m,r,\cdot)$-wreath with $m \geq 2$ and if $n=1$, (that is T is a $(1,k,(m,r,\cdot)\text{-wreath})$-wreath!), we call T a *1-wreath with parameters* $[k,m,r]$, see Figure 2.

If U is a $(2,s,\cdot)$-wreath and if $n=2$ we call T a *2-wreath with parameters* $[k,s]$, and finally if U is a 2-wreath with parameters $[r,s]$ and if $n=2$ we call T a *2-wreath with parameters* [k,r,s], see Figure 2.

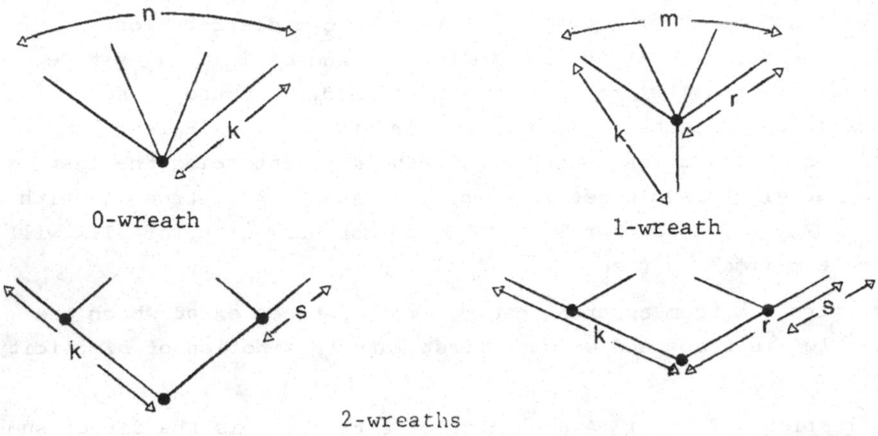

Figure 2.

We shall show that the i-wreaths, for i=0,1,2, are the building blocks for special maximal u-trees.

(c) A *simple tree* is an i-wreath, i=0,1 or 2. A *semisimple tree* is a direct sum of pairwise nonisomorphic simple trees. (Clearly each component of a semisimple tree is simple).

Theorem 1. *Let* $\underline{u} = (u_1, \ldots, u_h)$ *be a Hales' distribution of exponent* $h \geq 1$. *Then there is a maximal* $\underline{u}$*-tree* T *which is semisimple and has at most one component which is a 2-wreath. Further we may choose* T *such that there is a partition* (I_0, I_1) *of* $\{1,\ldots,h\}$ *and for each* $i \in I_0$ *an integer* w_i *with* $0 \leq w_i \leq u_i$, *such that the following hold.*

(a) For each $i \in I_1$, T *has a (unique) component which is a 1-wreath with parameters* $[j_i, u_i+1, i]$ *for some* $j_i \in I_0$, $j_i > i$.

(b) For each $i \in I_0$ *with* $w_i > 0$, T *has a (unique) component which is a 0-wreath with parameters* $[w_i, i]$, *and* $u_i - w_i$ *components which are 1-wreaths of exponent* i. *The non-zero* w_i *increases strictly as* i *increases.*

(c) For each $i \in I_0$ *with* $w_i = 0$ *which is not a parameter of a 2-wreath, there are* u_i *components which are 1-wreaths of exponent* i.

(d) If T *has a component which is a 2-wreath with parameters* [k,s] *or* [k,r,s], *then the parameters all lie in* I_0, $u_s=2$, $u_r=1$, *and there are* u_k-2, u_k-1 *components which are 1-wreaths of exponent* k *in the first and second cases respectively.*

Remark. Not all maximal $\underline{u}$-trees are semisimple, for example if $\underline{u} = (1,1,1)$ there are five maximal $\underline{u}$-trees of which three are semisimple, see Figure 3.

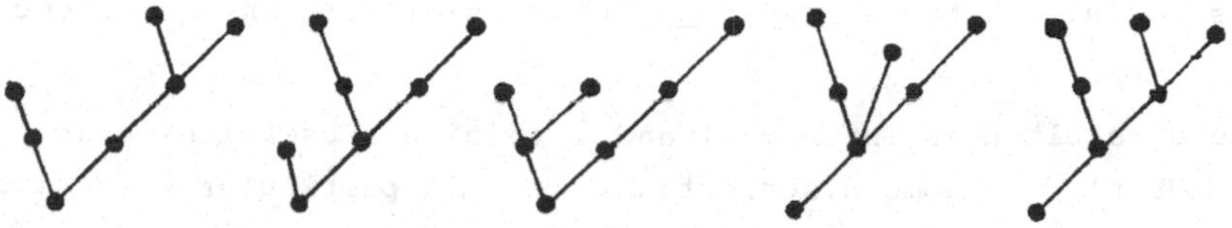

Figure 3.

The next result often provides us with a method of simplifying the determination of $UB(\underline{u})$. We use it repeatedly in the rest of the paper. (It is of interest to compare it with Lemma 4.5 of [2].)

Theorem 2. *Let* $\underline{u} = (\underline{w}_1, \underline{w}_2)$ *be a Hales' distribution of exponent* $h \geq 1$ *where* $\underline{w}_1 = (u_1,\ldots,u_t)$, $\underline{w}_2 = (u_{t+1},\ldots,u_h)$, $0 \leq t < h$, *and* $\underline{w}_1$ *is assumed to be vacuous if* $t=0$. *Then* $UB(\underline{u}) = UB(\underline{u}')$ *where* $\underline{u}' = (\underline{w}_1, 0, \underline{w}_2)$. *Further there is a* 1-1 *correspondence* ϕ *between the sets of maximal semisimple* $\underline{u}$*-trees and maximal semisimple* $\underline{u}'$*-trees: if* T *is a maximal semisimple* $\underline{u}$*-tree, then* $\phi(T)$ *is the tree obtained by replacing each component* C *of* T *by a component* C' *corresponding to* C *defined as follows.*

(a) If C *has exponent at most* t *then* C' = C.

(b) If C *is a 0-wreath with parameters* [n,k], $k > t$, *then* C' *is a 0-wreath with parameters* [n,k+1].

(c) If C *is a 1-wreath with parameters* [k,m,r] *then* C' *is a 1-wreath with parameters* [k+1,m,r+1] *if* $r > t$ *and* [k+1,m,r] *if* $r \leq t < k$.

(d) If C *is a 2-wreath with parameters* [k,s] *then* C' *is a 2-wreath with parameters* [k+1,s+1] *if* $s > t$ *and* [k+1,s] *if* $s \leq t < k$.

(e) If C *is a 2-wreath with parameters* [k,r,s] *then* C' *is a 2-wreath with parameters* [k+1,r+1,s+1] *if* $s > t$, [k+1,r+1,s] *if* $s \leq t < r$, *and* [k+1,r,s] *if* $r \leq t < k$.

Remarks. 1. In the proof of Theorem 2 an explicit definition of ϕ^{-1} is given.

2. In terms of height distributions, the first part of Theorem 2

states:

Theorem 2'. *Let* $\underline{v} = (v_0,\ldots,v_h)$ *be a height distribution of exponent* $h \geq 1$, *and let* t *be an integer,* $0 \leq t \leq h-1$. *Then* $UB(\underline{v}) = UB(\underline{v}')$ *where* $\underline{v}' = (v_0,\ldots,v_t,v_t,v_{t+1},\ldots,v_h)$

Definition 1.4. If $\underline{u}$ and $\underline{u}'$ are related as in Theorem 2 then we say that $\underline{u} \sim \underline{u}'$ and $\underline{u}' \sim \underline{u}$. If $\underline{u}$ and $\underline{u}^*$ are Hales' distributions for which there exist $\underline{u} = \underline{u}_1 \sim \underline{u}_2 \sim \ldots \sim \underline{u}_k = \underline{u}^*$ we say that $\underline{u}$ is *equivalent* to $\underline{u}^*$, $\underline{u} \equiv \underline{u}^*$; this is clearly an equivalence relation.

The next result uses Theorems 1 and 2 to give a useful step in evaluating $UB(\underline{u})$ for some distributions $\underline{u}$. In particular we derive $UB(\underline{u})$ in the Corollary for all $\underline{u}$ in which the nonzero entries are strictly increasing.

Theorem 3. *Suppose that in the Hales' distribution* $\underline{u} = (u_1,\ldots,u_h)$, $h \geq 1$, u_h *is the unique maximal entry. Then:*

(a) If $\underline{u}$ *is equivalent to* $\underline{u}^* = (1,1,\ldots,1,2,3)$, *where there are* $2t+1$ *1's for some* $t \geq 0$, *then* $UB(\underline{u}) = 2^{t+4}$ *and every maximal semisimple* $\underline{u}^*$*-tree satisfying the conditions of Theorem 1 is the direct sum of a 2-wreath with parameters* $[h,h-1]$ *and* $t+1$ *2-wreaths with parameters* $[k,2,r]$ *for distinct* $k > r$ *in* $\{1,\ldots,h\}$, *(see Figure 4).*

(b) Otherwise, $UB(\underline{u}) = u_h!\, UB(u_1,\ldots,u_{h-1})$ *and there is a maximal semisimple* $\underline{u}$*-tree as in Theorem 1 having a 0-wreath with parameters* $[u_h,h]$ *as a component.*

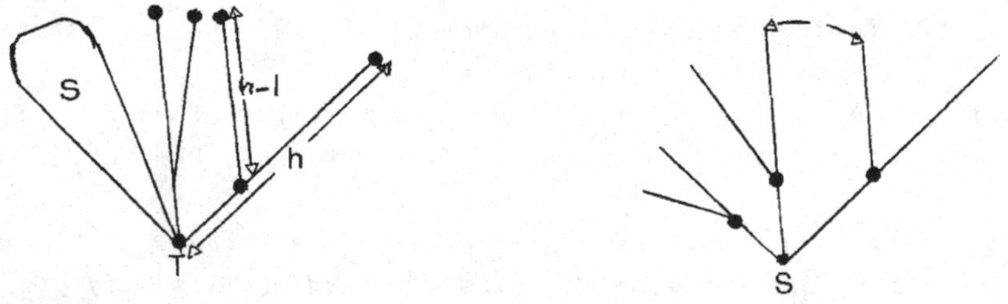

Figure 4.

Corollary. *Suppose that in the Hales' distribution* $\underline{u} = (u_1,\ldots,u_h)$ *of exponent* $h \geq 1$ *the nonzero entries are strictly increasing. Then*

(a) if the first three non-zero entries are $1,2,3$, *then*

$$UB(\underline{u}) = 4/3 \prod_{1\leq i\leq h} u_i!$$

(b) otherwise

$$UB(\underline{u}) = \prod_{1\le i\le h} u_i! \ .$$

This corollary follows immediately from Theorem 3. There remains to solve the problem:

Problem. Find an efficient algorithm for evaluating $UB(\underline{u})$ in terms of the parameter u_i.

We may assume by Corollary 2 that the non-zero entries are not strictly increasing, and hence by Theorem 1 that the maximal semi-simple $\underline{u}$-tree has at least one 1-wreath or 2-wreath as a component. A preliminary question is

Question. For what Hales' distributions $\underline{u}$ do all maximal semi-simple $\underline{u}$-trees (as in Theorem 1) have a 2-wreath as a component?

The next section contains a short discussion of automorphism groups of trees. This is followed by some preliminary results and then the proofs of the results stated in this section.

2. AUTOMORPHISM GROUPS OF TREES

Perhaps it is not out of place to give a brief discussion of the automorphism group of a rooted tree. If a rooted tree T has components $C_1,\ldots,C_r$ then $\mathrm{Aut}\ T$ is the direct product of $\mathrm{Aut}\ C_i$, $i=1,\ldots,r$. If T has just one component, so that T is an (n,k,U)-wreath for some positive integers n and k, and bud U, then $\mathrm{Aut}\ T$ is the wreath product $(\mathrm{Aut}\ U)\ \mathrm{wr}\ S_n$ so that $|\mathrm{Aut}\ T| = |\mathrm{Aut}\ U|^n.n!$. Thus $\mathrm{Aut}\ T$ is obtained by forming direct products and wreath products of the automorphism groups of smaller trees. The "building blocks" are symmetric groups and the simplest rooted trees whose automorphism groups are symmetric groups are the 0-wreaths and 1-wreaths. Theorem 1 shows that for any Hales' distribution $\underline{u}$, we can always find a maximal $\underline{u}$-tree whose automorphism group is a direct product of groups with either all factors or all but one factors being symmetric groups, and possibly one factor being a dihedral group of order 8, $D_8 \simeq S_2\ \mathrm{wr}\ S_2$.

3. PRELIMINARY RESULTS

This section contains some combinatorial results which will be used later.

Lemma 3.1. *Let* $\underline{u}$ *be a Hales' distribution and* T *a maximal* $\underline{u}$*-tree. If* D *is a direct sum of components of* T *then* D *is a maximal* $\underline{u}(D)$*-tree.*

Proof. We note that $UB(\underline{u}) = |Aut\ T| = \Pi|Aut\ C|$ where the product is over all components of T. If D' is any $\underline{u}(D)$-tree, then replacing D by D' in T gives another $\underline{u}$-tree T' with $|Aut\ T'| = UB(u).|Aut\ D'|/|Aut\ D| \le UB(\underline{u})$ so that $|Aut\ D'| \le |Aut\ D|$. Thus D is a maximal $\underline{u}(D)$-tree.

Lemma 3.2. *Let* a,b,c,d *be non-negative integers such that* $a+b = c+d$ *and* $d < b \le c$. *Then* $a!b! \le c!d!$ *with equality if and only if* $b=c$.

Proof. Let $b = d+x$, where $x > 0$, and $c = b+y$ where $y \ge 0$, so that $a = d+y$ and $c!d!/a!b!$ is 1 if $y=0$ and is $\prod_{1\le i\le y} (d+x+i)/(d+i) > 1$ if $y > 0$.

Corollary 3.3. *Let* m *and* n *be positive integers with* $n \ge 2$. *Then* $(m!)^n < (n(m-1))!$ *if and only if* $m+n \ge 6$.

Proof. The result is easily checked for $m+n \le 6$. Assume inductively that N is an integer greater than 6 and that $(m!)^n < (n(m-1))!$ whenever $6 \le m+n \le N-1$. Let $m+n = N$. Then $(m!)^n < ((n-1)(m-1))!m!$, by induction, which is $\le (n(m-1))!$ by Lemma 3.2 with $b = (n-1)(m-1)$, $c = n(m-1)$, $d=1$. By induction $(m!)^n < (n(m-1))!$ whenever $m+n \ge 6$ and the result follows.

Lemma 3.4. *Let* T *be the direct sum of* $n \ge 2$ *1-wreaths with parameters* $[k,m,r]$ *where* $m \ge 2$. *Then* T *is a maximal* $\underline{u}(T)$*-tree if and only if* $m=2$ *and* n *is* 2 *or* 3.

If $n=m=2$ *then* T *is a 2-wreath with parameters* $[k,r]$.

If $m=2$, $n=3$, *then the direct sum of a 0-wreath with parameters* $[2,k]$ *and a 1-wreath with parameter* $[k,4,r]$ *is also a maximal* $\underline{u}(T)$*-tree*.

Proof. Now $|Aut\ T| = (m!)^n n!$. If T^* is the direct sum of two 0-wreaths with parameters $[n(m-1),r]$ and $[k,n]$, then $\underline{u}(T) = \underline{u}(T^*)$ and $|Aut\ T^*| = (n(m-1))!n!$ which by Corollary 3.3 is greater than $|Aut\ T|$ if $n+m \ge 6$. Thus for T to be maximal we must have $n+m \le 5$. If $n=2$, $m=3$ then the direct sum S of a chain of length k and a 1-wreath with parameters $[k,5,r]$ has $|Aut\ S| = 5! > |Aut\ T|$. Thus if T is maximal then $m=2$ and $n=2$ or 3, and conversely for these values T is maximal. Finally if $n=3$, $m=2$ then the direct sum of two chains of length k and a 1-wreath with parameters $[k,4,r]$ also has automorphism group of order $48 = |Aut\ T|$.

Lemma 3.5. *Let* T *be the direct sum of trees* $T_1,\dots,T_n$, $n \ge 1$, *where for each* $i=1,\dots,n$, T_i *is a 2-wreath with parameters* $[k_i,s_i]$

or $[k_i, r_i, s_i]$. *Then* T *is a maximal* $\underline{u}(T)$*-tree if and only if* $n=1$.

Proof. Since it is easily checked that 2-wreaths are maximal, we assume that T is maximal and prove that $n=1$. First we show that the T_i are pairwise non-isomorphic. By Lemma 3.1 each component C of T is a maximal $\underline{u}(C)$-tree. Suppose that C is the direct sum of t 2-wreaths with parameters, $[k,s]$, that is C is the direct sum of 2t 1-wreaths with parameters $[k,2,s]$, so that $|\mathrm{Aut}\ C| = 2^{2t}(2t)!$. Let C' be the direct sum of two 0-wreaths with parameters $[2t,k]$ and $[2t,s]$. Then $\underline{u}(C) = \underline{u}(C')$ and $|\mathrm{Aut}\ C'| = (2t)!^2$. If $t > 1$ then by Corollary 3.3 $|\mathrm{Aut}\ C| < |\mathrm{Aut}\ C'|$, a contradiction to the maximality of C. Hence $t=1$. Suppose now that C is the direct sum of t 2-wreaths with parameters $[k,r,s]$ so that $|\mathrm{Aut}\ C| = 8^t t!$. Let C' be the direct sum of t 1-wreaths with parameters $[k,2,s]$ and t 1-wreaths with parameters $[r,2,s]$. Then $\underline{u}(C) = \underline{u}(C')$ and $|\mathrm{Aut}\ C'| = (2^t t!)^2$. It follows from Corollary 3.3 and the maximality of C that $t \leq 3$. Let C" be the direct sum of t chains of length k, (t-1) chains of length r and a 1-wreath with parameters $[r,2t+1,s]$. Then if t is 2 or 3, $|\mathrm{Aut}\ C''| > |\mathrm{Aut}\ C|$ and $\underline{u}(C) = \underline{u}(C'')$, which is a contradiction. Thus in either case C is a single 2-wreath, and hence $|\mathrm{Aut}\ T| = |\mathrm{Aut}\ T_1| \ldots |\mathrm{Aut}\ T_n|$.

To complete the proof that $n=1$ it is sufficient to show that the direct sum T of two nonisomorphic 2-wreaths is never maximal. The proof is clearest if we consider four cases separately, namely when the 2-wreaths have parameters

(i) $[k,s]$ and $[k',s']$
(ii) $[k,s]$ and $[k',r',s']$
(iii) $[k,r,s]$ and $[k',s']$
(iv) $[k,r,s]$ and $[k',r',s']$,

and we assume that $k \geq k'$. Then $|\mathrm{Aut}\ T| = 64$ in all cases we obtain a $\underline{u}$-tree T' with $|\mathrm{Aut}\ T'| = 72$. This will complete the proof. We take T' to be the direct sum of trees S_1, S_2, and S_3 where S_1 is a 1-wreath with parameters $[k,3,s']$, S_2 is a 1-wreath with parameters $[k,3,s]$ in cases (i) and (ii) and $[r,3,s]$ in cases (iii) and (iv), and S_3 is the direct sum of two chains of length k' in cases (i) and (iii) and a 1-wreath with parameters $[k',2,r']$ in cases (ii) and (iv).

4. PROOF OF THEOREM 1.

The proof is by induction on the number $N = \left(\sum_{1\leq i\leq h} iu_i\right) + 1$ of vertices. If $N=2$ the result is trivially true so assume inductively that $N > 2$ and the result is true for all Hales' distributions on less than N vertices. Let T be a maximal $\underline{u}$-tree. We show first that we may assume that T is a direct sum of i-wreaths, $i=0,1,2$. If T has components $C_1,\ldots,C_r$ where $r \geq 2$ then by induction and Lemma 3.1 we may assume that each C_i is semisimple. If T has only one component so that T is an (n,k,U)-wreath for some n,k,U, then either U is a single vertex as that T is a 1-wreath, or by induction we may assume that U is semisimple. If U has more than one component let C be a component of maximal exponent, say $U = C \oplus D$. Then we replace T by the direct sum T' of n copies of D and an (n,k,C')-wreath where $C' = C$ if C is not a 1-wreath and C' is the bud of C if C is a 1-wreath: T' is a $\underline{u}$-tree, $|\mathrm{Aut}\ T'| \geq |\mathrm{Aut}\ T|$ and as T is maximal so is T'. If U has only one component then, by the definition of a bud U is either a 0-wreath or a 2-wreath and T is a direct sum of 1-wreaths or 2-wreaths respectively.

Thus we may assume that T is a direct sum of i-wreaths, $i=0,1$ or 2. Let C be a component of T. Then using Lemmas 3.4 and 3.5 we may assume that C is an i-wreath, $i=0,1$, or 2. Thus T is semisimple and by Lemma 3.5 at most one component is a 2-wreath.

Suppose that T has a component C which is a 1-wreath with parameters $[k,m,i]$. We show that T may be chosen such that for all such C, $m = u_i+1$. Suppose that i is maximal such that $m < u_i+1$. Then there is a chain of length i in another component D of T where D is one of

(i) a 0-wreath with parameters $[t,i]$, $t \geq 2$.

(ii) a 1-wreath with parameters $[r,s,t]$, $s \geq 2$, where $i=r$ or $i=t$.

(iii) a 2-wreath with parameters $[r,s]$ or $[r,s,t]$ and $i \in \{r,s,t\}$.

We note that $E = C \oplus D$ is a maximal $\underline{u}(E)$-tree by Lemma 3.1. In case (i) let E' be a 1-wreath with parameters $[k,m+t,i]$. Then $\underline{u}(E) = \underline{u}(E')$ and $|\mathrm{Aut}\ E'| = (m+t)! > m!t! = |\mathrm{Aut}\ E|$ by Lemma 3.2, which is a contradiction. In case (ii) if $i=t$ let E' be the direct sum of a 1-wreath with parameters $[k,m+s-1,i]$ and a chain of length s. Then $\underline{u}(E) = \underline{u}(E')$ and $|\mathrm{Aut}\ E'| = (m+s-1)! > m!s! = |\mathrm{Aut}\ E|$ by Lemma 3.2, a contradiction. In case (ii) if $i=r$, let E' be the direct sum of a 1-wreath with parameters $[k,s,t]$ and a 0-wreath with parameters $[m,i]$: then $\underline{u}(E) = \underline{u}(E')$ and $|\mathrm{Aut}\ E| = |\mathrm{Aut}\ E'| = m!s!$. Replacing E by E' (and using Lemma 3.4 if necessary) we get a new maximal $\underline{u}$-tree which is semisimple and such that the bud exponent t of a 1-wreath with parameters $[r,s,t]$ and $s < u_t+1$, is less than i. In case (iv) in the case $[r,s]$ or $[r,s,i]$ there is a tree E' with a 1-wreath with parameters $[k,m+2,i]$ as component and $\underline{u}(E') = \underline{u}(E)$, $|\mathrm{Aut}\ E'| \geq (m+2)! > 8(m!) = |\mathrm{Aut}\ E|$, contradiction. In the case of parameters $[r,s,t]$ where i is r or s there is a tree E' with $\underline{u}(E) = \underline{u}(E')$ with components two 1-wreaths with parameters $[k,m+1,i]$ and $[j,3,t]$ where $\{i,j\} = \{r,s\}$; $|\mathrm{Aut}\ E'| = (m+1)!6 > |\mathrm{Aut}\ E|$, a contradiction.

Thus (if necessary by repeating the replacement process of case (ii)) we may assume that all 1-wreaths have parameters $[k,u_i+1,i]$ for some k,i.

Next let T have a component C which is a 0-wreath with parameters $[w,i]$, $w > 0$. We show that i cannot be a parameter of a 2-wreath. Suppose that T has a component D which is a 2-wreath with parameters $[r,s]$ or $[r,s,t]$ and $i \in \{r,s,t\}$. Let $E = C \oplus D$; then $|\mathrm{Aut}\ E| = w!\,8$. In the case of $[r,s]$ or $[r,s,i]$ there is a $\underline{u}(E)$-tree E' with $|\mathrm{Aut}\ E'| = (w+2)!2 > |\mathrm{Aut}\ E|$, contradiction. In the case of $[r,s,t]$, $i \in \{r,s\}$ there is a $\underline{u}(E)$-tree E' with $|\mathrm{Aut}\ E'| = (w+1)!\,6 > |\mathrm{Aut}\ E|$, contradiction. Thus i is not a parameter of a 2-wreath, so T has $u_i - w$ components which are 1-wreaths of exponent i.

Now suppose that T has a component C which is a 2-wreath with parameters $[r,t]$ or $[r,s,t]$. If $u_t > 2$ there must be a component $D \neq C$ which is a 1-wreath with parameters $[t,u_i+1,i]$ for some $i < t$. Let $E = D \oplus C$; so $|\mathrm{Aut}\ E| = (u_i+1)!8$. The direct sum E' of a 1-wreath with parameters $[r,u_i+1,i]$ and a 1-wreath with parameters $[j,4,t]$, where j is r or s in the first or second case respectively, is a $\underline{u}(E)$-tree with $|\mathrm{Aut}\ E'| = (u_i+1)!4! > |\mathrm{Aut}\ E|$, contradiction. Hence $u_t = 2$. In the case of parameters $[r,s,t]$ assume that $u_s > 1$. Then there must be a component $D \neq C$ which is a 1-wreath with parameters $[s,u_i+1,i]$ for some $i < s$. Again $E = D \oplus C$ has $|\mathrm{Aut}\ E| = (u_i+1)!8$ and the direct sum E' of a 1-wreath with parameters $[r,u_i+1,i]$ and a 2-wreath with parameters $[s,t]$ is a $\underline{u}(E)$-tree with $|\mathrm{Aut}\ E'| = |\mathrm{Aut}\ E|$. So we can always assume that $u_s = 1$ in this case.

Now let I_1 be the set of $i \in \{1,\ldots,h\}$ which are the exponents of buds in the components of T which are 1-wreaths and let I_0 be the complement of I_1 in $\{1,\ldots,h\}$. For $i \in I_0$ let $w_i = 0$ if there is no 0-wreath of exponent i as component, and $w_i = w$ if T has a 0-wreath C_i with parameters $[w,i]$ as a component. If w_i, w_j are nonzero, $i < j$, and $w_i \geq w_j$, then $|\mathrm{Aut}\ C_i \oplus C_j| = w_i!w_j!$, and the direct sum E of a 1-wreath with parameters $[j, w_i+1, i]$ and a 0-wreath with parameters $[j, w_j-1]$ is a $\underline{u}(C_i \oplus C_j)$-tree with $|\mathrm{Aut}\ E| = (w_i+1)!(w_j-1)! > w_i!w_j!$ by Lemma 3.2, a contradiction. Thus $w_i < w_j$ and the proof of Theorem 1 is complete.

5. PROOF OF THEOREM 2

Clearly $\phi(T)$ is a $\underline{u}'$-tree, is semisimple, and $UB(\underline{u}) = |\mathrm{Aut}\ T| = |\mathrm{Aut}\ \phi(T)| \leq UB(\underline{u}')$. We define a map ψ from the set of maximal semisimple $\underline{u}'$-trees to the set of semisimple $\underline{u}$-trees: for a maximal semisimple $\underline{u}'$-tree T', $\psi(T')$ is the tree obtained from T' by replacing each component C' of T' by a component C corresponding to C' defined as follows.

(a) If C' has exponent at most $t+1$ then $C = C'$.

(b) If C' is a 0-wreath with parameters $[n,k]$, $k > t+1$ then C is a 0-wreath with parameters $[n,k-1]$.

(c) If C' is a 1-wreath with parameters $[k,m,r]$ then C is a 1-wreath with parameters $[k-1,m,r-1]$ if $r > t+1$, and $[k-1,m,r]$ if $r \leq t+1 < k$.

(d) If C' is a 2-wreath with parameters $[k,s]$ then C is a 2-wreath with parameters $[k-1,s-1]$ if $s > t+1$, and $[k-1,s]$ if $s \leq t+1 < k$.

(e) If C' is a 2-wreath with parameters $[k,r,s]$ then C is a 2-wreath with parameters $[k-1,r-1,s-1]$ if $s > t+1$, $[k-1,r-1,s]$ if $s \leq t+1 < r$, and $[k-1,r,s]$ if $r \leq t+1 < k$.

Clearly $\psi(T')$ is a semisimple $\underline{u}$-tree and $UB(\underline{u}') = |\mathrm{Aut}\ T'| = |\mathrm{Aut}\ \psi(T')| \leq UB(\underline{u})$. It follows that $UB(\underline{u}) = UB(\underline{u}')$, that $\phi(T)$ is a maximal semisimple $\underline{u}'$-tree, and in fact $\psi = \phi^{-1}$.

6. PROOF OF THEOREM 3

The proof is by induction on the exponent h of $\underline{u}$. If h is 1 or 2 then considering the possible sorts of maximal semisimple trees given by Theorem 1, part (b) is true. Thus assume that $h \geq 3$ and that the result is true for all Hales' distributions of exponent less than h. Let $\underline{u}$ be a Hales' distribution of exponent h in which $u_h = x$ is the unique maximal entry. By induction and Theorem 2 we may assume that the result is true for all Hales' distributions of exponent h which have at least one zero entry, and hence that all the entries u_i of $\underline{u}$ are positive. Let T be a maximal semisimple $\underline{u}$-tree satisfying the conclusions of Theorem 1, and let D be the direct sum of all the components of T of exponent h. Suppose first that $D \neq T$. Then T has a component C of exponent say $t < h$ and by Theorem 1 no chains of length t are in D so that $\underline{u}(D)$ has t^{th} entry 0. Thus Theorem 3 is true for $\underline{u}(D)$. If (b) is true for $\underline{u}(D)$ we can assume that D has a 0-wreath with parameters [x,h] as a component so that (b) is true for $\underline{u}$. If (a) is true for u(D), then (since all components of D have exponent h) $\underline{u}(D) = (w_1,\ldots,w_h) \equiv (1,2,3)$ and D is the direct sum of a 2-wreath with parameters [h,s] and a 1-wreath with parameters [h,2,r], where $w_h = u_h = 3$, $w_r = 1$, $w_s = 2$, $1 \le r < s < h$. As all u_i, $i \le h-1$, are at most 2, the component C is either a 0-wreath with parameters [1,t] or [2,t] or a 1-wreath with parameters [t,2,q] or [t,3,q] for some q. Let $E = D \oplus C$; then E is a maximal $\underline{u}(E)$-tree. If C is a single chain of length t then the direct sum E' of 3 chains of length h, 2 chains of length s and a 1-wreath with parameters [t,2,r] if $t > r$ or [r,2,t] if $t < r$ is a $\underline{u}(E)$-tree with $|\text{Aut } E'| = 24 > 16 = |\text{Aut } E|$ which is a contradiction. If C is the direct sum of two chains of length h then the direct sum E' of 3 chains of length h, 1 chain of length r, and a 2-wreath with parameters [t,s] if $t > s$ or [s,t] if $t < s$ is a $\underline{u}(E)$-tree with $|\text{Aut } E'| = 48 > 32 = |\text{Aut } E|$ contradiction. If C is a 1-wreath with parameters [t,3,q] then the direct sum E' of 3 chains of length h, a 2-wreath with parameters [q,s] if $q > s$ or [s,q] if $q < s$, and a 1-wreath with parameters [r,2,t] if $r > t$ or [t,2,r] if $r < t$ is a $\underline{u}(E)$-tree with $|\text{Aut } E'| = 96 = |\text{Aut } E|$, so that E' is a maximal $\underline{u}(E)$-tree and replacing E by E' in T part (b) is true. Thus we may assume that all components C of exponent say $t < h$ are 1-wreaths with parameters [t,2,q] for some q. If $s \neq h-1$ then there is a component C of exponent $t = h-1$ which is a 1-wreath with parameters [h-1,2,q] for some q. The direct sum E of 3 chains of length h, a 1-wreath with parameters [h-1,3,s] and a 1-wreath with parameters [r,2,q] if $r > q$ or [q,2,r] if $r < q$ is a $\underline{u}(C \oplus D)$-tree and $|\text{Aut } E| = 72 > 32 = |\text{Aut } C \oplus D|$, contradiction. Thus $s = h-1$. If $u_t = 2$ for some $t < h-1$ then there are components C, C' which are 1-wreaths with parameters [t,2,q], and [t,2,q'] for some $q > q'$.

The direct sum E of 3 chains of length h, 1 chain of length r, a 2-wreath with parameters $[h-1,t]$ and a 1-wreath with parameters $[q,2,q']$ is a $\underline{u}(D \oplus C \oplus C')$-tree with $|\mathrm{Aut}\ E| = 96 > 64 = |\mathrm{Aut}\ D \oplus C \oplus C'|$, which is a contradiction. Thus for all $t < h-1$, $u_t = 1$, and as all components of T not in D are 1-wreaths it follows that $\underline{u} = (1,\dots,1,2,3)$ has an odd number say $2t+1$ entries equal to 1 and that $UB(\underline{u}) = |\mathrm{Aut}\ T| = 2^t.|\mathrm{Aut}\ D| = 2^{t+4}$. Thus part (a) is true.

We may now assume that $D = T$, that is, all components of T have exponent h, and we assume that T has more than one component. One of the following holds:

(1) T has $h-1$ components $C_1,\dots,C_{h-1}$ such that C_i is a 1-wreath with parameters $[h,u_i+1,i]$, and if, $x \ge h$, a further component C which is a 0-wreath with parameters $[h,x-h+1]$. Hence $x \ge h-1$.

(2) T has a component C which is a 2-wreath with parameters $[h,s]$ and $h-2$ components $\{C_i \mid 1 \le i < h,\ i \ne s\}$ such that C_i is a 1-wreath with parameters $[h,u_i+1,i]$. Here $x = h$.

(3) T has a component C which is a 2-wreath with parameters $[h,r,s]$ and $h-3$ components $\{C_i \mid 1 \le i < h,\ i \ne r,s\}$ such that C_i is a 1-wreath with parameters $[h,u_i+1,i]$. Here $x = h-2$ and $h \ge 5$, (for $x > u_s = 2$ and so the $h-3$ C_i contain at least 2 chains of length h).

In all cases let T' be the direct sum of the C_i so that $T = T' \oplus C$, and $|\mathrm{Aut}\ T| = \prod_i (u_i+1)!\,|\mathrm{Aut}\ C|$. Define a vector $\underline{y} = (y_1,\dots,y_q)$ by:

in case (1), $y_i = u_i$ for $1 \le i \le q=h-1$,

in case (2), $y_i = u_i$ for $1 \le i < s$, $y_i = u_{i+1}$ for $s \le i \le q=h-2$,

in case (3), $y_i = u_i$ for $1 \le i < s$, $y_i = u_{i+1}$ for $s \le i \le r-2$, $y_i = u_{i+2}$ for $r-1 \le i \le q=h-3$.

Then $\sum_{i\ge 2} y_i \ge 1$ except in case (2) with $h=3$, so apart from that case there is an I, $1 \le I \le q-1$, maximal such that $I \le \sum_{i \ge I+1} y_i = Y$ say. Then $Y \le I + y_{I+1}$ by maximality of I.

Suppose first that case (1) holds. Let S be the direct sum of I 1-wreaths, namely y_q 1-wreaths with parameters $[y_q,y_i+1,i]$ for $1 \le i \le y_q,\dots,y_{I+2}$ 1-wreaths with parameters $[y_{I+2},y_i+1,i]$ for $(\sum_{j \ge I+3} y_j) + 1 \le i \le \sum_{j\ge I+2} y_j$, $y_{I+1} - (Y-I)$ 1-wreaths with parameters

$[y_{I+1}, y_i+1, i]$ for $Y-y_{I+1}+1 \le i \le I$, and also $Y-I$ chains of length $I+1$, and x chains of length h. Then S is a u-tree and $|\mathrm{Aut}\ S| = \prod_{i\le I}(y_i+1)!(Y-I)!x!$. Thus $|\mathrm{Aut}\ T|/|\mathrm{Aut}\ S| = \prod_{i\ge I+1}(y_i+1)!$ $(x-h+1)/(Y-I)!x!$. Now using Lemma 3.2, since $\sum_{i\ge I+1}(y_i+1)+(x-h+1)=(Y-I)+x$ and $Y-I < y_{I+1}+1 \le x$, $(Y-I)!\ x! \ge (\sum_{i\ge I+2}(y_i+1)+x-h+1)!(y_{I+1}+1)!$ with equality if and only if $y_{I+1} = x-1$. Also $\prod_{i\ge I+2}(y_i+1)!(x-h+1)! \le (\sum_{i\ge I+2}(y_i+1)+x-h+1)!$ with equality if and only if $I = h-2$. Thus $|\mathrm{Aut}\ T| \le |\mathrm{Aut}\ S|$, and by maximality of T equality holds so that S is a maximal semisimple u-tree and part (b) is true in case (1).

Suppose next that case (2) is true. Suppose first that $h=3$ so that $x=3=u_3$, $u_s=2$, and u_r is 1 or 2 where $\{r,s\} = \{1,2\}$; thus $|\mathrm{Aut}\ T| = (u_r+1)!8$. If $u_r=2$ then the direct sum S of 3 chains of length 3 and a 2-wreath with parameters $[2,1]$ is a u-tree with $|\mathrm{Aut}\ S| = 48 = |\mathrm{Aut}\ T|$ so that part (b) is true. So assume that $u_r=1$. If $r=2$ then the direct sum S of 3 chains of length 3 and a 1-wreath with parameters $[2,3,1]$ is a u-tree with $|\mathrm{Aut}\ S| = 36 > |\mathrm{Aut}\ T|$, contradiction. Thus $r=1$, $\underline{u} = (1,2,3)$ and part (a) is true.

Now assume that $h \ge 4$. In this case let S be the direct sum of the I 1-wreaths described in the argument for case (1), $Y-I$ chains of length d where $d = I+1$ if $I+1 < s$ or $I+2$ if $I+1 \ge s$ 2 chains of length s, and x chains of length h. Then S is a u-tree, and $|\mathrm{Aut}\ T|/|\mathrm{Aut}\ S| = \prod_{i\ge I+1}(y_i+1)!8/(Y-I)!x!2$. Using Lemma 3.2, since $(\sum_{i\ge I+1}(y_i+1))+1 = Y+h-3-I+1 = (Y-I)+x$ and $Y-I < y_{I+1}+1 \le x$, $(Y-I)!x! \ge (y_{I+1}+1)!\ (\sum_{i\ge I+2}(y_i+1)+2)!$ with equality if and only if $y_{I+1} = x-1$. If $I \le h-4$ then $(\sum_{i\ge I+2}(y_i+1)+2)! \ge 12\{(\sum_{i\ge I+2}(y_i+1))!\}$ $\ge 12\prod_{i\ge I+2}(y_i+1)!$ and it follows that $|\mathrm{Aut}\ T| < |\mathrm{Aut}\ S|$, contradiction. Thus $I = h-3$, so by definition of I, $x-1 \ge y_{h-2} = Y \ge I = h-3 = x-3$, and $|\mathrm{Aut}\ T|/|\mathrm{Aut}\ S| = (y_{h-2}+1)!4/(Y-I)!x!$. It follows that $y_{h-2} \ne x-3$, and if $y_{h-2} = x-2$ then $x=h=4$, and $\underline{u}$ is $(y_1,2,2,4)$ or $(2,y_1,2,4)$, and $|\mathrm{Aut}\ T| = (y_1+1)!48$. In this case the direct sum S' of 4 chains of length 4, and two 1-wreaths with parameters, $[3,y_1+1,i]$ and $[3,3,j]$ where (i,j) is $(1,2)$ or $(2,1)$ respectively, is a u-tree with $|\mathrm{Aut}\ S'| = (y_1+1)!144 > |\mathrm{Aut}\ T|$, a contradiction. Thus $y_{h-2} = x-1$ so that $Y-I = 2$. In this case replacing the $Y-I = 2$ chains of length d and 2 chains of length s in S by a 2-wreath with parameters $[d,s]$ or $[s,d]$ gives a new u-tree S' with $|\mathrm{Aut}\ S'| = 2|\mathrm{Aut}\ S|$ so that $|\mathrm{Aut}\ S'| = |\mathrm{Aut}\ T|$ and part (b) is true.

TABLE 3

	R_1	R_2	R_3	R_4	R_5	R_6	R_7
1	21	24	53	52	61	51	71
2	31	34	42	62	32	72	22
3	41	64	73	43	43	33	44
4	54	74	63	14	14	11	13

Applying the definition in §5 of incidence between fixed elements in our partially transitive plane constructible from D_3 we obtain the following incidences for the Q_i and K_j

K_1	K_2	K_3	K_4	K_5	K_6	K_7
1	1	1	3	2	2	3
6	4	2	4	5	4	5
7	5	3	7	7	6	6

where Q_i is denoted by i for convenience. These incidences yield precisely the incidences of the plane $\bar{\pi}$ of order 2 upon which we took $G = PSL(3,2)$ to be acting in its natural manner. The fact that we have retrieved $\bar{\pi}$ essentially verifies planarity conditions (a) and (c). As a specimen verification of planarity condition (b) take $a = (12)(56)$, $i=1$ and $j=2$. From the last table we have $EL_{11} = ER_{21}$, $EL_{21} = ER_{24}$. Also $EL_{11}a = EL_{21}$ is not hard to verify. So subgroup R_2 works in condition (b). The only other possibility is $k=3$. ($EL_{12} = ER_{31}$ and $EL_{22} = ER_{34}$ from the table and the other cases $EL_{13} = ER_{41}$, $EL_{23} = ER_{64}$ and $EL_{14} = ER_{54}$, $EL_{24} = ER_{74}$ don't work.) But $EL_{12}a \neq EL_{22}$ since, for example, $b = (132)(576) \in EL_{12}$ and $ba = (23)(67) \notin EL_{22}$.

The work that we have done considerably clarifies what the planarity conditions mean and how to work with them. This may prove of significance in any work on planar partial difference sets of types $(4,m)$ and $(5,m)$ because the planarity conditions in those cases are quite similar. There are however, infinitely many planes of each of these types known.

Johnson and Ostrom [9] and independently Walker [16] have established that if π is a translation plane of order 16 which is tangentially transitive relative to a Fano subplane, then π is the unique known translation plane with this property. Also, Jha [4] has shown that a translation plane π that is tangentially transitive

ON PARTIALLY TRANSITIVE PLANES OF HUGHES TYPE (6,M)

ALAN RAHILLY AND DAVID SEARBY

1. INTRODUCTION

In his paper [3] Hughes considers the following type of projective plane π: π has a collineation group G which acts sharply transitively on the set of points of π not incident with any line of the fixed substructure $\pi(G) = \pi'$ of G and on the set of lines of π not incident with any point of π'. The elements of these two sets Hughes calls "ordinary points" and "ordinary lines" respectively. The remaining elements (points and lines) of π divide naturally into those of π' (fixed elements) and those incident with a unique element of π' (tangent elements). Hughes calls such a plane "partially transitive". The fixed substructure π' is easily shown to be one of the following types:

(0) π' is empty,

(1a) π' consists of an incident point/line pair,

(1b) π' consists of a non-incident point/line pair,

(2) π' consists of two points Q_0 and Q_1 and two lines K_0 and K_1, where $K_0 = Q_0Q_1$, and Q_0 is on K_1,

(3) π' consists of three non-collinear points Q_i, $i=0,1,2$, and the three lines $K_0 = Q_1Q_2$, $K_1 = Q_0Q_2$, and $K_2 = Q_0Q_1$,

(4,m) π' consists of m points $(m \geq 3)$ Q_i, $i=1,2,\dots,m$ on a line K_0, a point Q_0 not on K_0, and the $m+1$ lines K_0, $K_i = Q_0Q_i$,

(5,m) π' consists of $m+1$ points $(m \geq 2)$ Q_i, $i=0,1,\dots,m$ on a line K_0 and $m+1$ lines K_i, $i=0,1,\dots,m$ through Q_0,

(6,m) π' is a non-degenerate subplane of order m with points Q_i, lines K_i, $i=0,1,\dots,m^2+m$.

It is the aim of this paper to investigate partially transitive planes of type (6,m) and in particular to establish the following theorem:

<u>Theorem</u>. *There are no planes of Hughes type* (6,p) *where* p *is a prime greater than* 3.

In [3] Hughes mentions that he knows of no examples of planes of type (6,m). Recently a plane of order 16 has been constructed independently by two authors (Lorimer [5] and Rahilly [11]). It has been pointed out by Ostrom and Johnson (see [9]) that this plane is partially transitive of Hughes type (6,2). The partially transitive group G of this plane is isomorphic to PSL(3,2) and acts regularly on the set of 168 ordinary points and on the set of 168 ordinary lines with respect to a subplane of order 2.

Hughes [3] has shown that a partially transitive group of whatever type contains subsets which satisfy a difference condition with respect to certain subgroups. He calls these sets "partial difference sets". Such partial difference sets satisfy certain extra conditions which we shall call "the planarity conditions". Hughes showed that a partially transitive plane can be reconstructed from a partial difference set in its partially transitive group in an analogous manner to the way in which a cyclic projective plane can be reconstructed from one of its cyclic difference sets. Given a group G of an appropriate order containing a collection of subgroups of appropriate orders and a partial difference set relative to this collection of subgroups satisfying the planarity conditions a projective plane can be constructed on which G acts partially transitively relative to its fixed substructure π'. In the case of Hughes type (6,m) the group G has order $(m^4-m)(m^4-m^2)$ for some $m \geq 2$. The relevant subgroups are m^2+m+1 in number, are mutually group-disjoint and are each of order m^4-m^2. The partial difference sets contain m^4-m^2-m elements. Furthermore, the partially transitive plane π of Hughes type (6,m) is of order m^4 and the partially transitive group G elementwise fixes a subplane π' of π of order m. In the next section we shall recapitulate these results of Hughes in more detail. We remark that in §6 we describe an exhaustive computer search for planar partial difference sets of type (6,2) in PSL(3,2). In this case the relevant subgroups are a collection of 7 conjugate and mutually group-disjoint subgroups of PSL(3,2) isomorphic to A_4 and a partial difference set must contain 10 elements.

2. PRELIMINARY RESULTS

Following Hughes [3] let P_0 be an arbitrary ordinary point, and J_0 an arbitrary ordinary line of the partially transitive plane π. Let D be the set of all $d \in G$ such that P_0d is on J_0. Denote the tangent line Q_iP_0 by L_i and the tangent point $K_i \cap J_0$ by R_i. Also, let $\mathcal{L}_i$ be the stabilizer of L_i in G and $\mathcal{R}_i$ be the stabilizer of R_i in G. Then it is possible to establish the following results:

Lemma 1. (Hughes [3], Lemma 2). *If* Q_i *is not on* K_j *then* L_i *and* R_j *are conjugates in* G*; thus, if there is a line of* π' *which does not contain either* Q_i *or* Q_j*, then* L_i *and* L_j *are conjugates in* G.

Theorem 1. (Hughes [3], Theorem 5).

(a) If $g \in G$ *and* $g \notin L_i$ *for any* i, *then* $g = d_1 d_2^{-1}$ *for a unique ordered pair* $d_1, d_2 \in D$*; if* $g \in L_i$ *for some* $i, g \neq 1$*, then* $g \neq d_1 d_2^{-1}$ *for any* $d_1, d_2 \in D$.

(b) If $g \in G$, $g \notin R_i$ *for any* $i, g = d_1^{-1} d_2$ *for a unique ordered pair* $d_1, d_2 \in D$*; if* $g \in R_i$ *for some* $i, g =/ 1$, *then* $g \neq d_1^{-1} d_2$ *for any* $d_1, d_2 \in D$.

(c) $R_i \cap R_j = L_i \cap L_j = 1$ *if* $i \neq j$.

The set D we shall refer to as a *partial difference set* with respect to the L_i and R_j. Clearly, every partially transitive plane gives rise to a partial difference set in a group. Also, as Hughes shows in [3], a partial difference set satisfying certain extra "planarity" conditions on the subgroups L_i and R_j will give rise to a partially transitive plane. The extra conditions vary somewhat for the various types of partially transitive plane. In the case of Hughes type (6,m) we shall investigate those conditions more closely in §5.

From now on we shall restrict attention to partially transitive planes of type (6,m). Simple counting arguments show that if n is the order of π then there are $(n-m)(n-m^2)$ ordinary points, $n-m^2$ ordinary lines through each tangent point, $n-m^2$ ordinary points on each tangent line, and $n-m-m^2$ ordinary points on each ordinary line. It follows that the order of G is $(n-m)(n-m^2)$, the order of the L_i and R_j is $n-m^2$ for all i,j, and the order of D is $n-m-m^2$.

Theorem 3. (Hughes [3], Theorem 22). *The order of the normalizer of* R_i *is* $(m^2-m)(n-m^2)$ *and* $q = (n-m)/(m^2-m)$ *is an integer. Any pair of the* q *distinct conjugates of* R_i *intersect in the identity. For each conjugate of* R_i *there is a subplane of* π *of order* m^2 *containing* π'*, which is fixed element-wise by each element of the conjugate. Distinct conjugates fix distinct subplanes.*

Theorem 4. (Hughes [3], Theorem 23). *In type* (6,m), $n=m^4$.

It should be noted that in the case of type (6,m), Lemma 1 implies that the set $\{L_i\} \cup \{R_j\}$ is a set of conjugate subgroups of G. Hughes,s argument to establish Theorem 4 actually shows that every conjugate of R_i is in $\{R_j\}$. Consequently, $\{L_i\} = \{R_j\}$.

Note that Theorems 3 and 4 imply that $q = m^2+m+1$, that the subplane π_i fixed elementwise by R_i is a Baer subplane of π (that is, the elements of R_i are Baer collineations) and that the order of each R_i is m^4-m^2. It follows that each subgroup R_i acts tangentially transitively (Jha [4]) relative to its fixed Baer subplane π_i. Note also that π' is a Baer subplane of each of the m^2+m+1 subplanes π_i and that $\pi_i \cap \pi_j = \pi'$ for all $i \neq j$.

Finally, we shall need the following theorem on permutation groups.

Theorem 5. (Tsuzuku [15], Theorem 2). *Let* G *be a doubly transitive group acting on a set* Ω *of* $1+p+p^2$ *points, where* p *is prime. If the order of* G *is divisible by* p^3 *exactly, then*
(a) $p = 2$, *and* G *is the alternating group* A_7 *of degree* 7, *or*
(b) Ω *is identified with the desarguesian plane over the field of* p *elements, and* G *is isomorphic to a collineation group of the plane which contains the projective special linear group* PSL(3,p).

3. GENERALIZED HALL PLANES AND PLANES OF TYPE (6,m)

A *generalized Hall plane* is a plane π with the following properties:

(1) π is a translation plane, and

(2) π possesses a Baer subplane π_0 and a collineation group $G(\pi_0)$ which acts tangentially transitively on π_0.

If π is a finite generalized Hall plane of order greater than four then there is a unique translation axis ℓ_∞ (say) and ℓ_∞ is contained in π_0. Furthermore, it is not difficult to show that π_0 is desarguesian (see [12], Theorem 1, corollary).

A finite generalized Hall plane may be (Hall) coordinatized over a quadrangle O,I,X,Y in π_0, such that $XY = \ell_\infty$, by a (right) quasifield F with the properties

(a) F is a right vector space of dimension two over the subfield F_0 coordinatizing π_0,

(b) multiplication on the right by an element of F_0 is scalar multiplication in the vector space of (a), and multiplication on the right for any $z \in F\backslash F_0$ is given by

$$(z\alpha+\beta)z = z(f(\alpha)+h(\beta)) + g(\alpha) + k(\beta)$$

for all $\alpha, \beta \in F_0$, where f,g,h and k are endomorphisms of $(F_0,+)$ satisfying

(i) $h(1) = 1$, $k(1) = 0$,

(ii) $h \in \mathrm{Aut}(F_0,+)$,

(iii) $M_\lambda = g + (k-m_\lambda)h^{-1}(m_\lambda-f) \in \mathrm{Aut}(F_0,+)$ for all $\lambda \in F_0$, where m_λ denotes the field endomorphism $m_\lambda : F_0 \to F_0 : x \to \lambda x$ of $(F_0,+)$.

Such a quasifield F is called a *generalized Hall system* and the endomorphisms f,g,h and k the *defining functions* of F. Given endomorphisms f,g,h and k of a finite field F_0 satisfying conditions (i), (ii) and (iii) we can construct a generalized Hall system and hence a finite generalized Hall plane in an obvious way.

In [11] a generalized Hall system of order 16 is constructed with the defining functions

$$f = \begin{bmatrix} 0 & 0 \\ 0 & 1 \end{bmatrix}, \quad g = \begin{bmatrix} 1 & 0 \\ 0 & 1 \end{bmatrix} = h, \quad k = \begin{bmatrix} 1 & 0 \\ 0 & 0 \end{bmatrix}$$

being endomorphisms of $(F_0,+)$, where $F_0 = GF(4)$ the matrices being defined with respect to the basis $\{t,1\}$ of the vector space $GF(4)$ over $GF(2)$, where $t^2 = t+1$. Recently Johnson and Ostiem [9], p. 17, have pointed out that the plane coordinatized by this generalized Hall system is of Hughes type $(6,2)$. This plane and its dual are the only known planes of Hughes type $(6,m)$. For an interesting alternative construction of this plane see Lorimer [5].

Suppose π is a finite translation plane which is of Hughes type $(6,m)$. Our earlier remarks indicate that π will be tangentially transitive relative to m^2+m+1 Baer subplanes π_i of π which mutually intersect precisely on the fixed substructure π' of the partially transitive group. Now either all lines are a translation axis in which case π is $PG(2,4)$ (the only finite desarguesian plane which is tangentially transitive with respect to a Baer subplane. But $PG(2,4)$ is not of type $(6,m)$. So there is a unique translation axis which must belong to π' and also to each of the Baer subplanes π_i. Clearly π is a finite generalized Hall plane and the subplanes π_i and π' are desarguesian.

<u>Theorem 6</u>. *If π is a translation plane of Hughes type* $(6,m)$, *then π is of order* 16 *and is the generalized Hall plane previously defined.*

<u>Proof</u>. Consider the Baer subplane π_0 of π fixed elementwise by the subgroup R_0 of G. The subgroup of R_1 which fixes π_0 induces a group of collineations on π_0 which is tangentially transitive with respect to the subplane π' of π_0. But π_0 is a desarguesian plane and so must be of order four. So π is of order 16 and it then follows, by Johnson and Ostrom [9] (Theorems 2.5 and 4.1) that π is the generalized Hall plane defined above.

Remarks. (i) A plane π of Hughes type $(6,m)$ is of order m^4 and 6 acts tangentially transitively relative to the subplane π' of order m. If π is a translation plane, then it follows from a result of Jha [5] that π is a generalized Hall plane of order 16.

(ii) If π is a finite generalized Hall plane of order greater than 16 with translation axis ℓ_∞ and tangentially transitive Baer subplane π_0, then the point set $\pi_0 \cap \ell_\infty$ is fixed by all collineations of π (see Rahilly [11]).

(iii) Either of the results (i) and (ii) can be used to provide an alternative proof that a translation plane of Hughes type $(6,m)$ is a generalized Hall plane of order 16. The results of Johnson and Ostrom [9] used in the proof of Theorem 6 can then be applied to obtain Theorem 6.

4. THE NON-EXISTENCE OF PLANES OF TYPE $(6,p)$, $p > 3$.

In this section we shall consider planes of type $(6,p)$, where p is a prime number, and we shall prove:

Theorem 7. *If* π *is a projective plane of type* $(6,p)$, *where* p *is a prime, then* $p=2$ *or* 3, *and* G *is respectively* $PSL(3,2)$ *and* $PSL(3,3)$.

Proof. In this case the group G is of order $(p^4-p)(p^4-p^2)$. Let π_i denote the Baer subplane fixed pointwise by R_i, and Ω the set $\{\pi_i | i=0,\dots,p^2+p\}$. When π_i is considered as an element of Ω it will be denoted by i. Further, if $K < G$ we shall denote by $\bar{K}$ the permutation group induced on Ω, and ϕ will denote the natural homomorphism from G onto $\bar{G}$.

The proof of the theorem is divided into eight parts.

Part 1. $\bar{G}$ is doubly transitive on Ω.

R_i is regular on the set of points $\ell - \pi_i$ of π, where ℓ is any line of π'. Clearly, $\bar{R}_i$ is transitive on $\Omega - \{i\}$ and fixes i. It follows that $\bar{G}$ is doubly transitive.

Part 2. There is a unique subgroup $C_{i,j}$ of order p in R_i which fixes π_i, where $i \neq j$; furthermore, $C_{i,j} \triangleleft G_{i,j}$ (the stabilizer of π_i and π_j in G).

The order of the subgroup of R_i fixing π_j is $p(p-1)$. Such a group must have a unique Sylow p-subgroup $C_{i,j}$ of order p. Also, for any $x \in G_{i,j}$, $xC_{i,j}x^{-1}$ is a subgroup of R_i of order p fixing π_j, and so $xC_{i,j}x^{-1} = C_{i,j}$ for all $x \in G_{i,j}$.

Part 3. There is no element of order p in $\ker(\phi) \cap R_i$.

Suppose x_i is such an element. Then, since R_i and R_j are conjugate in G, there is an element $x_j \in R_j$ of order p in R_j $(i \neq j)$. Consider the group J generated by x_i and x_j. Clearly $\langle x_i \rangle \triangleleft J$ and $\langle x_j \rangle \triangleleft J$ and $\langle x_i \rangle \cap \langle x_j \rangle = 1$, so that J is elementary abelian of order p^2. Now J fixes each $\pi_k \in \Omega$ and each subgroup of order p of J fixes at most one subplane π_k pointwise. Since J possesses exactly $p+1$ subgroups of order p there is some subplane π_k which is not fixed pointwise by any element of J. Clearly, J permutes the points of π_k in orbits of length p^2. But $\ell \cap \pi_k$ is a union of orbits of J for any line ℓ of π', and $|\ell \cap \pi_k - \pi'| = p^2-p$. Thus we have a contradiction and so there is no element of order p in $\ker(\phi) \cap R_i$, for each i.

Part 4. $|\bar{G}_{i,j}|$ is divisible by p.

There is a unique subgroup $C_{i,j}$ of order p of R_i which fixes π_j, $i \neq j$. Also $C_{i,j} \cap \ker(\phi) = 1$ by Part 3.

Part 5. $|G|$ is divisible by p^3 exactly.

Now $|\bar{G}| = (p^2+p+1)(p^2+p)|\bar{G}_{i,j}|$. We show that $|\bar{G}_{i,j}|$ is divisible by p^2 exactly. To do this it is sufficient to show (Part 4) that $|\bar{G}_{i,j}|$ is not divisible by p exactly. Consider ϕ_i, the natural homomorphism of $\ker(\phi)$ onto π_i. Now $\mathrm{im}(\phi_i)$ is a group of collineations of π_i fixing π' pointwise and so $|\mathrm{im}(\phi_i)| \leq p(p-1)$. Also $\ker(\phi_i)$ fixes π_i pointwise and fixes π_j, $i \neq j$, and so $|\ker(\phi_i)| \,\big|\, p(p-1)$. But there is no element of order p in $\ker(\phi_i)$ (Part 3) and so $|\ker(\phi_i)| \,\big|\, p-1$. We conclude that $|\ker(\phi)|$ is divisible by p exactly or not by p at all. It follows that $|G_{i,j}|$ is divisible by at most p^2. But $G_{i,j}$ has $C = C_{i,j} \times C_{j,i}$ as a (normal) subgroup of order p^2. So $|G_{i,j}|$ is divisible by p^2 exactly and C is a normal Sylow p-subgroup of $G_{i,j}$. Now $G_{1,n}$ (the stabilizer of π_1 and π_n in G) is conjugate to $G_{i,j}$ in G since $\bar{G}$ is doubly, transitive on Ω. So, $G_{1,n}$ has a unique subgroup of order p^2. Now $C_{j,i}$ normalizes $C_{i,j}$ and so permutes the Baer subplanes π_k fixed by $C_{i,j}$. But $C_{i,j}$ and $C_{j,i}$ both fix π_i and π_j and so they share at least $p+1$ fixed subplanes π_k and so C fixes at least $p+1$ Baer subplanes in Ω.

Now denote by Γ the subset of Ω which consists of precisely those Baer subplanes fixed by C. We have that C and $C_{1,n} \times C_{n,1}$ are of order p^2 and fix π_1 and π_n, if $\pi_1, \pi_n \in \Gamma$. It follows that $C = C_{1,n} \times C_{n,1}$. Each subgroup $C_{r,s}$ of C fixes π_1 pointwise.

Clearly C possesses at least $p+1$ such subgroups (one for each $r \in \Gamma$ But C possesses exactly $p+1$ subgroups of order p. It follows that $C_{r,s} = C_{r,t}$ for all $s,t \in \Gamma$ and, denoting $C_{r,s}$ by C_r, we have $C = \bigcup_{r\in\Gamma} C_r$. Thus each non-identity element of C is in one of the subgroups R_r. But this means $\ker(\phi) \cap G = 1$ and so $C \cong \bar{C}$. But $\bar{C} < \bar{G}_{i,j}$ and so $|\bar{G}_{i,j}|$ is divisible by p^2, which is a contradiction.

Part 6. Ω may be identified with the desarguesian plane σ of order p and $\bar{G}$ is isomorphic to a collineation group of σ which contains $PSL(3,p)$.

This part follows immediately from Part 5 and Theorem 5.

Part 7. $\bar{R}_i$ is a normal subgroup of order $p^2(p^2-1)$ of $\bar{G}_i = \phi(G_i)$ which is transitive on $\Omega - \{i\}$, where G_i is the stabilizer of π_i in G.

If $x_i \in \ker(\phi) \cap R_i$ and $x_i \neq 1$, then there is $x_k \in \ker(\phi) \cap R_k$, $x_k \neq 1$, for all k, because R_i and R_k are conjugate in G. Since $x_k \neq x_i$ if $k \neq i$ we would then have $|\ker(\phi)| > p^2+p+1$, which is impossible. Hence we have $\ker(\phi) \cap R_i = 1$. Now R_i is normal in the stabilizer G_i of R_i in G and so $\bar{R}_i$ is normal in $\bar{G}_i$.

Part 8. The prime number p is either 2 or 3.

We establish this by showing that a collineation group $\bar{G}$ of σ containing $PSL(3,p)$ possesses a subgroup $\bar{R}_i$ as in the statement of Part 7 only if $p=2$ or 3.

Consider σ homogeneously coordinatized by $GF(p)$ in such a way that i has coordinates $\begin{pmatrix}0\\0\\1\end{pmatrix}$. The elements of the group $\bar{G}_i$ can be represented by non-singular matrices of the form $\begin{pmatrix}a&b&0\\c&d&0\\e&f&1\end{pmatrix}$ over $GF(p)$. Now consider the homomorphism $\chi:\bar{G}_i \to GL(2,p):\begin{pmatrix}a&b&0\\c&d&0\\e&f&1\end{pmatrix} \to \begin{pmatrix}a&b\\c&d\end{pmatrix}$. Note that $SL(2,p)$ is a subgroup of $\chi(\bar{G}_i)$. The group $\bar{N}_i = \ker(x) \cap \bar{R}_i$ is a normal subgroup of $\bar{R}_i$ and so $\bar{R}_i$ permutes the fixed points of $\bar{N}_i$. But $\bar{R}_i$ is transitive on the set $\Omega - \{i\}$ and so the set of fixed points of $\bar{N}_i$ is Ω or simply $\{i\}$. In the former case $\chi(\bar{R}_i)$ is a subgroup of $GL(2,p)$ of order $p^2(p^2-1)$, which is impossible. So $\bar{N}_i$ fixes only i.

Now $\bar{N}_i$ consists of matrices of the form $\begin{pmatrix}1&0&0\\0&1&0\\e&f&1\end{pmatrix}$. The group of all such matrices is elementary abelian of order p^2. Consequently $\bar{N}_i \cong C_p$ (the cyclic group of order p) or $C_p \times C_p$. In the former case, if $\begin{pmatrix}1&0&0\\0&1&0\\e&f&1\end{pmatrix} \in \bar{N}_i - \{1\}$, then $\bar{N}_i$ fixes all points $\begin{pmatrix}x\\y\\z\end{pmatrix}$ such that $ex + fy = 0$. Since $\bar{N}_i$ fixes only i we have that $\bar{N}_i \cong C_p \times C_p$ and so $|\chi(\bar{R}_i)| = p^2-1$.

Thus $\chi(\bar{R}_i)$ is a normal subgroup of order p^2-1 of the subgroup $\chi(\bar{G}_i)$ of $GL(2,p)$ containing $SL(2,p)$. But $GL(2,p)$ has such a subgroup only if $p=2$ or 3 ($GL(2,2)$ is isomorphic to the symmetric group on three letters and has a normal subgroup of order 3, and $GL(2,3)$ is of order 48 and has a normal quaternion subgroup of order 8).

If $p=2$ or 3 then $\bar{G}$ which is respectively $PSL(3,2)$ or $PSL(3,3)$ is clearly isomorphic to G and this completes the proof of the theorem.

Note that the proof of Theorem 7 shows that if $p=2$, then each R_i is isomorphic to the group of 3×3 matrices over $GF(2)$ of the form $\left[\begin{array}{c|c} A & \underset{\sim}{0} \\ \hline \underset{\sim}{b}^T & 1 \end{array}\right]$, where $\underset{\sim}{0}$ is the 2×2 zero column vector over $GF(2)$, $\underset{\sim}{b}$ is a 2×2 column vector over $GF(2)$ and $A \in \{\binom{10}{01}, \binom{11}{10}, \binom{01}{11}\}$. Thus each R_i is isomorphic to the subgroup of collineations of $PG(2,2)$ consisting of the elations fixing a particular point P and all collineations of order 3 fixing P. So each R_i is isomorphic to A_4 (the alternating group on four letters). The collineation group of $PG(2,2)$ is, of course, $PSL(3,2)$. Note that the involutions and elements of order 3 form a complex of 7 group-disjoint subgroups each isomorphic to A_4. To find all planes of Hughes type $(6,2)$ it is sufficient to determine all partial difference systems (G,R_i,D), where $G = PSL(3,2)$ and the R_i are the 7 subgroups of $PSL(3,2)$ just mentioned. We shall turn to this problem in §6.

5. PARTIAL DIFFERENCE SETS OF TYPE (6,m)

We shall now turn to a consideration of the partial difference sets a plane of Hughes type $(6,m)$ gives rise to. A type $(6,m)$ partial difference set D in a group G is a set of distinct left and right coset representatives in G for each of the subgroups R_i of G, that is, D is a partial transversal of the right and left coset decomposition in G of each R_i. A coset of R_i in G which contains no element of D will be called an *extra coset* relative to D. Since the order of G is $(m^4-m)(m^4-m^2)$, the order of each R_i is m^4-m^2 and the order of D is m^4-m^2-m there are m^2 extra left (and right) cosets of each R_i relative to D.

In [3] Hughes does not explicitly state the planarity conditions for a partial difference set of type $(6,m)$. However on p. 673 he remarks that they are obtained by a slight modification of the planarity conditions for a partial difference set of type (4,m) or (5,m) which he had discussed previously. We shall state the planarity conditions in a somewhat different style to the one adopted by Hughes in the cases he does explicitly state (for example, type (4,m)) in order to emphasise that the planarity conditions essentially connect the left and right coset partitions of the subgroups R_i and, in particular, the extra left and right cosets of the R_i relative to D. In the planarity conditions we shall, contrary to Hughes [3] and our practice so far, refer to subgroup R_i, $i=1,\dots,m^2+m+1$, rather than R_i, $i=0,\dots,m^2+m$. The planarity conditions for type (6,m) are:

(a) Each of the extra left cosets relative to D is an extra right coset relative to D. Furthermore the m^2 R_j's for which there is an extra right coset equal to an extra left coset of a given R_i are all different.

(b) Each R_j is a partial transversal of the right coset decomposition of R_i in G, where $i \neq j$. Further, if $|R_i a \cap R_j| = 0$ then there is a unique subgroup R_k, a unique extra left coset gR_i and a unique extra left coset hR_j such that $gR_i = R_k g$, $hR_j = R_k h$ and $(gR_i)a = hR_j$.

(c) For each R_i, R_k, where $i \neq k$, there is a unique R_j such that there is no extra right coset of R_i nor of R_k equal to an extra left coset of R_j.

Remarks. (i) Clearly each of the conditions in (a) implies its "dual" statement obtained by interchanging the words "left" and "right".

(ii) Hughes' planarity conditions for type (4,m) (p. 666 of [3]), modified as he suggests on p. 673 of [3], so as to be suitable for type (6,m), come down to the first part of (a) and to (b). Now type (6,m) differs from all the other Hughes types in that, should we wish to construct a partially transitive plane from a type (6,m) partial difference set, the incidence between the fixed elements Q_i and K_j does not follow purely from the subscripts. In fact we need to define the incidences between these elements and ensure that a subplane of order m results. The second part of (a) and also (c) are included to ensure this.

We shall refer to a group G together with appropriate subgroups R_i and a partial difference set D satisfying conditions (a), (b) and (c) as a *partial difference system of type (6,m)* and denote it by (G,R_i,D). We shall also refer to the partial difference set D in a partial difference system (G, R_i, D) as a *planar partial difference set.*

To construct a projective plane from a partial difference system of type $(6,m)$ one proceeds as follows: Points are elements a of G (denoted by (a)) or right cosets of R_i in G (dencted by $(R_i a)$ or Q_i, where $i=1,\dots,m^2+m+1$. Lines are right "pseudocosets" Db of D in G (denoted by $[Db]$) or right cosets of R_j in G (denoted by $[R_j b]$) or K_j, where $j=1,\dots,m^2+m+1$. Incidence is specified as in the following table and below it.

	(a)	$(R_i a)$	Q_i
[Db]	$a \in Db$	$b \in R_i a$	*
$[R_j b]$	$a \in R_j b$	see below	i=j
K_j	*	i=j	see below

Incidence between $[R_j b]$'s and $(R_i a)$'s is defined as follows: Firstly, $(R_i a)$ is incident with $[R_j b]$ only if there is an extra left coset gR_i relative to D which is identical to an extra right coset $R_j g$ relative to D. Further, under this circumstance $(R_i a)$ is incident with $[R_j b]$ if and only if $ga \in R_j b$.

Non-incidence (and hence incidence) between the Q_i's and K_j's is defined as follows: K_j is non-incident with Q_i if and only if there is an extra left coset of R_j and an extra right coset of R_i which are identical.

The specification of incidence between $(R_i a)$ and $[R_j b]$ can easily be shown to be independent of the coset representative g. It is a lengthy process to verify that points, lines and incidence thus specified yields a partially transitive projective plane of Hughes type $(6,m)$ with G as the partially transitive group and π' consisting of the Q_i and K_j. It is only necessary to verify that there are m^8+m^4+1 points, m^4+1 points on each line and each pair of points is on a unique line (Dembowski [2], p. 138). This can be done using the sort of arguments Hughes ([3], p. 661) applies to the case of type 2. Of course, each of the planarity conditions must be applied at some stage of this verification.

Before proceeding to consider partial difference systems in PSL(3,2) we explain the significance of planarity conditions (a) and (c). Conditions (c) and the second part of (a) are included in order to be able to show that the Q_i and K_j form a plane of order m with the previously specified incidence. Clearly there are m^2+m+1 points Q_i. The second part of condition (a) yields that $m+1$ of the Q_i lie on a particular K_j and condition (c) yields that each distinct pair of points Q_i, Q_k lies on a unique line K_j. By Dembowski [2], p. 138, this means that the Q_i and K_j form a plane π' of order m.

6. PLANAR PARTIAL DIFFERENCE SETS IN PSL(3,2)

In this section we shall use the partially transitive translation plane π of type (6,2) one of whose coordinate systems was given in §3 to construct a planar partial difference set of type (6,2) in PSL(3,2). Any subset $\{0,1,x,x+1\}$ of this coordinate system constitutes a subfield of order four and coordinatizes a tangentially transitive Baer subplane of π. We shall denote these seven subplanes as follows:

Subplane	Coordinate field
π_1	$\{0,1,t,t+1\}$
π_2	$\{0,1,z,z+1\}$
π_3	$\{0,1,z+t,z+t+1\}$
π_4	$\{0,1,zt,zt+1\}$
π_5	$\{0,1,zt+t,zt+t+1\}$
π_6	$\{0,1,z(t+1),z(t+1)+1\}$
π_7	$\{0,1,z(t+1)+t,z(t+1)+t+1\}$

We shall denote the elementwise stabilizer of π_i by R_i for each i. Each R_i is isomorphic to the group of linear transformations $x \to xa+b$ of GF(4) which is isomorphic to A_4. The group G generated by the R_i is the type (6,2) partially transitive group on π and is isomorphic to PSL(3,2). The action of G on $\Omega = \{\pi_i | i=1,2,\ldots,7\}$ is precisely the natural action of PSL(3,2) on the plane $\bar{\pi}$ of order 2

1	1	1	2	2	3	3
2	4	6	4	5	4	5
3	5	7	6	7	7	6

where we have dropped the symbol "π" for convenience. The point (t,z) of π is an ordinary point and the line $y = x(t+1)+z+1$ is an ordinary line of π containing (t,z). In fact, the ordinary points on this line are:

$(t,z), (t+1,z+t+1), (z,zt+1), (z+1,zt+t), (zt+1,t)$
$(z+t,zt), (z+t+1,zt+t+1), (z(t+1)+1,zt+z+t), (zt+t+1,t+1)$
$(z(t+1)+t,zt+z)$.

Choosing the point P_0 of §2 to be (t,z) and the line J_0 to be $y = x(t+1)+z+1$ we obtain the following difference set of type $(6,2)$ in $PSL(3,2)$ where we write the partial difference set elements as permutations on the plane $\bar{\pi}$ except for the identity element which we denote by 1. Our partial difference set D_1 (following the order of the ordinary points just listed) is:

$1, (23)(4756), (1245736), (1254637), (35)(1472), (1376524),$
$(1367425), (27)(1643), (34)(1562), (26)(1753).$

At this stage we mention a lemma which is a natural analogue of a well-known result on the group difference sets studies by Bruck [1]. We shall state the lemma without proof and refer the reader to Dembowski [2], p. 13.

<u>Lemma 2</u>. *If* (G,R_i,D) *is a partial difference system then so is* (G,R_i,xDy) *for all* $x,y \in G$ *and the planes constructed from them are identical.*

We also note that the set D^{-1} of inverses of elements in a planar partial difference set D of type $(6,m)$ is a planar partial difference set of type $(6,m)$. The partial difference set D^{-1} arises out of the dual plane π^d of the plane π arising out of D which must also be a partially transitive plane of type $(6,m)$. To see this let the point P_0 in π play the role in π^d of the line J_0 and the line J_0 in π play the role in π^d of the point P_0.

If we take the difference set D^{-1} and construct the set $D_2 = D^{-1}e$ where $e = (23)(4756)$ we find that D_2 is:

$(23)(4756), 1, (1627453), (1726543), (37)(1256),$
$(1435762), (1534672), (24)(1375), (36)(1247), (25)(1364).$

Note that D_2 is a planar partial difference set of type $(6,2)$ arising from the dual plane of the translation plane π and that 1 and $e = (23)(4756)$ both belong to D_1 and D_2.

Let us now turn to $G = PSL(3,2)$ acting in its natural way on the plane $\bar{\pi}$ above with its 7 group-disjoint conjugate subgroups R_i isomorphic to A_4. We shall denote $\bigcup_{i=1}^{7} R_i$ by Γ. A partial difference set of type $(6,2)$ contains 10 elements and all planes of Hughes type $(6,2)$ can be obtained by finding all partial difference systems (G,R_i,D). Now suppose D is any partial difference set of type $(6,2)$. Then there is a unique pair of elements d_1 and d_2 of D such that $d_1 d_2^{-1} = e = (23)(4756)$, and consequently $D' = Dd_2^{-1}$ is a partial difference set of type $(6,2)$ which contains 1 and e. It is thus sufficient to construct all partial difference sets of type $(6,2)$ containing these two elements.

The following approach can be used to construct a partial difference set D of type $(6,2)$: Choose a first element d_1 of D, then a second element $d_2 \notin \Gamma d_1$ and then a third $d_3 \notin \Gamma d_1 \cup \Gamma d_2$, and so on. Suppose we have selected $d_1, d_2, \ldots, d_k$ in this manner then $d_k d_i^{-1} \notin \Gamma$ for any $i < k$ and so $d_i d_k^{-1} \notin \Gamma$. Also, if $d_i^{-1} d_k \in \Gamma$, then $d_k \in d_i \Gamma = \Gamma d_i$, a contradiction. So $d_i^{-1} d_k$ and $d_k^{-1} d_i \notin \Gamma$ for any $i < k$. Thus our method guarantees us a set all of whose non-trivial differences lie outside Γ. When such a set of 10 is reached retain those whose differences fill $G-\Gamma$. It is then necessary to sort them into equivalence classes coming from the one plane using Lemma 2 and then to check that the planarity conditions are satisfied.

An algorithm based on the above method (choosing 1 and e as the first two elements) was devised by the second author, coded into FORTRAN and run on a CDC 7600 computer. Fourteen planar partial difference sets of type $(6,2)$ were obtained two of which are D_1 and D_2 listed above. Six of the remaining twelve are of the form xD_1y and six of the form xD_2y. We thus have the result:

Theorem 8. *There are precisely two partially transitive projective planes of Hughes type* $(6,2)$.

The following set D_3 containing 1 and e is also a planar partial difference set of type $(6,2)$ arising out of the translation plane π:

1, (23)(4756), (1467352), (1576342), (1376524), (1245736),
(17)(2543), (47)(1625), (26)(1753), (27)(1643).

The subgroup R_1 of G consists of the following twelve elements:

1, (45)(67), (23)(45), (23)(67), (246)(357), (247)(356), (256)(347)
(257)(346), (264)(375), (265)(374), (275)(364), (274)(365).

The subgroups R_2, R_3, R_4, R_5, R_6 and R_7 are in order $g^{-6}R_1g^6$, $g^{-4}R_1g^4$, $g^{-5}R_1g^5$, $g^{-1}R_1g$, $g^{-3}R_1g^3$, $g^{-2}R_1g^2$, where $g = (1243675)$. The extra left and right cosets of the R_i relative to D_3 are listed (using a single coset representative) in the tables below. We shall later use ER_{ij} to indicate the extra right coset of R_i in column j and similarly EL_{ij} will indicate the extra left coset of R_i in column j.

TABLE 1

Extra Right Cosets

	1	2	3	4
R_1	(156)(374)	(154)(376)	(167)(345)	(165)(347)
R_2	(12)(4576)	(35)(1472)	(1637542)	(45)(67)
R_3	(132)(576)	(35)(1472)	(36)(1742)	(23)(45)
R_4	(35)(1472)	(152)(347)	(1637542)	(36)(1742)
R_5	(12)(56)	(132)(456)	(35)(1472)	(1534672)
R_6	(132)(456)	(1534672)	(1754362)	(265)(374)
R_7	(12)(56)	(1534672)	(162)(374)	(274)(365)

TABLE 2

Extra Left Cosets

	1	2	3	4
R_1	(12)(4576)	(132)(576)	(35)(1472)	(1534672)
R_2	(45)(67)	(23)(45)	(265)(374)	(274)(365)
R_3	(35)(1472)	(152)(347)	(162)(374)	(1754362)
R_4	(132)(456)	(1534672)	(1637542)	(154)(376)
R_5	(132)(456)	(35)(1472)	(1637542)	(165)(347)
R_6	(12)(56)	(1534672)	(36)(1742)	(156)(374)
R_7	(12)(56)	(35)(1472)	(36)(1742)	(167)(345)

Since D_3 is a planar partial difference set of type (6,2) each extra right coset is an extra left coset and *vice versa* from planarity condition (a). We can use Tables 1 and 2 to decide which extra right coset equals which extra left coset since the same coset representatives appear in both tables. We can then draw up the following table which contains the extra right coset subscripts. Rows are labelled by extra left coset second subscripts.

TABLE 3

	R_1	R_2	R_3	R_4	R_5	R_6	R_7
1	21	24	53	52	61	51	71
2	31	34	42	62	32	72	22
3	41	64	73	43	43	33	44
4	54	74	63	14	14	11	13

Applying the definition in §5 of incidence between fixed elements in our partially transitive plane constructible from D_3 we obtain the following incidences for the Q_i and K_j

K_1	K_2	K_3	K_4	K_5	K_6	K_7
1	1	1	3	2	2	3
6	4	2	4	5	4	5
7	5	3	7	7	6	6

where Q_i is denoted by i for convenience. These incidences yield precisely the incidences of the plane $\bar{\pi}$ of order 2 upon which we took $G = PSL(3,2)$ to be acting in its natural manner. The fact that we have retrieved $\bar{\pi}$ essentially verifies planarity conditions (a) and (c). As a specimen verification of planarity condition (b) take $a = (12)(56)$, $i=1$ and $j=2$. From the last table we have $EL_{11} = ER_{21}$, $EL_{21} = ER_{24}$. Also $EL_{11}a = EL_{21}$ is not hard to verify. So subgroup R_2 works in condition (b). The only other possibility is $k=3$. ($EL_{12} = ER_{31}$ and $EL_{22} = ER_{34}$ from the table and the other cases $EL_{13} = ER_{41}$, $EL_{23} = ER_{64}$ and $EL_{14} = ER_{54}$, $EL_{24} = ER_{74}$ don't work.) But $EL_{12}a \neq EL_{22}$ since, for example, $b = (132)(576) \in EL_{12}$ and $ba = (23)(67) \notin EL_{22}$.

The work that we have done considerably clarifies what the planarity conditions mean and how to work with them. This may prove of significance in any work on planar partial difference sets of types (4,m) and (5,m) because the planarity conditions in those cases are quite similar. There are however, infinitely many planes of each of these types known.

Johnson and Ostrom [9] and independently Walker [16] have established that if π is a translation plane of order 16 which is tangentially transitive relative to a Fano subplane, then π is the unique known translation plane with this property. Also, Jha [4] has shown that a translation plane π that is tangentially transitive relative to a Fano subplane π' is of order 16 and necessarily possesses a Baer subplane π_1 such that $\pi' \subseteq \pi_1$ and π is tangentially transitive relative to π_1.

Now suppose π is a projective plane of order m^4 which is tangentially transitive relative to a subplane π' of order m. The tangentially transitive group G acts transitively on the m^4-m points in $\ell-\pi'$ for any line of π'. Since m^4-m is even G must contain an involution which necessarily fixes π'. Such an involution must be a Baer collineation of π fixing a Baer subplane π_1 of π elementwise. It follows that there are m^2+m+1 Baer subplanes π_i such that $\pi_i \cap \pi_j = \pi'$, where $i \neq j$. Now if the elementwise stabilizer R_1 of π_1 in G is tangentially transitive on π_1 the same will be true for each π_i. Now the groups R_1 and R_2 generate a group of order at least $(m^4-m^2)^2$ since $|R_1| = |R_2| = m^4-m^2$ and $R_1 \cap R_2 = 1$. By Johnson and Ostrom [9], Lemma 2.10, G will act semiregularly on the set of ordinary points and on the set of ordinary lines relative to π' which are $(m^4-m)(m^4-m^2)$ in number. But $(m^4-m^2)^2 > \frac{1}{2}(m^4-m)(m^4-m^2)$ for all $m \geq 2$. So G must act regularly on the set of ordinary points and on the set of ordinary lines relative to π'. This means that π is of Hughes type $(6,m)$. Thus we have

Theorem 9. *Let* π *be a plane of order* m^4 *that possesses a group* G *which is tangentially transitive relative to a subplane* π' *of order* m. *Suppose that the* m^2+m+1 *Baer subplanes* π_i *containing* π' *which we have shown to be present have respective elementwise stabilizers* R_i *in the tangentially transitive group* G *which are tangentially transitive on their respective* π_i. *Then* π *is a partially transitive plane of Hughes type* $(6,m)$ *and* G *is the partially transitive group. Furthermore, if* m *is a prime, then* $m=2$ *and* $G = PSL(3,2)$ *or* $m=3$ *and* $G = PSL(3,3)$. *If* $m=2$ *then* π *is the translation plane of Hughes type* $(6,2)$ *or its dual.*

7. FURTHER COMMENTS.

The first author originally constructed the translation plane π of Hughes type $(6,2)$ in his Ph.D. thesis [11]. It is the smallest of a class of planes of order 2^{4r}, $r \geq 1$. For the construction of this class see Rahilly [13]. Independently Lorimer [5] gave an alternative construction for π. Neither Lorimer or the first author were originally aware that their planes were isomorphic nor that they were of Hughes type $(6,2)$. These facts were established by Johnson and Ostrom who communicated them to the first author prior to the publication of their [9]. The first author began to investigate planes of Hughes type $(6,m)$ in 1975 and presented the substance of our first four sections in a talk at the Fourth Australian Conference on Combinatorial Mathematics in Adelaide, August, 1975. At that conference Cheryl Praeger and the first author began a collaboration which resulted in the paper [10] which contains results about planes of Hughes types $(4,m)$, $(5,m)$ and $(6,m)$. The main result on planes of Hughes type $(6,m)$ in [10] is:

Theorem 10. *If* π *is a partially transitive projective plane of Hughes type* $(6,m)$ *and* $R_i^\Omega \cap R_j^\Omega = 1$ *for all* $i \neq j$, *where* R_i^Ω *is the group induces by* R_i *on* $\Omega = \{\pi_i | i=1,2,\ldots,m^2+m+1\}$, *then* $m=2$ *and* $G \cong G^\Omega \cong PSL(3,2)$ *or* $m=3$ *and* $G \cong G^\Omega \cong PSL(3,3)$.

To prove Theorem 10 Praeger and Rahilly applied a stronger result due to O'Nan [8] on doubly transitive groups in place of Tsuzuku's result (Theorem 5). More recently Lorimer [7] has looked more closely at the information contained in the proof of O'Nan's result which is relevant to the action of a partially transitive group of type $(6,m)$ and proved the complete result originally conjectured by the first author, namely

Theorem 11. *If* π *is a partially transitive projective plane of Hughes type* $(6,m)$, *then* $m=2$ *and* $G \cong PSL(3,2)$ *or* $m=3$ *and* $G \cong PSL(3,3)$.

The collaboration of the present authors on partial difference sets of type $(6,2)$ occurred in 1976 when the first author was visiting the Istituto di Geometria in Bologna. At that time the first author obtained some other results on planes of Hughes types $(4,m)$, $(5,m)$ and $(6,m)$, in particular, a simple classification of finite tangentially transitive planes involving these types. For these results see [14].

Theorem 7 of this paper has, of course, been superseded by the results of Praeger/Rahilly and Lorimer. However, the proof we offer here contains the germs of the ideas used by these authors in settling more general cases.

The significance of Theorem 7 is that it was the initial evidence which led to the first author's general conjecture (see [10], p. 96) settled by Lorimer [7]. Lorimer's result, of course implies that the restriction that m be a prime in Theorem 9 can be omitted. It is also significant here in that it makes quite clear in which group (PSL(3,2)) an exhaustive search for planar partial difference sets can be carried out. Our results also throw considerable light on the undecided $(6,3)$ case. The partial difference set approach requires the construction of a set of 69 elements in $PSL(3,3)$ with the usual difference property relative to a complex of 13 group-disjoint and conjugate subgroups of order 72 isomorphic to the semi-direct product of $C_3 \times C_3$ and Q_8. A plane of Hughes type $(6,3)$ can not be a translation plane by Theorem 6. It will possess 13 non-desarguesian Baer subplanes π_i such that $\pi_i \cap \pi_j = \pi'$, $i \neq j$, where π' is a subplane of order 3. We note that the three known non-desarguesian planes of order 9 are tangentially transitive relative to a Baer subplane and each may be coordinatized by a nearfield of order 9, the multiplicative group of which is isomorphic to Q_8. Results of the first author ([14], Theorems 10 and 11) imply that a plane of Hughes type $(6,3)$ must be in Levy-Barlotti class I.1.

REFERENCES

[1] R.H. Bruck, Difference sets in a finite group, *Trans. Amer. Soc.* 78, (1955), 464-481.

[2] Peter Dembowski, *Finite Geometries*, Springer-Verlag, Berlin, Heidelberg, New York, 1968.

[3] D.R. Hughes, Partial difference sets, *Amer. J. Math.* 78, (1956), 650-674.

[4] V. Jha, On tangentially transitive translation planes and related systems, *Geom. Ded.* 4, (1975), 457-483.

[5] P. Lorimer, A projective plane of order 16, *J. Combinatorial Theory* (A), 16, (1974), 334-347.

[6] P. Lorimer, On projective planes of type (6,m), *Mathematical Proceedings of the Cambridge Philosophical Society*, 88, (1980), 199-204.

[7] P. Lorimer, Correction to "On projective planes of type (6,m)", *Mathematical Proceedings of the Cambridge Philosophical Society*, to appear.

[8] M.E. O'Nan, Normal structure of the one-point stabilizer of a doubly transitive permutation group. I. *Trans. Amer. Math. Soc.* 214, (1975), 1-42.

[9] T.G. Ostrom and N.L. Johnson, Tangentially transitive planes of order 16, *Journal of Geometry* 10, (1977), 146-163.

[10] Cheryl E. Praeger and Alan Rahilly, On partially transitive projective planes of certain Hughes types, *Group Theory*, Proc. of a Miniconference, A.N.U., Lect. Notes in Math. Vol. 573, Springer-Verlag, (1977), 85-111.

[11] A.J. Rahilly, *Finite generalized Hall planes and their collineation groups*, Ph.D. Thesis, University of Sydney, 1973.

[12] Alan Rahilly, Generalized Hall planes of even order, *Pacific J. Math.*, 55 No. 2, (1974), 543-551.

[13] Alan Rahilly, Some translation planes with elations which are not translations, *Combinatorial Mathematics III: Proceedings of the Third Australian Conference*, ed. A.P. Street and W.D. Wallis, Lecture Notes in Mathematics, Vol. 452, Springer-Verlag, (1975), 197-209.

[14] Alan Rahilly, On tangentially transitive projective planes, *Geom. Ded.* to appear.

[15] T. Tsuzuku, On doubly transitive permutation groups of degree $1+p+p^2$ where p is a prime number, *J. Algebra*, 8, (1968), 143-147.

[16] M. Walker, A note on tangentially transitive affine planes, *Bull. London Math. Soc.* 8, (1976), 273-277.

EMBEDDING INCOMPLETE IDEMPOTENT LATIN SQUARES

C.A. RODGER

We provide necessary and sufficient conditions to ensure the embedding of an incomplete idempotent latin square of size n *on the symbols* $\sigma_1, \sigma_2, \ldots, \sigma_t$ *in an idempotent latin square of size* t, *for* $t = 2n$. *The value of* $t = 2n$ *fills the gap between the unsolved problem when* $t < 2n$, *where substantial complications inherent in these values of* t *arise, and the previously proven results when* $t > 2n$.

1. INTRODUCTION AND DEFINITIONS

A *partial latin square* on the t symbols $\sigma_1, \sigma_2, \ldots, \sigma_t$ of size n is an $n \times n$ matrix, each of the cells of which may be empty or contain one of the symbols $\sigma_1, \ldots, \sigma_t$, where no symbol occurs more than once in any row or in any column. An *incomplete latin square* on t symbols of size n is a partial latin square in which there are no empty cells; it is a *latin square* of size n if $t = n$. A partial latin square is said to be *idempotent* if, for each i, $1 \le i \le n$, the cell (i,i) contains the symbol σ_i. Let $N_R(\sigma_j)$ represent the number of times that the symbol σ_j occurs in the partial latin square R.

The object of this paper is to prove the following theorem.

Theorem 1. *An incomplete idempotent latin square* R *of size* n, $n > 16$ *on the symbols* $\sigma_1, \ldots, \sigma_{2n}$ *can be embedded in an idempotent latin square of size* $2n$ *if and only if*

$$N_R(\sigma_j) \ge 1 \quad \textit{for} \quad 1 \le j \le 2n.$$

This problem has quite a long history. The problem of embedding a partial latin square of size n on n symbols in a latin square of size t for all $t \ge 2n$ was settled by Evans [7] in 1960. In 1971, Lindner [12] showed that a partial idempotent latin square R of size n on n symbols could be embedded in an idempotent latin square of size t, where t is finite. This was improved by Hilton [9] in 1973 who solved this problem when $t = 4(n+k)$, $k=0,1,2,\ldots$ or $t \ge 8n+1$ and he then posed the conjecture that this result should be true for all $t \ge 2n+1$. Andersen [2] has recently improved Hilton's result to all $t \ge 4n$, $t \ne 4n+1$ and in 1982, Andersen, Hilton and Rodger [6] finally obtained the best possible value of $t \ge 2n+1$; if t is to be lowered any further, then necessary conditions must be introduced.

The problem that then arose was to show that an incomplete idempotent latin square of size n on t symbols could be embedded in an idempotent latin square of size t for all $t \geq 2n+1$. Clearly this result would imply that of Andersen, Hilton and Rodger [6].

The problem of requiring a latin square to be idempotent was generalized by Andersen, Häggkvist, Hilton and Poucher [4] and then added to by Andersen and Hilton [3] to consider a generally prescribed diagonal; this resulted in the following theorem. The *diagonal of* T *outside* S, where S is an $r \times s$ (say $s \geq r$) incomplete latin rectangle embedded in a latin square T of size t, is the cells $(r+1,s+1),(r+2,s+2),\ldots,(t-s+r,t)$ of T.

Theorem 2. (L.D. Andersen et. al. [3] and [4]). *Let* $t \geq s \geq r > 0$. *Let* S *be an* $r \times s$ *incomplete latin rectangle on the symbols* $\sigma_1,\ldots,\sigma_t$. *Let* f *be a non-negative integral valued function on the symbols* $\sigma_1,\ldots,\sigma_t$ *such that*

$$\sum_{j=1}^{t} f(\sigma_j) \leq \begin{cases} t-s-1 & \text{if } r = s, \\ t-s & \text{if } r \neq s. \end{cases}$$

Then S *can be embedded in a latin square* T *of size* t *in which* σ_j $(1 \leq j \leq t)$ *occurs at least* $f(\sigma_j)$ *times on the diagonal of* T *outside* S *if and only if*

$$N_S(\sigma_j) \geq r+s-t+f(\sigma_j) \quad \textit{for all } j,\ 1 \leq j \leq t.$$

The requirement that $\sum_{j=1}^{t} f(\sigma_j) \leq t-s-1$ when $r = s$ cannot be improved to include the case when $\sum_{j=1}^{t} f(\sigma_j) = t-s$ using the method of proof for the above theorem. This case has now been proved using different methods (see C.A. Rodger [13]); by setting $f(\sigma_j) = 1$ for $n+1 \leq j \leq t$ and zero otherwise, this result also proves the idempotent embedding problem for incomplete latin squares for all $t \geq 2n+1$.

An added difficulty arises when considering the embedding of an incomplete latin square of size n in a latin square of size t where the diagonal has been prescribed and $t < 2n+1$ or when considering the embedding of an incomplete idempotent latin square of size n in an idempotent latin square of size t where $t < 2n$. In these cases it can be shown that the arrangement of the symbols within the given incomplete latin square R can determine whether or not R is embeddable, and so numerical conditions on the symbols are no longer sufficient. For discussion and results concerning this problem, see Andersen [1],

Andersen, Hilton and Rodger [5] and Hilton and Rodger [10].

We shall need the following theorems to prove Theorem 1.

Theorem 3. (D. König, [11]). *A bypartite multigraph with maximum degree* Δ *can be properly edge-coloured with* Δ *colours.*

Theorem 4. (M. Hall, [8]). *A* $t \times r$ *incomplete latin rectangle on the symbols* $\sigma_1, \ldots, \sigma_t$ *can be embedded in a latin square of size* t.

2. PROOF OF THEOREM 1

Necessity. Since R is idempotent, clearly $N_R(\sigma_j) \geq 1$ for each j, $1 \leq j \leq n$. Let R be embedded in the idempotent latin square T, and let T be subdivided as indicated in Figure 1. Then for each j, $n+1 \leq j \leq 2n$, $N_B(\sigma_j) = n$, $N_A(\sigma_j) \leq n-1$ (since σ_j occurs in cell (j,j) of T), $N_T(\sigma_j) = 2n$ and $N_R(\sigma_j) = N_T(\sigma_j) - N_A(\sigma_j) - N_B(\sigma_j)$. Therefore $N_R(\sigma_j) \geq 1$.

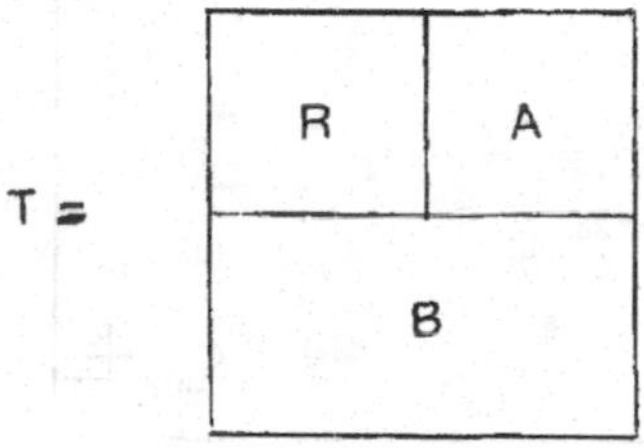

Figure 1.

Sufficiency. We shall begin by adding rows ρ_0 and possibly ρ_{-1} and columns c_0, c_{-1} and possibly c_{-2} to R either to form an incomplete latin rectangle R^+ of size $(n+1) \times (n+2)$ where σ_{2n} occurs in cells $(0,-1)$ and $(1,0)$ of R^+ and σ_{2n-1} occurs on cells $(0,0)$ and $(1,-1)$ of R^+, or to form an incomplete latin triangle R' of size $(n+2) \times (n+3)$ where σ_{2n} occurs in cells $(-1,-1)$ and $(0,0)$, σ_{2n-1} occurs in cells $(-1,0)$ and $(0,-1)$ and σ_{2n-2} occurs in cell $(-1,-2)$. In addition, we shall ensure that

$$(1) \qquad N_{R^+}(\sigma_j) \geq \begin{cases} 3 & \text{for } 1 \leq j \leq n \text{ and } 2n-1 \leq j \leq 2n, \\ 4 & \text{for } n+1 \leq j \leq 2n-2, \end{cases}$$

and

$$(2) \qquad N_{R'}(\sigma_j) \geq \begin{cases} 5 & \text{for } 1 \leq j \leq n \text{ and } 2n-1 \leq j \leq 2n, \\ 6 & \text{for } n+1 \leq j \leq 2n-2. \end{cases}$$

We can then apply Theorem 2 to R^+ or R' as appropriate with $f(\sigma_j) = 1$ for $n+1 \le j \le 2n-2$ or $n+1 \le j \le 2n-3$ respectively to form a latin square T^*. Finally T^* will be slightly altered to produce the required latin square T (see Figure 2).

We shall begin forming R^+ or R' from R, considering three cases in turn.

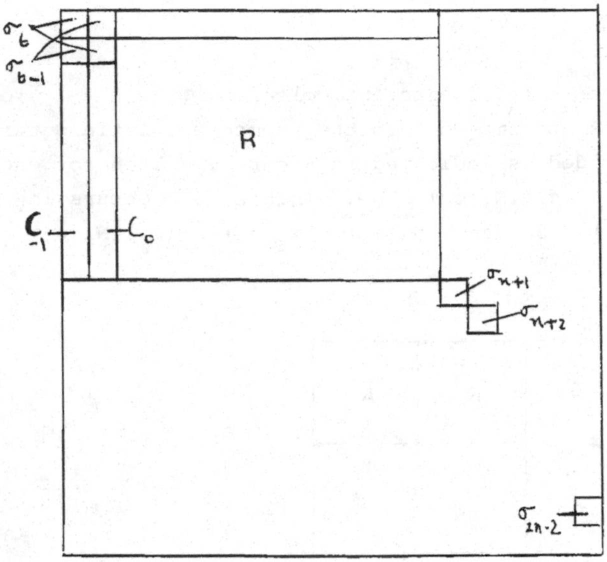

Figure 2. The latin square T^*, formed from R^+.

Case 1. Suppose that R contains at most one symbol σ_j, where $N_R(\sigma_j) \le 3$ and $n+1 \le j \le 2n$. We shall consider two subcases in turn.

Case 1a. Suppose that each row of R contains $n-1$ of the symbols $\sigma_{n+1},\ldots,\sigma_{2n}$. We shall assume that if there exists a symbol σ_j with $n+1 \le j \le 2n$ and $N_R(\sigma_j) = 1$ (in case 1 we have assumed there is at most one such symbol) then $j = n+1$ and subject to this, that $N_R(\sigma_j) \le N_R(\sigma_{j-1})$ for $n+2 \le j \le 2n$. We shall now form R' from R.

Add two rows, ρ_0 and ρ_{-1} and three columns, c_0, c_{-1} and c_{-2} to R. Place σ_{2n} in cells $(-1,-1)$ and $(0,0)$, if $N_R(\sigma_{2n}) = 2$ then place σ_{2n} in a cell in c_{-2}, place σ_{2n-1} in cells $(-1,0)$ and $(0,-1)$, place σ_{2n-2} and σ_{2n-3} in cells $(-1,-2)$ and $(0,-2)$ respectively and if $N_R(\sigma_{n+1}) = 1$ then place σ_{n+1} in each of ρ_0,

ρ_{-1}, c_0, c_{-1} and c_{-2}. The remaining empty cells can easily be filled using the symbols $\sigma_1,\ldots,\sigma_n$ so that each such symbol occurs at least four times in the added rows and columns (for $n \geq 8$). Call the resulting incomplete latin rectangle R'; it is not hard to check that all symbols σ_j, $1 \leq j \leq 2n$ satisfy (2).

Case 1b. Suppose that there exists a row of R, say the first, which contains at most $n-2$ of the symbols $\sigma_{n+1},\ldots,\sigma_{2n}$; we can assume that σ_{2n-1} and σ_{2n} do not occur in row 1 of R. We shall now form R^+ from R.

Add one row, ρ_0 and two columns, c_0 and c_{-1} to R. Place σ_{2n} in cells $(0,-1)$ and $(1,0)$, place σ_{2n-1} in cells $(0,0)$ and $(1,-1)$ and if there exists a symbol σ_j with $n+1 \leq j \leq 2n-2$ and $N_R(\sigma_j) \leq 3$ (in case 1 there is at most one such symbol) then place it in $4 - N_R(\sigma_j)$ cells in the added row and columns. At this stage there remains at least $3n-5$ empty cells. A symbol that occurs at most twice in R could be placed in any except at most 6 of these empty cells; therefore the symbols $\sigma_1,\ldots,\sigma_n$ can be placed in empty cells in the added row and columns so that each such symbol occurs at least 3 times in the resulting partial latin rectangle, since this requires filling at most $2n$ cells and since $2n \geq 3n-5-6$ for $n \geq 11$ So all symbols occur at least 3 times in the resulting partial latin rectangle and the symbols σ_j where $n+1 \leq j \leq 2n-2$ each occurs at least 4 times (recalling the assumption for case 1). However some cells may still be empty.

The remaining empty cells can be filled one by one until at most two cells remain empty in each of ρ_0, c_0 and c_{-1}. Suppose ρ_0 contains two empty cells. One can be filled immediately unless the symbols occurring in the columns of R corresponding to the empty cells contain different symbols to those occurring in ρ_0. Since $N_R(\sigma_{2n}) \geq 1$, let σ_{2n} occur in, say column x of R. Remove the symbol from cell $(0,x)$, place it in one of the two empty cells and now cell $(0,x)$ can be filled with some symbol. Suppose ρ_0 contains one empty cell, say $(0,y)$. This can be filled unless c_y and ρ_0 have exactly one symbol in common. By the assumption of case 1, either $N_R(\sigma_{2n}) \geq 4$ or $N_R(\sigma_{2n-1}) \geq 4$ (say the latter), and so σ_{2n} occurs in a column of R, say c_z, where the symbol in cell $(0,z)$ does not occur in c_y (assuming cell $(0,y)$ is still empty). Remove the symbol from $(0,z)$ and place it in $(0,y)$. Since ρ_0 contains $n+1$ symbols, each of which occurs at least once in R, at least one column of R, say c_w, contains at least two symbols that occur in ρ_0.

Remove the symbol from cell $(0,w)$, place it in $(0,z)$ and complete the filling of ρ_0 by placing some symbol in $(0,w)$.

Providing there are at least two empty cells in $c_0 \cup c_{-1}$ similar arguments to that used to fill ρ_0 can be used to fill a cell in $c_0 \cup c_{-1}$. Therefore we shall suppose that c_0 has been filled and that c_{-1} contains one empty cell. Let R^- be the $(n-1) \times n$ incomplete latin rectangle formed by deleting the first row from R. As before, since $N_{R^-}(\sigma_{2n}) \geq 4$ or $N_{R^-}(\sigma_{2n-1}) \geq 4$ (say the latter), we can "move" the empty cell to a row which contains σ_{2n-1}, say row x. The cell $(x,-1)$ can be filled unless σ_{2n-1} is the only symbol occurring in both c_{-1} and row x of $R^- \cup c_0$. If a row of $R^- \cup c_0$, say row y, contains at least two symbols that also occur in c_{-1}, then we can move the symbol from cell $(y,-1)$ to $(x,-1)$ and fill the resulting vacancy in $(y,-1)$. Finally, if each row of $R^- \cup c_0$ has exactly one symbol in common with the symbols in c_{-1} then since $N_{R^-}(\sigma_{2n-1}) \geq 4$, at least 4 symbols in c_{-1} do not occur in $R^- \cup c_0$; this is impossible since clearly $\sigma_2,\ldots,\sigma_n$ occur in R^- and by the assumption for case 1, so do at least $n-1$ of $\sigma_{n+1},\ldots,\sigma_{2n}$.

Let the resulting incomplete latin rectangle be R^+. All symbols in R^+ satisfy (1).

<u>Case 2</u>. Suppose that R contains an $n \times (n-1)$ or an $(n-1) \times n$ incomplete latin rectangle (by considering the transpose of R if necessary, we may assume it to be $n \times (n-1)$) on n symbols which include $n-1$ of the symbols $\sigma_1,\ldots,\sigma_n$, excluding, say σ_n, and one of $\sigma_{n+1},\ldots,\sigma_{2n-2}$, say σ_{n+1}. Clearly $N_R(\sigma_{n+1}) = n-1$ and so is missing from one row of R, say the first. Since $N_R(\sigma_{2n-1}) = 1$ and $N_R(\sigma_{2n}) = 1$, either σ_{2n-1} or σ_{2n} does not occur in the first row of R (say σ_{2n-1}). Relabel σ_{n+1} and σ_{2n} by σ_{2n} and σ_{n+1} respectively; clearly once T is formed, σ_{n+1} and σ_{2n} can be swapped back and rows and columns $n+1$ and $2n$ interchanged to obtain the required idempotent embedding. Place σ_{2n} in cells $(0,-1)$ and $(1,0)$ and place σ_{2n-1} in cells $(0,0)$ and $(1,-1)$. R^+ can now be formed by placing σ_1 in cell $(0,n)$ and placing each of $\sigma_n,\ldots,\sigma_{2n-2}$ in one cell in each of ρ_0, c_0 and c_{-1}. Then it is easy to check that R^+ satisfies (1).

<u>Case 3</u>. Suppose that R contains at least two symbols σ_j where $n+1 \leq j \leq 2n$ and $N_R(\sigma_j) \leq 3$; we shall assume that $N_R(\sigma_j) \leq N_R(\sigma_{j-1})$ for $n+2 \leq j \leq 2n$ and so $N_R(\sigma_j) \leq 3$ for $2n-1 \leq j \leq 2n$. Suppose also that R contains no $(n-1) \times n$ and no $n \times (n-1)$ incomplete latin rectangle on $n-1$ of the symbols $\sigma_1,\ldots,\sigma_n$ and one of

$\sigma_{n+1},\ldots,\sigma_{2n-2}$. By considering the transpose of R if necessary, we may assume that there exist two distinct rows of R, say ρ_a and ρ_b which contain the symbols σ_{2n-1} and σ_{2n} respectively. Also, since $N_R(\sigma_{2n-1}) + N_R(\sigma_{2n}) \le 6$ and $n \ge 7$, there exists a row of R, say the first, which is missing both σ_{2n-1} and σ_{2n} (so $a > 1$ and $b > 1$).

Add one row, ρ_0 and two columns, c_0 and c_{-1} to R. Place σ_{2n} in cells $(0,-1)$ and $(1,0)$ and place σ_{2n-1} in cells $(0,0)$ and $(1,-1)$. We shall now complete the filling of ρ_0, c_0 and c_{-1} so that in the resulting incomplete latin rectangle, R^* all symbols satisfy (2).

To complete the filling of ρ_0 we proceed as follows. Form a bipartite graph on the vertex sets $\{c_1,\ldots,c_n,c^*\}$ and $\{\sigma_1',\ldots,\sigma_{2n-2}'\}$ by joining c_i to σ_j' if and only if σ_j is missing from column i of R. If $N_R(\sigma_j) = 1$ then σ_j' is adjacent to $n-1$ of $c_1,\ldots,c_n$ and so is adjacent to at least one of c_1 and c_2; delete one such edge. If σ_{2n-1} or σ_{2n} occurs in column i of R then c_i has degree at least $n-1$; in any such case remove edges (c_i,σ_j') for some j $(1 \le j \le 2n-2)$ until c_i has degree $n-2$ and add the edge (c^*,σ_j'), so the degree of σ_j' is unchanged. Notice that since $N_R(\sigma_{2n-1}) + N_R(\sigma_{2n}) \le 6$, at most 6 edges are joined to c^*. If $N_R(\sigma_j) = 3$ and $n+1 \le j \le 2n-2$ then join σ_j' to c^* with one edge. Finally, if we now find that c^* has degree greater than $n-3$ then remove edges joining c^* to σ_j' where $N_R(\sigma_j) = 3$ and $n+1 \le j \le 2n-2$ until c^* has degree $n-3$; notice that since $n-3 > 6$, the initial 6 edges joined to c^* need not be removed, and since $|\{\sigma_{n+1},\ldots,\sigma_{2n-2}\}| = n-2$, at most 7 edges need to be removed. Call the resulting bipartite graph G.

By the construction of G, $d_G(c^*) \le n-3$ and $d_G(\sigma_j') \le n-2$ for $1 \le j \le 2n-2$. Also $d_G(c_i) = n-2$ for $3 \le i \le n$ and $d_G(c_i) \le n-2$ for $1 \le i \le 2$. Furthermore, if $N_R(\sigma_j) \le 2$ then $d_G(\sigma_j') = n-2$ and if $N_R(\sigma_j) = 3$ and $n+1 \le j \le 2n-2$ then $d_G(\sigma_j') = n-2$ for all except at most 7 such vertices. Therefore G has maximum degree $n-2$ and so by Theorem 3 can be edge-coloured with $n-2$ colours. Let k be a colour which occurs on no edge incident with c^*. Place σ_j in cell $(0,i)$ if and only if (c_i,σ_j') is an edge on G that is coloured with k. Then all symbols corresponding to vertices of degree $n-2$ are joined to some vertex c_i $(1 \le i \le n)$ by an edge coloured k and so are placed in ρ_0. Similarly, if $d_G(c_i) = n-2$ the cell $(0,i)$ is filled by some symbol and so all cells except possibly for $(0,1)$ and $(0,2)$ of ρ_0 are filled. Therefore, if $N_R(\sigma_j) \le 2$ then σ_j is placed in ρ_0 and if $N_R(\sigma_j) = 3$ and $n+1 \le j \le 2n-2$ then all except at most 7 such symbols are placed in ρ_0.

If both the cells (0,1) and (0,2) are still empty then one can be filled immediately unless all symbols occurring in columns 1 and 2 of R do not occur in ρ_0. In this case, since $N_R(\sigma_{2n}) \ge \sigma_{2n}$ occurs in say column x of R; move the symbol from cell (0,x) to (0,2) and fill (0,x) with some other symbol. If (0,1) is the only empty cell, it can be filled unless c_1 and ρ_0 have exactly one symbol in common. Since $N_R(\sigma_j) \ge 1$ for $1 \le j \le 2n$ and since ρ_0 contains n+1 symbols, at least one column of R contains at least 2 symbols in common with ρ_0. If there are 2 such columns, say c_{n-1} and c_n, then the symbol in either (0,n-1) or (0,n) can be removed and placed in (0,1), the resulting vacancy in ρ_0 being filled by some symbol. Therefore either ρ_0 has been filled or exactly one column of R, say c_n, contains at least 2 symbols in common with ρ_0. Let σ_k occur in cell (0,n). If σ_k does not occur in column 1 of R then ρ_0 can be filled by moving σ_k to (0,1) and filling (0,n) with another symbol. Suppose σ_k occurs in column 1 of R. If σ_k does not occur in some column c_i of R for $2 \le i \le n-1$ then move the symbol from cell (0,i) to (0,1) (σ_k is now the only symbol in both column 1 of R and ρ_0), move σ_k from (0,n) to (0,i) and fill (0,n) with another symbol. Finally, if σ_k occurs in all columns c_i of R for $1 \le i \le n-1$ then as it is the only symbol in these columns of R and in ρ_0, the n-1 symbols not occurring in ρ_0 occur in each column c_i for $1 \le i \le n-1$; then however R contains an $n \times (n-1)$ incomplete latin rectangle which contains the symbols $\sigma_1,\ldots,\sigma_{n-1}$ (as R is idempotent) and excludes σ_n, σ_{2n-1} and σ_{2n} (as $N_R(\sigma_j) \le 3 < n-1$ for $2n-1 \le j \le 2n$), which was covered by Case 2.

This completes the filling of ρ_0. To summarize the position, ρ_0 has been filled in such a way that if $1 \le N_R(\sigma_j) \le 2$ and $1 \le j \le 2n-2$ then σ_j occurs in ρ_0 and if $N_R(\sigma_j) = 3$ and $n+1 \le j \le 2n-2$ then with up to 7 exceptions, σ_j has been placed in ρ_0.

To fill c_0 we proceed as follows. Form a bipartite graph on the vertex sets $\{\rho_2,\rho_3,\ldots,\rho_n,\rho^*\}$ and $\{\sigma_1',\ldots,\sigma_{2n-2}'\}$ by joining ρ_i to σ_j' if and only if σ_j does not occur in row i of R^-. If $N_{R^-}(\sigma_j) = 0$ then remove the edge (ρ_b,σ_j') (recall that σ_{2n} occurs in row b of R^-). If for some i, $2 \le i \le n$, ρ_i has degree at least n-1 (so σ_{2n-1} or σ_{2n} occurs in row i of R^-) then remove edges (ρ_i,σ_j') for some j, $1 \le j \le 2n-2$ until ρ_i has degree n-2 and add the edge (ρ^*,σ_j'). Let the resulting bipartite graph be G'.

Since $N_R(\sigma_{2n-1}) + N_R(\sigma_{2n}) \le 6$, $d_{G'}(\rho^*) \le 6 < n-2$ for $n \ge 9$. $d_{G'}(\rho_i) = n-2$ for $2 \le i \le n$, $i \ne b$ and $d_{G'}(\rho_b) \le n-2$. $d_{G'}(\sigma_j') \le n-2$ for $1 \le j \le 2n-2$ with $d_{G'}(\sigma_j') = n-2$ if $N_{R^-}(\sigma_j') \le 1$. Therefore G' has maximum degree n-2 and so by Theorem 3 can be edge-coloured with n-2 colours. Let k be a colour that occurs on no edge incident with ρ^*. Place σ_j in cell (i,0) if and only if the edge (ρ_i,σ_j') in G' is coloured with k. The cell (b,0) may still be empty (since $d_{G'}(\rho_b) \le n-2$), in which case since row b of R and c_0 contain at most 2n-1 different symbols (b,0) can now be filled with some symbol.

If $N_{R^-}(\sigma_j) \le 1$ then $d_{G'}(\sigma_j') = n-2$, so σ_j' is joined to c_i (for some i, $2 \le i \le n$) by an edge coloured k, so σ_j is placed in c_0. Thus, the only symbols that we have to ensure are placed in c_{-1} are those for which $N_{R\cup\rho_0\cup c_0}(\sigma_j) = 3$ and $n+1 \le j \le 2n-2$. Recall that if $N_R(\sigma_j) = 3$ and $n+1 \le j \le 2n-2$ then with up to 7 exceptions, σ_j occurs in ρ_0 and so $N_{R\cup\rho_0\cup c_0}(\rho_j) \ge 4$. Therefore in order that all symbols σ_j where $n+1 \le j \le 2n-2$ satisfy $N_{R\cup\rho_0\cup c_0\cup c_{-1}}(\sigma_j) \ge 4$, it suffices to fill c_{-1} so that if $N_{R^-\cup c_0}(\sigma_j) \le 2$ and $n+1 \le j \le 2n-2$ then σ_j is placed in c_{-1} and also so that at most 7 further symbols σ_j for which $N_{R^-\cup c_0}(\sigma_j) = 3$, $N_{R\cup\rho_0\cup c_0}(\sigma_j) = 3$ and $n+1 \le j \le 2n-2$ are placed in c_{-1}.

To fill c_{-1} as required we begin by forming a bipartite graph on the vertex sets $\{\rho_2,\ldots,\rho_n,\rho^*\}$ and $\{\sigma_1',\ldots,\sigma_{2n-2}'\}$, joining ρ_i to σ_j' if and only if σ_j does not occur in row i of $R^- \cup c_0$. If $N_{R^-\cup c_0}(\sigma_j) = 1$ then remove either the edge (ρ_a,σ_j') or the edge (ρ_b,σ_j') (recall that σ_{2n-1} and σ_{2n} occur in rows a and b of R^- respectively; notice that if $N_{R^-}(\sigma_j) = 0$ then σ_j is placed in c_0, so $N_{R^-\cup c_0}(\sigma_j) \ge 1$ for $1 \le j \le 2n-2$. If for any i, $2 \le i \le n$, ρ_i has degree at least n-2 (so σ_{2n-1} or σ_{2n} occurs in row i of R^-) then remove edges (ρ_i,σ_j') for some j, $1 \le j \le 2n-2$ until ρ_i has degree n-3 and add the edge (ρ^*,σ_j'). If $N_{R\cup\rho_0\cup c_0}(\sigma_j) = 3$, $N_{R^-\cup c_0}(\sigma_j) = 3$ and $n+1 \le j \le 2n-2$ (this is possible for at most 7 symbols) then join σ_j' to ρ^* with one edge. Let the resulting bipartite graph be G".

Now $d_{G''}(\rho^*) \le N_R(\sigma_{2n-1}) + N_R(\sigma_{2n}) + 7 \le 13 < n-3$ for $n > 16$, $d_{G''}(\rho_i) = n-3$ for $2 \le i \le n$, $i \notin \{a,b\}$, $d_{G''}(\rho_i) \le n-3$ for $i \in \{a,b\}$, $d_{G''}(\sigma_j') \le n-3$ for $1 \le j \le 2n-2$ and $d_{G''}(\sigma_j') = n-3$ if $N_{R^-\cup c_0}(\sigma_j) \le 2$

or if $N_{R \cup \rho_0 \cup c_0}(\sigma_j)=3$, $N_{R^- \cup c_0}(\sigma_j)=3$ and $n+1 \le j \le 2n-2$. Therefore G'' has maximum degree $n-3$ and so can be edge-coloured with $n-3$ colours. Let k be a colour on no edge incident with ρ^*. Place σ_j in cell $(i,-1)$ if and only if (ρ_i, σ_j) is an edge in G'' that is coloured k. Then all symbols that we required to occur in c_{-1} have now been placed in c_{-1} and c_{-1} has been filled except possibly for cells $(a,-1)$ and $(b,-1)$.

If c_{-1} contains 2 empty cells then one can be filled immediately, say $(b,-1)$, since ρ_b and c_{-1} contain at most $2n-1$ different symbols. If one cell in c_{-1} is still empty (say $(a,-1)$) then it can be filled unless ρ_a and c_{-1} only have σ_{2n-1} in common. Each symbol occurs at least once in $R^- \cup c_0$ (since if $N_{R^-}(\sigma_j) = 0$ then σ_j is placed in c_0) and so at least one row of $R^- \cup c_0$, say row x, contains at least 2 symbols that occur in c_{-1}. Take the symbol from cell $(x,-1)$, place it in $(a,-1)$ and fill $(x,-1)$ with another symbol, thus completing the filling of c_{-1}.

Call the resulting incomplete latin rectangle R^*. Then R^* satisfies (1). This completes case 3.

In all three cases we have now formed either R^* or R' from R so that (1) or (2) is satisfied respectively.

If R^* has been formed then apply Theorem 2 to R^* with $f(\sigma_j) = 1$ for $n+1 \le j \le 2n-2$ and $f(\sigma_j) = 0$ otherwise to form the latin square T^* of size $2n$. Form the incomplete idempotent latin square T' by deleting rows 0 and $2n-1$ and columns 0 and -1 from T^*.

If R' has been formed then apply Theorem 2 to R' with $f(\sigma_j) = 1$ for $n+1 \le j \le 2n-3$ and $f(\sigma_j) = 0$ otherwise to form the latin square T^* of size $2n$. Form the incomplete idempotent latin square T' from T^* by deleting rows 0 and $2n-2$ and columns 0 and -1 and by relabelling row -1 by row $2n-2$ and column -2 by column $2n-2$; since σ_{2n-2} occurs in cell $(-1,-2)$ of T^*, σ_{2n-2} occurs in cell $(2n-2,2n-2)$ of T' as required.

In any case, R has now been embedded in the incomplete idempotent latin square T' of size $2n-2$. T' has the further properties that $N_{T'}(\sigma_j) = 2n-3$ for $2n-1 \le j \le 2n$ and that σ_{2n-1} and σ_{2n} are missing from the same row of T' (row 1 if R^* was formed and row $2n-2$ if R' was formed).

We shall now add two different rows to T', filling cells (i,j) for $2n-1 \le i \le 2n$ and $1 \le j \le 2n-2$ so that σ_{2n} occurs in row $2n-1$ and σ_{2n-1} occurs in row $2n$. To do this, form a bipartite graph on the vertex sets $\{c_1,\ldots,c_{2n-2}\}$ and $\{\sigma'_1,\ldots,\sigma'_{2n}\}$ by joining c_i to σ'_j if and only if σ_j is missing from column i of T'. This graph, say G_0, has maximum degree 2 and since $N_{T'}(\sigma_j) = 2n-3$ for $2n-1 \le j \le 2n$, $d_{G_0}(\sigma'_j) = 1$ for $2n-1 \le j \le 2n$. Amalgamate the vertices σ'_{2n-1} and σ'_{2n} into a single vertex of degree 2 and edge-colour the resulting bipartite graph with 2 colours. This corresponds to an edge-colouring of G_0 with 2 colours in which the edge incident with σ'_{2n-1} and the edge incident with σ'_{2n} receive different colours. Use the edges of one colour to fill ρ_{2n-1} and the edges of the other colour to fill ρ_{2n}; do this in such a way that σ_{2n} and σ_{2n-1} are placed in ρ_{2n-1} and ρ_{2n} respectively.

Finally, complete the latin square T using Theorem 4. Since σ_{2n-1} and σ_{2n} are both missing from the same row of T' the final two columns of this latin square can be interchanged if necessary so that σ_{2n-1} and σ_{2n} occur on cells $(2n-1,2n-1)$ and $(2n,2n)$ respectively. Thus we obtain the required idempotent latin square.

REFERENCES

[1] L.D. Andersen, *Latin squares and their generalizations*, Ph.D. Thesis, Reading, 1979.

[2] L.D. Andersen, Embedding latin squares with prescribed diagonal, *Annals of Discrete Maths*. 15 (1982), 9-26.

[3] L.D. Andersen and A.J.W. Hilton, Thank Evans!, submitted.

[4] L.D. Andersen, R. Häggkvist, A.J.W. Hilton and W.B. Poucher, Embedding incomplete latin squares in a latin square whose diagonal is almost completely prescribed, *Europ. J. Combinatorics*, 1 (1980), 5-7.

[5] L.D. Andersen, A.J.W. Hilton and C.A. Rodgers, Small embeddings of incomplete idempotent latin squares,

[6] L.D. Andersen, A.J.W. Hilton and C.A. Rodger, A solution to the embedding problem for partial idempotent latin squares, *J. Lond. Maths. Soc.*, to appear.

[7] T. Evans, Embedding incomplete latin squares, *Amer. Math. Monthly*, 67, (1960), 958-961.

[8] M. Hall Jr., An existence theorem for latin squares, *Bull. Amer. Math. Soc.*, 51 (1945), 387-388.

[9] A.J.W. Hilton, Embedding an incomplete diagonal latin square in a complete diagonal latin square, *J. Combinatorial Theory A*, 15, (1973), 121-128.

[10] A.J.W. Hilton and C.A. Rodger, Edge-colouring regular bipartite graphs, *Annals of Discrete Maths.*, 13 (1982), 139-158.

[11] D. König, *Theorie der endlichen und unendlichen graphen*, Chelsea Pub. Co., New York, 1950.

[12] C.C. Lindner, Embedding partial idempotent latin squares, *J. Combinatorial Theory A*, 10, (1971), 240-245.

[13] C.A. Rodger, *Embedding problems for latin squares*, Ph.D. Thesis, Reading, 1982.

THE COMPLETION OF PARTIAL F-SQUARES

BOHDAN SMETANIUK

F-squares were introduced into mathematics and statistics as a generalization of the concept of latin square. In this paper we consider the problem of completing partial F-squares. We show the problem of completing a partial F-square can be transformed into the problem of completing a partial latin square by showing that each partial F-square is a partial latin square with some entries set equal to one another. The now proved "Evans Conjecture" on partial latin squares can then be used to prove that a partial $F(n;\lambda_1,\lambda_2,\ldots,\lambda_m)$ *square with at most* $n-1$ *cells occupied can be completed to an* $F(n;\lambda_1,\lambda_2,\ldots,\lambda_m)$ *square.*

1. INTRODUCTION

The reader is assumed to be familiar with the terms *latin square* and *partial latin square* (those who are not are referred to Dénes and Keedwell [2]). The frequency square is a generalization of the concept of latin square. The idea of such generalized latin squares was first used indirectly by Finney [4], [5] and [6]. Later Freeman [7] and Addelman [1] used the idea before Hedayat and Seiden [8] refined and formalized the concept of frequency square. The frequency sqaure (F-square for short) is defined by:

DEFINITION: Let $A = [a_{ij}]$ be an $n \times n$ matrix and let $\Sigma = \{s_1, s_2, \ldots, s_m\}$ $m \le n$, be the set of distinct elements of A. Suppose that for each i in the range $1 \le i \le m$, the element s_i occurs exactly λ_i times $(\lambda_i \ge 1)$ in each row and each column of A. Then we shall say that A is an *F-square* of order n on Σ with *frequency vector* $(\lambda_1, \lambda_2, \ldots, \lambda_m)$. We shall abbreviate this by saying that A is an $F(n;\lambda_1,\lambda_2,\ldots,\lambda_m)$ square on Σ.

Thus

$$\begin{pmatrix} 1 & 1 & 2 & 3 \\ 3 & 1 & 1 & 2 \\ 2 & 3 & 1 & 1 \\ 1 & 2 & 3 & 1 \end{pmatrix}$$

is an $F(4;2,1,1)$ square on $\{1,2,3\}$.

Accordingly we shall define a *partial* $F(n;\lambda_1,\lambda_2,\ldots,\lambda_m)$ square on the set $\Sigma = \{s_1,s_2,\ldots,s_m\}$ as an $n \times n$ array A in which:

(i) each cell is either empty or it contains an element of Σ,

(ii) the element s_i occurs no more than λ_i times in any row or column of A.

For example

$$A = \begin{bmatrix} \cdot & \cdot & \cdot & \cdot \\ \cdot & 1 & 1 & \cdot \\ \cdot & \cdot & 2 & \cdot \\ \cdot & \cdot & \cdot & \cdot \end{bmatrix}$$

is a partial $F(4;2,1,1)$ sqaure on $\{1,2,3\}$. A can also serve as an example of a partial $F(4;2,2)$ square on $\{1,2\}$ or a partial $F(4;3,1)$ square on $\{1,2\}$.

In this paper we show that there is a connection between completing partial F-squares and completing partial latin squares.

2. CONCEPTS

To assist us in analysing partial F-squares we now make the following definitions:

Definition: Let A be an $n \times n$ array and let m be an integer in the range $1 \le m \le n$. Suppose M is a set whose elements are cells of A. Then if M has the following properties:

(i) each row and column of A has at most m cells which are elements of M,

(ii) at least one row or column of A has exactly m cells which are elements of M, we shall say that M is a *partial m-region* of A. If each row and column of A has m cells which are elements of M we say that M is an *m-region* of A.

Example

(a) 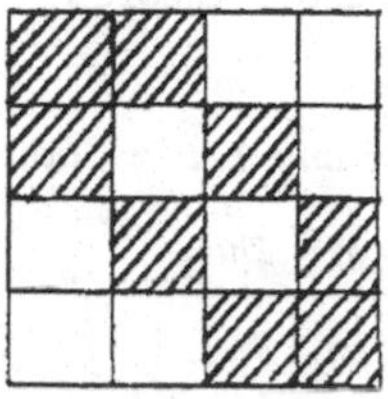 (b) 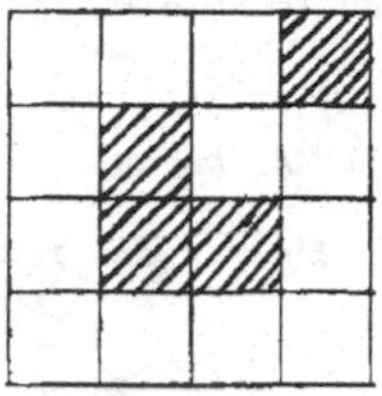

The shaded cells of the array (a) form a 2-region and of the array (b) form a partial 2-region.

Now suppose A is an $F(n;\lambda_1,\lambda_2,\ldots,\lambda_m)$ square on the set $\{s_1,s_2,\ldots,s_m\}$. Clearly the cells containing the element s_i constitute a λ_i-region of A. Similarly if B is a partial $F(n;\lambda_1,\lambda_2,\ldots,\lambda_m)$ square on $\{s_1,s_2,\ldots,s_m\}$ and if the element s_i occurs in B then the cells of B which contain s_i constitute a partial m_i-region of B where $1 \le m_i \le \lambda_i$.

3. DECOMPOSING PARTIAL M-REGIONS

Definition: A rectangular matrix with the property that all entries are either 0 or 1 is called a (0,1)-matrix.

Definition: A (0,1)-matrix with the property that no row or column contains more than one 1 shall be called a *Z-matrix*.

Definition: By *line* of a matrix we shall mean either a row or a column. A *line-sum* is the sum of all elements on a line.

The following theorem is to be found in Mirsky [9] (theorem 11.1.6):

Theorem 3.1. *Let A be a rectangular matrix whose elements are non-negative integers, and let m denote the maximum line-sum of A. Then*

$$A = Z_1 + \ldots + Z_m$$

where each Z_i *is a Z-matrix with at least one* 1.

Corollary 3.2 is an immediate consequence of the above theorem:

Corollary 3.2. *Let* A *be an* $n \times n$ *(0,1)-matrix and let* m *(where* $m \le n$*) be the maximum number of* 1*'s in any row or column of* A*. Then*

$$A = Z_1 + \ldots + Z_m$$

where each Z_i *is a Z-matrix.*

We can now show that if A is an $n \times n$ matrix and M is a partial m-region of A then M is the union of m pairwise disjoint partial 1-regions:

Theorem 3.3. *Let* A *be an* $n \times n$ *array and let* M *be a partial m-region of* A*. Then there exist partial 1-regions* $M_1, M_2, \ldots, M_m$ *such that* $M_1 \cup M_2 \cup \ldots \cup M_m = M$ *and* $M_i \cap M_j = \phi$ *for all* $i \neq j$.

Proof. Let B be the $n \times n$ (0,1)-matrix defined so that $b_{ij} = 1$ if and only if cell (i,j) of A is an element of M. The cells of B with entry 1 constitute a partial m-region of B. The result now follows from application of corollary 3.2.

4. CONNECTION BETWEEN PARTIAL F-SQUARES AND PARTIAL LATIN SQUARES.

We now use theorem 3.3 to show the connection between partial F-squares and partial latin squares.

Let A be a partial $F(n;\lambda_1,\lambda_2,\ldots,\lambda_m)$ square on $\Sigma = \{s_1,s_2,\ldots,s_m\}$ where $n > 1$. If the element s_i occurs in A then denote by M^i the partial

m_i-region of A constituted by those cells of A which contain s_i (clearly $m_i \le \lambda_i$). If M^i exists, then by theorem 3.3 there exist partial 1-regions $M_1^i, M_2^i, \ldots, M_{m_i}^i$ such that $M_1^i \cup M_2^i \cup \ldots \cup M_{m_i}^i = M_i$ and $M_\ell^i \cap M_k^i = \phi$ if $\ell \neq k$. We can define an $n \times n$ partial latin square B based on $\{1,2,\ldots,n\}$ as follows:

(i) if $1 \le i,j \le n$ and the cell (i,j) of A is empty, then the cell (i,j) of B is empty,

(ii) if $1 \le i,j \le n$ and the cell (i,j) of A contains s_ℓ then there exists an integer k in the range $1 \le k \le m_\ell$ such that cell (i,j) of A is an element of M_k^ℓ. Cell (i,j) of B contains the integer $\left(\left(\sum_{\alpha<\ell} \lambda_\alpha\right) + k\right)$.

We shall refer to B as a *derived partial latin square* of A. Note that A has at least one derived partial latin square.

Example

$$\begin{bmatrix} . & 1 & 4 & . & . \\ 1 & . & 3 & 2 & . \\ . & 2 & . & . & 1 \\ . & 4 & . & . & . \\ . & . & 1 & . & . \end{bmatrix}$$

is a derived partial latin square of the partial $F(5;2,3)$ square on $\{1,2\}$

$$\begin{bmatrix} . & 1 & 2 & . & . \\ 1 & . & 2 & 1 & . \\ . & 1 & . & . & 1 \\ . & 2 & . & . & . \\ . & . & 1 & . & . \end{bmatrix}$$

Theorem 4.1 now establishes a connection between completing partial frequency squares and completing partial latin squares.

<u>Theorem 4.1</u>. *Let* A *be a partial* $F(n;\lambda_1,\lambda_2,\ldots,\lambda_m)$ *square on* $\Sigma = \{s_1,s_2,\ldots,s_m\}$. *Then* A *is completable to an* $F(n;\lambda_1,\lambda_2,\ldots,\lambda_m)$ *square if and only if there exists a derived partial latin square* B *of* A *such that* B *is completable to a latin square.*

<u>Proof</u>. Suppose B is a derived partial latin square of A which is completable to a latin square C based on $\{1,2,\ldots,n\}$. We show A is completable to an $F(n;\lambda_1,\lambda_2,\ldots,\lambda_m)$ square. Let A' be the $F(n;\lambda_1,\lambda_2,\ldots,\lambda_m)$ square on Σ defined so that if $1 \le i,j \le n$ the cell (i,j) of A' contains

$$\begin{array}{lll} s_1 & \text{if} & 1 \le c_{i,j} \le \lambda_1 \\ s_2 & \text{if} & \lambda_1 < c_{i,j} \le \lambda_1 + \lambda_2 \\ \vdots & & \\ s_m & \text{if} & \sum_{\alpha=1}^{m-1} \lambda_\alpha < c_{i,j} \le \sum_{\alpha=1}^{m} \lambda_\alpha \end{array}$$

By our construction of A' it is clearly a completion of A.

Now suppose A is completable to an $F(n;\lambda_1,\lambda_2,\ldots,\lambda_m)$ square A' on Σ. We show that there exists a derived partial latin square B of A such that B is completable to a latin square C. For $\alpha = 1,\ldots,m$ define M^α to be the λ_α region of A' constituted by those cells of A' which contain s_α. By theorem 3.3, for each α in the range $1 \le \alpha \le m$ there exist 1-regions $M_1^\alpha, M_2^\alpha, \ldots, M_{\lambda_\alpha}^\alpha$ of A' such that $M_1^\alpha \cup M_2^\alpha \cup \ldots \cup M_{\lambda_\alpha}^\alpha = M^\alpha$ and $M_\ell^\alpha \cap M_k^\alpha = \phi$ if $\ell \ne k$. If $1 \le i,j \le n$ then there exists exactly one pair of integers (ℓ,k) such that cell (i,j) of A' is an element of M_k^ℓ. Let C be the latin square of order n based on $\{1,2,\ldots,n\}$ defined so that if $1 \le i,j \le n$ then cell (i,j) of C contains the integer $\left((\sum_{\alpha<\ell} \lambda_\alpha) + k\right)$ where M_k^ℓ is the 1-region which contains cell (i,j) of A'. Now define the $n \times n$ partial latin square B based on $\{1,2,\ldots,n\}$ as follows:

(i) if cell (i,j) of A is empty then cell (i,j) of B is empty,

(ii) if cell (i,j) of A is not empty then the cell (i,j) of B contains $c_{i,j}$.

B is then a derived partial latin square of A and C is a completion of B. □

5. A RESULT ON COMPLETING PARTIAL F-SQUARES

The following theorem on partial latin squares was known as the Evans Conjecture for many years (see Evans [3]). Recently it was shown to be true by Smetaniuk [10].

Theorem 5.1. *An* $n \times n$ *partial latin square with at most* $n-1$ *cells occupied can be completed to a latin square of order* n.

Using theorem 4.1 we can generalize the above theorem to obtain a result on completing partial F-squares:

Theorem 5.2. *Let* A *be a partial* $F(n;\lambda_1,\lambda_2,\ldots,\lambda_m)$ *square with at most* $n-1$ *cells occupied. Then* A *can be completed to an* $F(n;\lambda_1,\lambda_2,\ldots,\lambda_m)$ *square.*

Proof. It was shown at the start of the previous section that A has at least one derived partial latin square. Let B be a derived partial latin square of A. By construction B has as many occupied cells as A has and so by theorem 5.1 it is completable to a latin square. By theorem 4.1 A is thus completable to an $F(n;\lambda_1,\lambda_2,\ldots,\lambda_m)$ square.

ACKNOWLEDGEMENT

The author would like to thank Dr. Jennifer Seberry for her helpful suggestions during the writing of this paper.

REFERENCES

[1] S. Addelman, Equal and proportional frequency squares, *J. Amer. Statist. Assoc.*, 62 (1967), 226-240.

[2] J. Dénes and A.D. Keedwell, *Latin Squares and their Applicaitons*, English Universities Press Limited, London, 1974.

[3] T. Evans, Embedding incomplete latin squares, *Amer. Math. Monthly*, 67 (1960), 958-961.

[4] D.J. Finney, Some orthogonal properties of the 4×4 and 6×6 latin squares, *Ann. Eugenics*, 12 (1945), 213-219.

[5] D.J. Finney, Orthogonal partitions of the 5×5 latin squares, *Ann. Eugenics*, 13 (1946), 1-3.

[6] D.J. Finney, Orthogonal partitions of the 6×6 latin squares, *Ann. Eugenics*, 13 (1946), 184-196.

[7] G.H. Freeman, Some non-orthogonal partitions of the 4×4, 5×5 and 6×6 latin squares, *Ann. Math. Statist.* 37 (1966), 661-681.

[8] A. Hedayat and E. Seiden, F-square and orthogonal F-square designs: A generalization of latin square and orthogonal latin square designs, *Ann. Math. Statist.* 41 (1970), 2035-2044.

[9] L. Mirsky, *Transversal Theory*, Vol. 75 in Mathematics, Science and Engineering, Academic Press, Inc., New York and London, 1971.

[10] B. Smetaniuk, A new construction on latin squares - I: A proof of the Evans Conjecture, *Ars Combinatoria*, Vol. 11 (1981), 155-172.

BAER SUBSPACES IN THE N DIMENSIONAL PROJECTIVE SPACE

MARTA SVED

Baer subplanes are subplanes of order q *of a projective plane of order* q^2. *Their intersection configurations are well known. The concept of Baer subplanes is extended to* n *dimensions and two dimensional results are generalised to Baer subspaces of* $PG(n,q^2)$.

1. INTRODUCTION

Let S be the set of points of the projective geometry $PG(n,q^2)$ over $GF(q^2)$ of dimension n and order q^2 (q is a power of some prime). The points of S may be identified with ordered sets of $n+1$ homogeneous coordinates: $(x_1 x_2, \ldots, x_{n+1})$ *with* $x_i \in GF(q^2)$ $(i=1,\ldots,n+1)$ and not all x_i being equal to zero. A Baer *subspace* $PG(n,q)$ of $PG(n,q^2)$ can be obtained by selecting those points of S for which $x_i \in GF(q)$, $(i=1,\ldots,n+1)$ and *restricting* the geometry to this set. We will *denote* this set of points *by* B_0 and call it the *"real Baer subspace"*. Clearly, a change of coordinates will lead to *different subsets of* S *with geometries isomorphic to that of* B_0. The coordinates of all the points of S are determined by the choice of $n+2$ fundamental points: (10...0), (01...0), ... (00...1), (11...1). At the same time these points determine also B_0. Another set of $n+2$ points of S, no $n+1$ of them linearly dependent, may be chosen for fundamental points and determine thus another Baer subspace. The group of homographies Γ of $PG(n,q^2)$ is known to be transitive on ordered sets of $n+2$ points, no $n+1$ linearly dependent; hence the *Baer subspaces* B of S, are the homographical images of B_0.

The two dimensional case has been extensively studied.

The Baer subspaces of S, where S is $PG(2,q^2)$, are known as Baer subplanes. The real Baer plane B_0, is determined by the four fundamental points (100), (010), (001), (111). Any set of points A, B, C, D where the quadrilateral $ABCD$ is non degenerate (i.e. no three of the points A,B,C,D are collinear) will be represented as the homographic image of the fundamental points. The order of the group of homographies Γ of $PG(2,q^2)$ is

$$|\Gamma| = (q^4+q^2+1)(q^4+q^2)q^4(q^2-1)^2.$$

Each homography of Γ carries $\mathcal{B}_0$ into some Baer plane $\mathcal{B}$, and conversely every Baer plane, uniquely determined by some non-degenerate quadrilateral is the homographic image of $\mathcal{B}_0$.

Denote by Γ_0 the subgroup of Γ which *fixes* $\mathcal{B}_0$, i.e. takes the points of the *real* Baer plane into real points (with homogeneous coordinates in $GF(q)$). We have

$$|\Gamma_0| = (q^2+q+1)(q^2+q)q^2(q-1)^2.$$

Thus $PG(2,q^2)$ has

$$\frac{|\Gamma|}{|\Gamma_0|} = (q^2-q+1)q^3(q^2+1)(q+1) \quad \text{Baer subplanes.}$$

The possible intersection configurations of the Baer planes of a projective field-plane have been investigated by J. Colman [1], R.C. Bose, J.W. Freeman and D.G. Glynn [2], and the author [3].

The results established in these earlier investigations are listed in the following

(a) Two Baer subplanes may be disjoint or may intersect.

(b) The intersection of two Baer subplanes is a closed configuration; if two points are common to two subplanes, so is the line joining them, and if two subplanes share two lines, they share their intersection.

(c) The intersection of two distinct Baer subplanes has 0, 1, 2 or $q+1$ points common with any line of S. This means that if a line has two points in *common* with two Baer subplanes, then it contains 2 or $q+1$ points of the intersection.

(d) The number of points shared by two Baer subplanes is equal to the number of lines shared.

(e) The possible intersection configurations of two *distinct* Baer subplanes are

(1) the empty set;

(2) one point and one line, which may or may not be incident.

(3) two points and two lines, one of the two points being the intersection of the two lines and one of the lines being the line joining the two points;

(4) three points and three lines, forming a triangular configuration.

(5) $q+1$ points and $q+1$ lines, the $q+1$ points lying on one of the lines, the $q+1$ lines passing through one of the points.

(6) $q+2$ points and $q+2$ lines, one line being incident with $q+1$ of the points, the other lines being the joins of the remaining point with those $q+1$ points.

2. THE THREE DIMENSIONAL CASE. COMPUTING RESULTS.

A computer program, described in [3], based on Singer's Theorem [4] was adapted for higher dimensions. In particular, Baer subspaces of the three dimensional projective geometries PG(3,4), PG(3,9), PG(3,16) and PG(3,25) were investigated in some detail.

The program generated the points and the planes of S, the projective space of 3 dimensions of order q^2 $(q=2,3,4,5)$, and the "real" lines, i.e. the lines belonging in each case to $\mathcal{B}_0$. A collection of other Baer subspaces was found by applying homography groups (in particular, Singer groups) to $\mathcal{B}_0$. Computer search was used to study intersection configurations. Out of these computations some conjectures could be made, listed below.

(1) Each plane of S intersects each Baer subspace $\mathcal{B}$ in a plane or in a line of $\mathcal{B}$.

(2) The number of points shared by two Baer subspaces is equal to the number of planes shared by them.

(3) The points shared by two Baer subspaces form one of the following configurations:

(a) the empty set;
(b) one point;
(c) two points;
(d) three non collinear points;
(e) four non coplanar points;
(f) $q+1$ points of a line;
(g) $q+2$ points, $q+1$ on one line;
(h) $q+3$ points: $q+1$ on one line, and the line joining the other two skew to the first line;
(i) $2(q+1)$ points of two skew lines;
(j) q^2+q+1 points of one plane;
(k) q^2+q+2 points: q^2+q+1 points coplanar and one external.

In the attempt to prove these conjectures, a generalisation to n dimensions was made. In the following sections, theorems for n dimensions will be stated and proved. These settle conjectures (1) and (2). As for the conjectures listed in (3), no proof is given at this stage to establish the *existence* of all of them for the general case of $PG(3,q^2)$, though the computer search found examples for each for $q = 2,3,4,5$. However, a theorem, describing the only *possible* structures of the intersections of two Baer subgeometries in n dimensions will be stated and proved.

3. BASIC PROPERTIES OF n DIMENSIONAL BAER SUBGEOMETRIES

As stated in the Introduction, the points of S, the n dimensional projective space of order q^2 correspond to $(n+1)$-tuples of homogeneous coordinates belonging to $GF(q^2)$, those of $\mathcal{B}_0$, the *real* Baer subspace of order q to $(n+1)$-tuples of homogeneous coordinates belonging to $GF(q)$ and all the other Baer subspaces are homographic images of $\mathcal{B}_0$.

To determine the number of Baer subspaces in the geometry $PG(n,q^2)$, we proceed as in two dimensions, as seen in the Introduction. Let Γ be the group of homographies of S. Let Γ_0 be the subgroup of Γ fixing $\mathcal{B}_0$. Then

$$|\Gamma| = q^{n(n+1)} \prod_{i=2}^{n+1} (q^{2i}-1)$$

and

$$|\Gamma_0| = q^{\frac{n(n+1)}{2}} \prod_{i=2}^{n+1} (q^{i}-1).$$

Thus the number of Baer subspaces in S is

$$N = \frac{|\Gamma|}{\Gamma_0} = q^{\frac{(n+1)n}{2}} \prod_{i=2}^{n+1} (q^{i}+1). \tag{3.1}$$

Let now $\mathcal{B}$ be any of the Baer subspaces of S. A point of S not belonging to $\mathcal{B}$ is said to be *external* to $\mathcal{B}$. We can generalise in the following theorem a basic property of Baer subplanes.

<u>Theorem 1</u>. *Let* P *be a point of* S, *external to the Baer subspace* $\mathcal{B}$. *Then* P *lies on exactly one line of* $\mathcal{B}$.

<u>Proof</u>. P can lie on *at most* one line of $\mathcal{B}$, since two lines belonging to $\mathcal{B}$ intersect at a point of $\mathcal{B}$. Hence we must show that through each external point P we can find a line of $\mathcal{B}$. Equivalently, we show that S has no other points than those lying on lines of $\mathcal{B}$. To show this we count the points on the lines of $\mathcal{B}$, but external to $\mathcal{B}$.

The number of the lines of $\mathcal{B}$ is

$$L = \frac{(q^{n+1}-1)(q^n-1)}{(q-1)(q^2-1)},$$

(using the formula for the number of 1 dimensional subspaces of a projective space of order q and dimension n. See e.g. [5].)

Each of these lines has q^2-q points external to $\mathcal{B}$, and these cannot be common to two different lines, as seen before. Thus the number of the points external to $\mathcal{B}$, but incident with some line of $\mathcal{B}$ is

$$\text{(3.2)} \qquad \frac{(q^{n+1}-1)(q^n-1)}{(q-1)(q^2-1)}(q^2-q) = \frac{q(q^{n+1}-1)(q^n-1)}{q^2-1}.$$

On the other hand the number of points of S, external to $\mathcal{B}$ is

$$\text{(3.3)} \qquad \frac{(q^2)^{n+1}-1}{q^2-1} - \frac{q^{n+1}-1}{q-1} - \frac{q(q^{n+1}-1)(q^n-1)}{q^2-1}$$

Comparing the results (3.2) and (3.3) shows that each external point lies on a line of $\mathcal{B}$. □

In the two dimensional case it is also true that each line of the projective plane $PG(2,q^2)$ has at least one point in common with any of its Baer subplanes. If the line does not belong to the Baer subplane, then it has exactly one point in common with it. (Clearly, if the line has 2 points in common with the Baer subplane, then it belongs to the subplane, and so it must have $q+1$ points in common with the Baer subplane.)

In dimensions higher than 2, a line does not necessarily intersect a Baer subgeometry $\mathcal{B}$. In fact we can show that through each point external to $\mathcal{B}$, the number of lines skew to $\mathcal{B}$ is

$$\text{(3.4)} \qquad L_S = q^3\,\frac{(q^{n-1}-1)(q^{n-2}-1)}{q^2-1} > 0 \quad \text{for } n > 2.$$

To prove (3.4) we count the number of lines through the external point P, intersecting $\mathcal{B}$. Exactly one of these lines contains $q+1$ points of $\mathcal{B}$, hence the remaining points of $\mathcal{B}$,

$$\frac{q^{n+1}-1}{q-1} - (q+1) = q^2\,\frac{q^{n-1}-1}{q-1} \quad \text{in number,}$$

are each on exactly one line through P. Since the total number of lines through P is equal to the number of points of some hyperplane of S external to P and hence is

$$\frac{q^{2n}-1}{q^2-1},$$

We obtain for the number of lines through P skew to B

$$\frac{q^{2n}-1}{q^2-1} - q^2\,\frac{q^{n-1}-1}{q-1} - 1.$$

Simplifying this, we obtain (3.4).

In the two dimensional situation the lines of S can be regarded as hyperplanes in $PG(2,q^2)$. Hence it is appropriate in the general situation of n dimensions to look at the intersections of the hyperplanes of S and B. Here we have the situation summarised in the following

Theorem 2. *The intersection of a hyperplane of* S *with a Baer subgeometry is either a hyperplane of the subgeometry, (of dimension* $n-1$*), or a "hyperline" of dimension* $n-2$.

Proof. Clearly, for any pair of points of the intersection the joining line belongs to both: the hyperplane H of S, and the subgeometry, B. Thus all the points of that line which also belong to B are in the intersection, ensuring that the intersection is a subspace of B. It is clear that some intersections are hyperplanes of B, since the hyperplanes of B extend uniquely into hyperplanes of S. However, there are also hyperplanes of S intersecting B in lower dimensional subspaces, since the number of the hyperplanes of B is less than the number of the hyperplanes of S.

We show first that the intersection of a hyperplane of S with B is not empty. The number of lines of B is given by

$$\frac{(q^{n+1}-1)(q^n-1)}{(q-1)\;(q^2-1)} \text{ as seen before.}$$

From dimensional considerations it follows that each line intersects the hyperplane in at least one point.

Comparing the number of points of the hyperplane H with the number of lines of B, we find that

$$\frac{(q^{n+1}-1)(q^n-1)}{(q-1)\;(q^2-1)} - \frac{q^{2n}-1}{q^2-1} = \frac{q(q^n-1)(q^{n-1}-1)}{(q-1)(q^2-1)} > 0\,.$$

This means that some points of H must be common to at least two lines of B and hence must be *internal* points of B.

In order to determine the dimension of the possible intersections of some hyperplane with B, we count the *incidences* of *points of* H

with *lines of* $\mathcal{B}$.

Let x be the number of points and y the number of lines of $H \cap \mathcal{B}$.

Then $\frac{q^{2n}-1}{q^2-1} - x$ points of H *do not belong* to $\mathcal{B}$ and so by Theorem 1, each of these points accounts for just one incidence of the type to be counted.

Similarly, $\frac{(q^{n+1}-1)(q^n-1)}{(q-1)(q^2-1)} - y$ lines of $\mathcal{B}$ *do not belong* to H and so these lines intersect H in just one point, and so each counts for exactly one incidence.

Each of the x points in $H \cap \mathcal{B}$ has $\frac{q^n-1}{q-1}$ lines of $\mathcal{B}$ passing through it, and each of the y lines of $H \cap \mathcal{B}$ has q^2+1 points of H incident with it.

Thus the incidence equation becomes

$$(3.5) \qquad x\,\frac{q^n-1}{q-1} + \left(\frac{q^{2n}-1}{q^2-1} - x\right) = y(q^2+1) + \left(\frac{(q^{n+1}-1)(q^n-1)}{(q-1)(q^2-1)} - y\right)$$

Simplifying and dividing by q, we obtain

$$(3.6) \qquad x\,\frac{q^{n-1}-1}{q-1} - qy = \frac{(q^n-1)(q^{n-1}-1)}{(q-1)(q^2-1)}$$

Denote now the dimension of $H \cap \mathcal{B}$ by d. Since $H \cap \mathcal{B}$ is a proper subspace of $\mathcal{B}$, $d < n$.

Then

$$(3.7) \qquad x = \frac{q^{d+1}-1}{q-1} \quad \text{and} \quad y = \frac{(q^{d+1}-1)(q^d-1)}{(q-1)(q^2-1)} .$$

Substituting (3.7) into (3.6), we obtain

$$\frac{(q^{d+1}-1)(q^{n-1}-1)}{(q-1)^2} - q\,\frac{(q^{d+1}-1)(q^d-1)}{(q-1)(q^2-1)} = \frac{(q^n-1)(q^{n-1}-1)}{(q-1)(q^2-1)} ,$$

whence

$$(3.8) \quad (q+1)(q^{d+1}-1)(q^{n-1}-1) - (q^{d+1}-1)(q^{d+1}-q) = (q^n-1)(q^{n-1}-1) .$$

Setting $t = q^{d+1}$ in (3.8) and simplifying we obtain the quadratic

$$t^2 - t(q^n+q^{n-1}) + q^{2n-1} = 0.$$

This has two solutions for t: $t = q^n$ or q^{n-1}, hence $d = n-1$ *or* $n-2$. These are the only values possible for the dimension of $H \cap \mathcal{B}$.

The solution $d = n-1$ gives the hyperplanes as expected and the only other possibility is a subspace of dimension $n-2$ (called "hyperline"). □

Theorem 2 may be regarded as the dual of Theorem 1.

In this light the case of Baer planes can be reviewed. The hyperplanes are the *lines* of $PG(2,q^2)$ which intersect any Baer subplane in a "hyperplane" of the Baer plane, i.e. in a line, or in a subspace of dimension 0, i.e. in a point.

4. INTERSECTIONS OF BAER SUBGEOMETRIES

The following theorem generalises conjecture (2) in Section 2 and also the corresponding theorem on Baer subplanes, proved in [2].

Theorem 3. *The number of points of intersection of two Baer subspaces of* S *is equal to the number of hyperplanes shared by them.*

Note: "sharing" a hyperplane (or more generally a subspace) does not necessarily mean that the hyperplane is shared *pointwise* by the two subgeometries. It means merely that there exists a hyperplane of S which intersects each of the two Baer subspaces in one of their hyperplanes which however need not coincide pointwise.

Proof. Let B_0 and B_1 be the two Baer subgeometries considered and let the *number of points shared by them be* r, where $r \geq 0$.

Let k_i *be the number of hyperplanes belonging to* B_0 *which share* i *points with* B_1, $h_i \geq 0$.

Then

$$\sum_i h_i = \frac{q^{n+1}-1}{q-1} \text{ (= the number of hyperplanes of } B_0) \tag{4.1}$$

and

$$\sum_i ih_i = r\,\frac{q^n-1}{q-1}. \tag{4.2}$$

Equation (4.2) arises through counting incidences of points of $B_0 \cap B_1$ with hyperplanes of B_0. The right hand side is obtained by noting that through each point of B_0 there are $\frac{q^n-1}{q-1}$ hyperplanes of B_0.

Next we count the incidences of the points of $B_1 \backslash B_0$ and the hyperplanes of B_0.

Assume that x_i out of the h_i hyperplanes defined as above intersect B_1 in a hyperplane. Then, using Theorem 2, $h_i - x_i$ hyperplanes intersect B_1 in a hyperline.

The contribution of the h_i hyperplanes to the number of incidences to be counted is then

$$(4.3) \qquad x_i\left(\frac{q^n-1}{q-1}-i\right)+(h_i-x_i)\left(\frac{q^{n-1}-1}{q-1}-i\right).$$

On the other hand, the number of required incidences is given by the sum

$$\sum_{P\in\mathcal{B}_1\setminus\mathcal{B}_0} h_P$$

where h_P denotes the number of hyperplanes of $\mathcal{B}_0$ through some point P. The points P considered in the above sum are all external to $\mathcal{B}_0$. All hyperplanes of $\mathcal{B}_0$ going through a fixed external point P intersect in one line, the unique line ℓ_P of $\mathcal{B}_0$ through P. For let H_P be any hyperplane belonging to $\mathcal{B}_0$ through P, then by dimensional considerations, it is intersected by any line belonging to the projective geometry of $\mathcal{B}_0$. Hence ℓ_P intersects H_P in at least one point of $\mathcal{B}_0$ as well as in P and so ℓ_P belongs to H_P. Thus the number h_P is the same as the number of hyperplanes of $\mathcal{B}_0$ through a line of $\mathcal{B}_0$ and thus it is the same for all points P, external to $\mathcal{B}_0$. The number may be calculated by counting the incidences of the hyperplanes of $\mathcal{B}_0$ and points of S external to $\mathcal{B}_0$.

Each of the hyperplanes of $\mathcal{B}_0$ when extended into S contains $\frac{q^{2n}-1}{q^2-1}-\frac{q^2-1}{q-1}$ external points, while the total number of external points is

$$\frac{q^{2n+2}-1}{q^2-1}-\frac{q^{n+1}-1}{q-1},$$

hence

$$h_P\left(\frac{q^{2n+2}-1}{q^2-1}-\frac{q^{n+1}-1}{q-1}\right)=\frac{q^{n+1}-1}{q-1}\left(\frac{q^{2n}-1}{q^2-1}-\frac{q^n-1}{q-1}\right),$$

giving

$$h_P=\frac{q^{n-1}-1}{q-1}$$ (the number of points of an $n-2$ dimensional subspace of $\mathcal{B}_0$).

Thus

$$(4.4) \qquad \sum_{P\in\mathcal{B}_1\setminus\mathcal{B}_0} h_P=\frac{q^{n-1}-1}{q-1}\left(\frac{q^{n+1}-1}{q-1}-r\right).$$

Using (4.3) and (4.4) we obtain the required incidence equation:

$$(4.5) \qquad \sum_i\left(x_i\left(\frac{q^n-1}{q-1}-i\right)+(h_i-x_i)\left(\frac{q^{n-1}-1}{q-1}-i\right)\right)=\frac{q^{n-1}-1}{q-1}\left(\frac{q^{n+1}-1}{q-1}-r\right)$$

The left hand side of equation (4.5) simplifies to

$$\left(\frac{q^n-1}{q-1}-\frac{q^{n-1}-1}{q-1}\right)\sum_i x_i+\frac{q^{n-1}-1}{q-1}\sum_i h_i-\sum_i ih_i .$$

Here $\sum_i x_i = x$ *gives the number of hyperplanes shared by* B_i *and* B_1.

Furthermore we use (4.1) and (4.2), thereby obtaining

$$q^{n-1}x + \frac{q^{n-1}-1}{q-1}\,\frac{q^{n+1}-1}{q-1} - r\,\frac{q^{n}-1}{q-1} = \frac{q^{n-1}-1}{q-1}\left(\frac{q^{n+1}-1}{q-1} - r\right)$$

whence

$$q^{n-1}x = r\left(\frac{q^{n}-1}{q-1} - \frac{q^{n-1}-1}{q-1}\right) = rq^{n-1}$$

and so $x = r$ as claimed.

In particular if two Baer subgeometries are disjoint pointwise, they have no common hyperplane. □

Theorem 3 does not give any information about the possible number of common points of intersections of the two Baer subgeometries.

It is already clear from the known two dimensional results and the computer findings in three dimensions that in general, the intersection configurations of two Baer subspaces do not form necessarily subspaces of the two subgeometries. The geometrical nature of the possible intersections and the restrictions on the possible numbers of intersection points will be summarised in the following.

It will be first shown that the result for the intersections of two Baer subplanes of $PG(2,q^2)$ listed as Result (c) in the Introduction holds generally in the n dimensional case.

Theorem 4. *Let* P *and* Q *be two points shared by the Baer subgeometries* B_0 *and* B_1. *Then the line* ℓ, *determined by* P *and* Q *has either just the two points* P *and* Q *common to both spaces, or the two spaces share* $q+1$ *points of* ℓ.

Proof. Without loss of generality, the coordinates in S may be chosen so that B_0 is the *real* Baer subspace (i.e. the coordinates of the points of B_0 belong to $GF(q)$.) In particular, P and Q may be selected to be the fundamental points $(1,0,0,\ldots,0)$ and $(0,1,0,0,\ldots,0)$ ($n+1$ coordinates). Since P and Q are also in B_1, we may consider a homography taking B_0 to B_1 and leaving P and Q invariant. This has a matrix of form

$$\begin{bmatrix} \alpha & 0 & * & * \\ 0 & \beta & * & * \\ 0 & 0 & * & * \\ \vdots & \vdots & & \\ 0 & 0 & * & * \end{bmatrix}$$

Consider a point R in $(\ell \cap \mathcal{B}_0)$. Without loss of generality we may assume R is the point $(1,1,0,\ldots,0)$.

Then the image of R in $\mathcal{B}_1$ is

$$(\alpha,\beta,0,\ldots,0).$$

The image of R belongs to $\mathcal{B}_0$ if and only if $\frac{\alpha}{\beta} \in GF(q)$. We can show now that if T is any other point of $\mathcal{B}_0 \cap \ell$, then the image of T in $\mathcal{B}_1$ belongs to $\mathcal{B}_0$ if and only if R belongs to $\mathcal{B}_0$.

Indeed, if T is the point $(x,y,o,\ldots,0)$ where $x,y \in GF(q)$ and $x \neq 0$, $y \neq 0$, then the image of T is

$$(\alpha x,\beta y,0,\ldots,0) \text{ or equivalently}$$

$$(\tfrac{\alpha}{\beta}x,y,0,\ldots,0).$$

Clearly, the coordinates of this point belong to $GF(q)$ if and only if $\frac{\alpha}{\beta} \in GF(q)$. This means that the Baer subspaces $\mathcal{B}_0$ and $\mathcal{B}_1$ may share either $q+1$ points of the line ℓ or none other than P and Q, as stated in the theorem. □

The above theorem illustrates the difference between the two statements:

"A subspace of S belongs to $\mathcal{B}_0 \cap \mathcal{B}_1$"

or "A subspace of S belongs *pointwise* to $\mathcal{B}_0 \cap \mathcal{B}_1$".

Definition. A subspace of $\mathcal{B}_0$ is a *component* of the intersection $\mathcal{B}_0 \cap \mathcal{B}_1$ if

(1) *all its points* belong to $\mathcal{B}_0 \cap \mathcal{B}_1$,

(2) it is *maximal*, in the sense that it is not a subspace of some higher dimensional subspace belonging pointwise to $\mathcal{B}_0 \cap \mathcal{B}_1$.

Lemma 5. *Let S_d be a d dimensional subspace of order q belonging pointwise to $\mathcal{B}_0 \cap \mathcal{B}_1$. Let ℓ be a line intersecting S_d in P, and containing two points Q,R distinct from P, belonging to $\mathcal{B}_0 \cap \mathcal{B}_1$. Then the $d+1$ dimensional subspace S_{d+1}, spanned by S_d and ℓ, belongs pointwise to $\mathcal{B}_0 \cap \mathcal{B}_1$.*

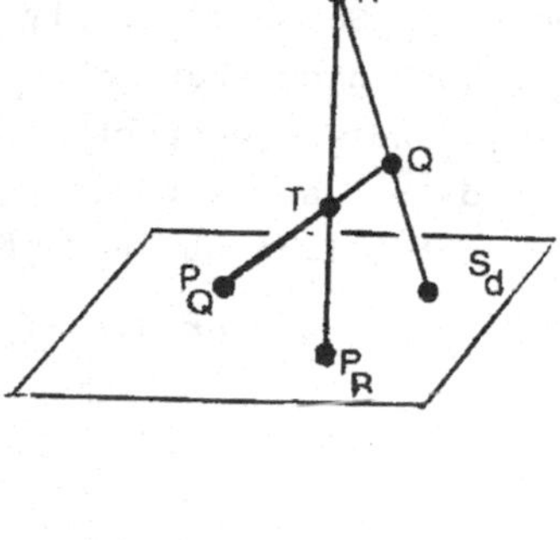

Proof. Let T be a point of $S_{d+1}\backslash(\ell \cup S_d)$. By dimensional considerations, the line QT intersects S_d in *one* point P_Q. So the line QP_Q belongs to $\mathcal{B}_0 \cap \mathcal{B}_1$, having two points in it. Similarly, T and R determine the line RP_R (P_R in S_d), belonging to $B_0 \cap B_1$. Thus the intersection of QP_Q and RP_R must belong to $B_0 \cap B_1$, hence T is a point of $B_0 \cap B_1$ as claimed. □

Lemma 6. *If two subspaces,* S_1 *and* S_2 *of dimensions* d_1 *and* d_2 *respectively belong pointwise to* $B_0 \cap B_1$ *and* $S_1 \cap S_2 \neq \phi$, S_1 *or* S_2, *then each is contained in a higher dimensional subspace which belongs pointwise to* $B_0 \cap B_1$.

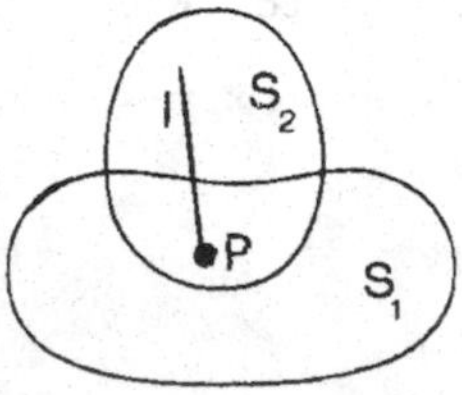

Proof. Suppose P is an intersection point of S_1 and S_2. Let ℓ be a line in S_2 through P, then by Lemma 5, the d_1+1 dimensional space, spanned by S_1 and ℓ belongs pointwise to $B_0 \cap B_1$. Similarly S_2 is a subspace of some d_2+1 dimensional subspace, belonging pointwise to $B_0 \cap B_1$. □

It follows from the definition of components, and Lemma 6, that the *components* of the intersection B_0 and B_1 are *pointwise skew*.

Lemma 7. *The space spanned by two components* S_1 *and* S_2 *of dimensions* d_1 *and* d_2 *respectively, does not contain any point of* $B_0 \cap B_1$ *other than those in* S_1 *and* S_2.

Proof. Let S_3 be the space spanned by S_1 and S_2. From the dimensional equation, the dimension of S_3 is

$$d_1+d_2+1 \quad \text{(since } S_1 \text{ and } S_2 \text{ are skew).}$$

Suppose that P is a point of $S_3 \setminus (S_1 \cup S_2)$ and P belongs to $B_0 \cap B_1$.

Let S_1' be the space spanned by S_1 and P. The dimension of S_1' is d_1+1. Similarly, the dimension of S_2', the space spanned by P and S_2 is d_2+1.

Applying the dimensional equation to S_1' and S_2' we find that the two spaces intersect in a *line* ℓ through P. Since S_1 is a hyperplane of S_1' and S_2 a hyperplane of S_2' it follows that the line ℓ intersects each, S_1 and S_2 (in a point). So we can apply Lemma 5 to line ℓ and S_1, also to ℓ and S_2 obtaining that both are subspaces of higher dimensional subspaces (S_1' and S_2' respectively) which belong pointwise to $B_0 \cap B_1$ and so S_1 and S_2 are not components. This contradiction shows that P cannot belong to $B_0 \cap B_1$. □

Lemma 8. *The dimension of the space spanned by* t *components of the intersection of* B_0 *and* B_1 *is*

$$d_1+d_2+\ldots+d_t+t-1.$$

There is no additional point of $\mathcal{B}_0 \cap \mathcal{B}_1$ *in this space other than the points of the* t *components.*

Proof. We proceed by induction from the previous lemma which settles the case $t=2$. Assume now that the proposition is true for t components: $S_1, S_2 \ldots S_t$ $(t \geq 2)$, and add the component S_{t+1}. Since S_{t+1} is skew to each of the first t components and by the inductive hypothesis the space S', spanned by $S_1, S_2, \ldots, S_t$ has no other point of $\mathcal{B}_0 \cap \mathcal{B}_1$, it follows that S_{t+1} is skew to S'. Furthermore, the space spanned by S' and S_{t+1} cannot contain any point of $\mathcal{B}_0 \cap \mathcal{B}_1$, external to $S' \cup S_{t+1}$. For suppose that P is such a point. Let S'' be the subspace spanned by S_{t+1} and P. By dimensional considerations S'' intersects S' in a point Q. The line $\ell = PQ$ intersects S_{t+1} in the point R. Since both P and R belong to $\mathcal{B}_0 \cap \mathcal{B}_1$, ℓ is in $\mathcal{B}_0 \cap \mathcal{B}_1$. S' is a subspace of $\mathcal{B}_0 \cap \mathcal{B}_1$ (though not pointwise). The intersection of S' and ℓ belongs to $\mathcal{B}_0 \cap \mathcal{B}_1$, hence $Q \in \mathcal{B}_0 \cap \mathcal{B}_1$. However, by the inductional hypothesis Q cannot belong to $S' - (S_1 \cup S_2 \ldots \cup S_t)$. Hence Q must belong to one of the components S_i $(i=1,2,\ldots,t)$. This, however contradicts Lemma 7.

If we denote the dimension of S_{t+1} by d_{t+1}, it follows from the inductive hypothesis and the dimensional equation that the dimension of the space spanned by

$$S' \text{ on } S_{t+1} \text{ is}$$

$$(d_1+d_2+\ldots+d_t+t-1) + d_{t+1} + 1$$

or

$$d_1+d_2+\ldots+d_t+d_{t+1}+(t+1)-1.$$

This completes the proof. □

Note: Isolated points of the intersection $\mathcal{B}_0 \cap \mathcal{B}_1$ are components of dimension 0.

We need one further lemma for obtaining bounds for r, the number of points in a possible intersection of $\mathcal{B}_0$ and $\mathcal{B}_1$.

Lemma 9. *Let* q *and* m *be integers greater than* 1 *and the set* $\{r_1, r_2, \ldots, r_k\}$ *a non trivial partition of* m, *i.e.*

$$r_1 + r_2 + \ldots + r_k = m$$

where $1 \leq r_1 \leq r_2 \leq \ldots \leq r_k$ *and* $k > 1$. *Then*

$$\sum_{i=1}^{k} q^{r_i} \leq q^m . \tag{4.6}$$

The inequality is strict except for the case

$$q = m = 2.$$

<u>Proof</u>. When $m=2$, the only non-trivial partition is

$$r_1 = r_2 = 1.$$

In this case

$$\sum_{i=1}^{2} q^{r_i} = 2q \begin{cases} < q^2 & \text{for } q > 2 \\ = q^2 & \text{for } q = 2. \end{cases}$$

We proceed by induction assuming that (4.6) is valid for all $m \le n$. Let

$$\sum_{i=1}^{k} r_i = n+1.$$

Then

$$\sum_{i=1}^{k} q^{r_i} = q^{r_1} + \sum_{i=2}^{k} q^{r_i},$$

where

$$\sum_{i=2}^{k} r_i = n+1-r_1 \le n \quad \text{since} \quad r_1 \ge 1.$$

By the inductive hypothesis

$$\sum_{i=2}^{k} q^{r_i} \le q^n$$

and so

$$\sum_{i=1}^{k} q^{r_i} \le q^{r_1} + q^n \quad \text{where} \quad 0 < n-r_1 < n .$$

We have $q^{r_1} + q^n = q^{r_1} \quad (1+q^{n-r_1}) < q^{r_1}.q^{n-r_1+1} = q^{n+1}$ for all $q > 1$. Thus for all $q > 1$ and $m > 2$ and $\sum_{i=1}^{k} r_i = m \; (r_i \ge 1)$

$$\sum_{i=1}^{k} q^{r_i} < q^m. \quad \square$$

The structure of the intersections of two distinct Baer subgeometries can now be characterised in the following.

<u>Theorem 10</u>. *(a) If the intersection of two Baer subgeometries* $\mathcal{B}_0$ *and* $\mathcal{B}_1$ *is non-empty, then it is the union of* t *pairwise skew components* $(t \ge 1)$, *of dimensions* $d_1, d_2, \ldots, d_t$ *respectively. The dimension of the space spanned by these components is*

$$d_1 + d_2 + \ldots + d_t + t - 1 \le n.$$

(b) If r *denotes the number of points common to* $\mathcal{B}_0$ *and* $\mathcal{B}_1$, *then*

$$(4.7) \qquad 0 \le r = \sum_{i=1}^{t} \frac{q^{d_i+1}-1}{q-1} \le \frac{q^n-1}{q-1} + 1.$$

<u>Proof</u>. Part (a) is a restatement of Lemma 8, noting also that the dimension of the space spanned by the components of the intersection cannot exceed the dimension of $\mathcal{B}_0$ (or $\mathcal{B}_1$).

In part (b) the number of intersection points is expressed as the sum of the numbers of points in the components. The only thing remaining to be proved is the inequality in (4.7).

We discuss first three cases;

(i) The components are a hyperplane H_0 of B_0 and a point P not belonging to H_0. In section (5) we will show that such intersections always exist. In this case clearly

$$r = \frac{q^n-1}{q-1} + 1,$$

hence the upper bound stated in (4.7) is reached in this case.

(ii) The components are t linearly independent points where $t \le n+1$. Write $\frac{q^n-1}{q-1} + 1 = q^{n-1} + q^{n-2} + \ldots + 1 + 1 > n+1$ since $q > 1$ and $n > 1$.

Hence in this case the inequality is strict.

(iii) $t=1$, i.e. the intersection is a single subspace, of dimension at most $n-1$. In this case the inequality is obviously strict.

We next show that in the remaining cases where $t > 1$ and for $\max\{d_1, d_2, \ldots, d_t\} = d_t$ we have

$$1 \le d_t \le n-2$$

the strict inequality

$$\sum_{i=1}^{t} \frac{q^{d_i+1}-1}{q-1} < \frac{q^n-1}{q-1} + 1 \quad \text{is valid.}$$

Write $r_i = d_i + 1 \quad (i=1,\ldots,t)$.

The inequality to be proved becomes

$$\sum_{i=1}^{t} q^{r_i} < q^n - 1 + q - 1 + t.$$

Since $q \ge 2$ and $t > 1$, it suffices to show that

$$\sum_{i=1}^{t} q^{r_i} \le q^n$$

provided that

$$\sum_{i=1}^{t} r_i = \sum_{i=1}^{t} d_i + t \le n+1$$

and

$$2 \le r_t \le n-1.$$

Write

(4.8) $$\sum_{i=1}^{t} q^{r_i} = \sum_{i=1}^{t-1} q^{r_i} + q^{r_t}.$$

Since

$$\sum_{i=1}^{t-1} r_i = \sum_{i=1}^{t} r_i - r_t \le n+1-2 = n-1,$$

we have by Lemma 9 that

$$\sum_{i=1}^{t-1} q^{r_i} \leq q^{n-1} ,$$

also $q^{r_t} \leq q^{n-1}$ since $r_t \leq n-1$.

Hence we have for the right hand side of (4.8)

$$\sum_{i=1}^{t-1} q^{r_i} + q^{r_t} \leq 2q^{n-1} \leq q^n \quad \text{since} \quad q \geq 2.$$

This completes the proof. □

5. OUTLINE FOR FURTHER INVESTIGATIONS

Theorem 10 settles in general the possible structures of the intersections of two Baer subspaces. However, existence questions are left open. The conjectures in Section 2 list 11 types of intersections in the three dimensional case. Theorem 10 confirms that there are no other possible configurations. Also the computer results show examples for each type of intersection in all the geometries investigated: q=2,3,4,5. At this stage it remains a conjecture that all the configurations permitted by Theorem 10 actually exist for all values of q $(q = p^h)$.

It is easy to confirm the existence of the configurations listed under (3)(k) in Section 2. In fact, we show that for any dimension n we can find Baer subspaces intersecting in $\frac{q^n-1}{q-1} + 1$ points: all points of a hyperplane of each subspace and a point external to it.

Consider again the real Baer subspace $\mathcal{B}_0$. The homography with the $(n+1) \times (n+1)$ matrix

$$\begin{bmatrix} 1 & 0 & \dots & 0 & 0 \\ 0 & 1 & \dots & 0 & 0 \\ \cdot & \cdot & & & \cdot \\ \cdot & \cdot & & & \cdot \\ \cdot & \cdot & & & \cdot \\ 0 & 0 & \dots & 1 & 0 \\ 0 & 0 & \dots & 0 & \alpha \end{bmatrix}$$

fixes all points of the hyperplane $x_{n+1} = 0$, as well as the point $(0,0,\dots,1)$. If $\alpha \notin GF(q)$, then $\mathcal{B}_0$ does not map into itself, e.g. the images of $(1,0,0,\dots,1)(0,1,0,\dots,1)$ etc. do not belong to $\mathcal{B}_0$. If $\mathcal{B}_1$ is the image induced by this homography then $\mathcal{B}_0$ and $\mathcal{B}_1$ intersect precisely in the hyperplane $x_{n+1} = 0$ and the point $(0,\dots,0,1)$.

A method was outlined in [3], whereby the Baer planes intersecting a fixed plane $\mathcal{B}_0$ in any permitted configuration could be counted.

There a *cluster* of Baer planes was defined as a set of Baer subplanes sharing pointwise some line and linewise some point. This concept can be extended to n dimensions and it is feasible that using it in a manner similar to the two dimensional case some more results can be extracted in higher dimensionsregarding the number of possible intersection types and thus their existence.

In two dimensions the intersection structures involve points and lines only. In higher dimensions, we have to consider subspaces of various dimensions, the survey of which is a complex task, since subspaces may be shared by two Baer subspaces, without any points, lines, etc. being shared. Even if the dimensions of the components are fixed according to Theorem 10, a complete characterisation can involve many variations, hence a classification presents a challenging task.

REFERENCES

[1] J. Cofman, Baer subplanes in finite projective and affine planes, *Can. J. Math.*, vol. XXIV, No. 1, (1972).

[2] R.C. Bose, J.W. Freeman and D.G. Glynn, On the intersection of two Baer subplanes in a finite projective plane, *Utilitas Mathematica*, 17, (1980), 65-77.

[3] M. Sved, On configurations of Baer subplanes of the projective plane over $GF(q^2)$, *Combinatorial Mathematics Proc., Brisbane Australia*, (1981), 423-443.

[4] Marshall Hall Jr., *Combinatorial Theory*, Blaesdel, 1967, 128-131.

[5] G.E. Andrews, The theory of partitions, *Enc. of Mathematics and its Applications*, 2, (1976), 212-213.

DISTRIBUTION OF LABELLED TREES BY DIAMETER

G. SZEKERES

A. Rényi and the author [1] have determined the asymptotic distribution according to height above their root of labelled and rooted trees of (large) order n. Following a (verbal) suggestion of R. Robinson we shall examine here similarly the distribution of labelled trees according to diameter. By diameter we mean the number of edges in a longest path.

An interesting feature of this distribution is that trees with odd or even diameter have very different enumerators, due to the fact that trees with even diameter $2k$ have a unique centre, that is a common midpoint of all paths of length $2k$, whereas trees with odd diameter $2k+1$ are bicentred, that is they have a unique common mid-edge of all paths of length $2k+1$. It is intuitively evident, and has been conjectured by Harary and Robinson (unpublished), that trees with odd diameters have the same asymptotic distribution as those with even diameters, and in particular that about half of all trees of order n should have an even (odd) diameter. The purpose of this note is to confirm these conjectures and to determine the precise form of the asymptotic distribution. Trees henceforth will always mean labelled trees.

Let $t(n,k)$ be the number of rooted trees with n nodes and height $\le k$ when the root itself is arbitrarily labelled, $\tau(n,k) = t(n,k)-t(n,k-1)$ the number of such trees with height equal to k. Let $\delta(n,k)$ be the number of (non-rooted) trees with n nodes and diameter k. Then according to Riordan [2] if $G_k(x)$, $k \ge 0$ is defined by

$$G_0(x)=x, \quad G_k(x) = x\exp(G_{k-1}(x)), \quad k \ge 1$$

and

$$H_0(x)=x, \quad H_k(x) = G_k(x)-G_{k-1}(x), \quad k \ge 1$$

then

$$(1) \qquad G_k(x) = \sum_{n=1}^{\infty} \frac{t(n,k)}{n!} x^n, \quad H_k(x) = \sum_{n=1}^{\infty} \frac{\tau(n,k)}{n!} x^n, \qquad k \ge 0.$$

Similarly the enumerator

$$(2) \qquad D_k(x) = \sum_{n=1}^{\infty} \frac{\delta(n,k)}{n!} x^n$$

is given by

$$(3.1)\qquad D_{2k+1}(x) = \tfrac{1}{2}H_k^2(x), \quad k \ge 0$$

and

$$(3.2)\qquad D_{2k}(x) = H_k(x) - H_{k-1}(x)G_{k-1}(x), \quad k \ge 1.$$

The first equation (3.1) is obtained by representing a tree with n nodes and diameter $2k+1$ as the union of two disjoint rooted trees of height k each, their roots being joined by an edge. The second equation (3.2) is obtained by representing a tree with diameter $2k$ as a rooted tree of height k and having at least two distinct branches from the root with height k. That is we remove those which have a branch of height $k-1$ on a stem emanating from the root and a second branch from the root with height $\le k-1$.

Let us briefly review the method employed in [1]. From (1) we obtain

$$(4)\qquad \frac{\tau(n,k)}{n!} = \frac{1}{2\pi i}\int_{C+} \frac{H_k(y)}{z^{n+1}}\,dy$$

where C is any circular path with centre 0. As shown in [1], the most favourable radius is e^{-1} and for appropriately small u, $G_k(e^{-1+iu})$ has an asymptotic expansion

$$(5)\qquad G_k(e^{-1+iu}) \simeq 1 - \frac{2}{k}\sigma\cot\sigma + \frac{2}{3}\frac{\sigma^2}{\sin^2\sigma}\frac{\log k}{k^2} - \frac{4\sigma^2}{3k^2} + \frac{\sigma^2}{k^2\sin^2\sigma}\left(C + \frac{2}{3}\log\frac{\sin\sigma}{\sigma}\right) + \dots$$

for a certain constant C, where $iu = 2\sigma^2/k^2$. (The corresponding formula (3.21) in [1] has some misprints). From here one obtains

$$(6)\qquad H_k(e^{-1+iu}) = G_k(e^{-1+iu}) - G_{k-1}(e^{-1+iu}) \simeq \frac{2}{k^2}\frac{\sigma^2}{\sin^2\sigma} + \dots$$

and from (4) by a simple application of the theorem of residues

$$(7)\qquad \frac{\tau(n,k)}{n!} \simeq -\frac{8e^n}{k^4}\sum_{p=1}^{\infty}\operatorname*{res}_{\sigma=p\pi}\left(\frac{\sigma^3}{\sin^2\sigma}e^{-\beta\sigma^2}\right)$$

where $\beta = 2n/k^2$. This with Stirling's formula gives the asymptotic probability distribution

$$(8)\qquad \frac{\tau(n,k)}{n^{n-1}} \simeq 2\left(\frac{2\pi}{n}\right)^{1/2}\sum_{p=1}^{\infty}(2p^4\pi^4\beta^3 - 3p^2\pi^2\beta^2)e^{-\beta\pi^2p^2}$$

which is the main result of [1]. More precisely (8) is valid uniformly in every fixed interval $0 < \varepsilon \le \beta \le \Omega$.

Consider now the distribution according to diameter. In the odd diameter case we get from (2) and (3.1)

$$\frac{\delta(n,2k+1)}{n!} = \frac{1}{4\pi i}\int_{C+} \frac{H_k^2(y)}{z^{n+1}}\,dy .$$

Now from (6) we obtain

$$(9)\qquad \tfrac{1}{2}H_k^2(e^{-1+iu}) \simeq \frac{2}{k^4}\,\frac{\sigma^4}{\sin^4\sigma} + \ldots$$

and by the same argument as (7) was obtained from (6),

$$\frac{\delta(n,2k+1)}{n!} \simeq -\frac{8e^n}{k^6}\sum_{p=1}^{\infty} \operatorname*{res}_{\sigma=p\pi} \left(\frac{\sigma^5}{\sin^4\sigma}e^{-\beta\sigma^2}\right).$$

This gives, by a straightforward evaluation of the residues,

$$(10)\quad p_n(2k+1) = \frac{\delta(n,2^{k+1})}{n^{n-2}} \simeq \frac{1}{3}\left(\frac{2\pi}{n}\right)^{\frac{1}{2}} \sum_{p=1}^{\infty} \{\beta^2(4b_p^4-36b_p^3+75b_p^2-30b_p) + \beta(4b_p^3-10b_p^2)\}e^{-b_p}$$

where for brevity we have written $b_p = p^2\pi^2\beta$. Since the total number of labelled trees of order n is n^{n-2} by Cayley's well known formula, this is the required distribution function.

Note that

$$(11)\quad \sum_{0\le k<\frac{n}{2}} p_n(2k+1) \simeq \frac{1}{3}\left(\frac{2\pi}{n}\right)^{\frac{1}{2}}\int_0^{n/2} \sum_{p=1}^{\infty} \{\beta^2(4b_p^4-36b_p^3+75b_p^2-30b_p) + \beta(4b_p^3-10b_p^2)\}e^{-b_p}dk$$

$$\simeq \frac{1}{3}\pi^{\frac{1}{2}}\int_0^{\infty}\sum_{p=1}^{\infty} \{\beta^{\frac{1}{2}}(4b_p^4-36b_p^3+75b_p^2-30b_p) + \beta^{-\frac{1}{2}}(4b_p^3-10b_p^2)\}e^{-b_p}d\beta$$

$$\simeq \frac{1}{3}\pi^{\frac{1}{2}}\lim_{\substack{\varepsilon=0\\ \Omega=\infty}}\left[\sum_{p=1}^{\infty}(-4\beta^{\frac{1}{2}}b_p^2-4\beta^{3/2}b_p^3+18\beta^{3/2}b_p^2-12\beta^{3/2}b_p)e^{-b_p}\right]_{\varepsilon}^{\Omega}$$

$$= \lim_{\beta=0}\frac{4}{3}(\beta\pi)^{\frac{1}{2}}\sum_{p=1}^{\infty} b_p^2 e^{-b_p} \simeq \frac{4}{3}\pi^{-\frac{1}{2}}\int_0^{\infty} u^4e^{-u^2}du$$

$$= \frac{1}{2}\,.$$

This shows that asymptotically one half of all trees has an odd diameter.

The case of even diameter is somewhat more troublesome, and we need higher order expansion terms in (6). According to (3.2) we need an expansion for

$$H_k(e^{-1+iu}) - G_{k-1}(e^{-1+iu})H_{k-1}(e^{-1+iu}).$$

To calculate G_{k-1} and H_{k-1} observe that for fixed u, σ is a function of k, indeed $\sigma(k-1) = \sigma(k)-\sigma(k)/k$. Therefore when calculating $G_{k-1}(e^{-1+iu})$, σ must be replaced by $\sigma-\sigma/k$, $\cot\sigma$ by

$$\cot(\sigma-\sigma/k) = \cot\sigma + \frac{\sigma}{k\sin^2\sigma} + \frac{\sigma^2}{k^2}\frac{\cot\sigma}{\sin^2\sigma} + \dots,$$

$\frac{1}{\sin^2\sigma}$ by $\frac{1}{\sin^2(\sigma-\sigma/k)} = \frac{1}{\sin^2\sigma}(1+\frac{2\sigma}{k}\cot\sigma+\frac{\sigma^2}{k^2}(1+3\cot^2\sigma)) + \dots,$

$\log\sin\sigma$ by $\log\sin(\sigma-\sigma/k) = \log\sigma - \frac{\sigma}{k}\cot\sigma + \dots .$

Using these expressions we get in turn

$$H_k(e^{-1+iu}) = G_k(e^{-1+iu}) - G_{k-1}(e^{-1+iu})$$
$$\simeq \frac{2}{k^2}\frac{\sigma^2}{\sin^2\sigma} - \frac{4}{3k^3}\frac{\sigma^3\cot\sigma}{\sin^2\sigma}(\log k + \log\frac{\sin\sigma}{\sigma} + \frac{3}{2}c - 2) + \dots,$$

$$H_{k-1}(e^{-1+iu}) \simeq H_k(e^{-1+iu}) + \frac{4}{k^3}\frac{\sigma^3\cot\sigma}{\sin^2\sigma} - \frac{3}{2}\frac{\sigma^4}{k^4\sin^2\sigma}$$
$$- \frac{4}{3k^4}\frac{\sigma^4(1+2\cos^2\sigma)}{\sin^4\sigma}(\log k + \log\frac{\sin\sigma}{\sigma} + \frac{3}{2}c - 4) + \dots,$$

$$H_{k-1}(e^{-1+iu})G_{k-1}(e^{-1+iu}) \simeq H_k(e^{-1+iu})G_{k-1}(e^{-1+iu})$$
$$+ \frac{4}{k^3}\frac{\sigma^3\cot\sigma}{\sin^2\sigma} + \frac{10}{3k^4}\frac{\sigma^4}{\sin^4\sigma}$$
$$- \frac{4}{3k^4}\frac{\sigma^4(1+2\cos^2\sigma)}{\sin^4\sigma}(\log k + \log\frac{\sin\sigma}{\sigma} + \frac{3}{2}c - 1)+ \dots$$

$$H_k(e^{-1+iu})(1-G_{k-1}(e^{-1+iu})) \simeq \frac{4}{k^3}\frac{\sigma^3\cot\sigma}{\sin^2\sigma} + \frac{16}{3k^4}\frac{\sigma^4}{\sin^4\sigma}$$
$$- \frac{4}{3k^4}\frac{\sigma^4(1+2\cos^2\sigma)}{\sin^4\sigma}(\log k + \log\frac{\sin\sigma}{\sigma} + \frac{3}{2}c - 1) + \dots$$

where $+\dots$ stands for the higher order asymptotic terms. Finally when $H_k-H_{k-1}G_{k-1} = H_k(1-G_{k-1}) + (H_k-H_{k-1})G_{k-1}$ is calculated, all expansion terms up to order $1/k^4$ cancel except a single term, namely

$$H_k(e^{-1+iu})-H_{k-1}(e^{-1+iu})G_{k-1}(e^{-1+iu}) \simeq \frac{2}{k^4}\frac{\sigma^4}{\sin^4\sigma}.$$

This is the same as the main asymptotic term for $\frac{1}{2}H_k^2(e^{-1+iu})$ in (9). It follows immediately that the asymptotic probability distribution function for even diameter trees is the same as for odd diameter trees. Therefore the probability (for large n) of a labelled tree with n nodes to have diameter $2\sqrt{2n/\beta}$ (uniformly for $0 < \varepsilon \le \beta \le \Omega$) is

$$(12)\quad p_n(k) \simeq \frac{1}{3}\left(\frac{2\pi}{n}\right)^{\frac{1}{2}} \sum_{p=1}^{\infty} \{\beta^2(4b_p^4-36b_p^3+75b_p^2-30b_p) + \beta(4b_p^3-10b_p^2)\}e^{-b_p}$$

where $b_p = p^2\pi^2\beta$.

From this expression we can calculate if we wish all moments of the distribution. In particular, the expectation value of the diameter of a random tree with n nodes is, by a calculation similar to that of (11),

$$(13)\quad \sum_k k p_n(k) \approx \frac{4}{3}(2\pi n)^{\frac{1}{2}} \int_0^\infty \sum_{p=1}^\infty \{(4b_p^4-36b_p^3+75b_p^2-30b_p) + \beta^{-1}(4b_p^3-10b_p^2)\}e^{-b_p} d\beta$$

$$= \frac{4}{3}(2\pi n)^{\frac{1}{2}} \lim_{\beta=0} \sum_{p=1}^\infty \{\beta(4b_p^3-20b_p^2+15b_p) + (4b_p^2-2b_p-2)\}e^{-b_p}$$

$$= \frac{4}{3}(2\pi n)^{\frac{1}{2}} \lim_{\beta=0} \sum_{p=1}^\infty (4b_p^2-2b_p-2)e^{-b_p}$$

$$= \frac{4}{3}(2\pi n)^{\frac{1}{2}} = 3.342171\sqrt{n}$$

by Euler-Maclaurin, since

$$\int_0^\infty (4u^4-2u^2-2)e^{-u^2} du = 0 .$$

In comparison the expectation value of the height of a random tree is $(2\pi n)^{\frac{1}{2}} = 2.506628\sqrt{n}$, according to [1], formula (4.7).

The maximum of the distribution occurs where $\frac{d}{d\beta} p_n(k) = 0$, i.e.

$$\sum_{p=1}^\infty \{\beta(-4b_p^5+60b_p^4-255b_p^3+330b_p^2-90b_p) + (-4b_p^4+26b_p^3-30b_p^2)\}e^{-b_p} = 0.$$

Numerical evaluation gives

$$(14)\qquad \beta(\max) = .78051168$$

i.e.

$$(15)\qquad d(\max) = 3.20151315\sqrt{n}$$

for the diameter with maximal probability. In comparison the maximum of the distribution by height occurs at

$$h(\max) = 2.31515436\sqrt{n}$$

according to [1], (3.32).

It would be interesting to prove some of these results, in particular the (asymptotic) equiprobability of even and odd diameter more directly, without the use of asymptotic analysis.

REFERENCES

[1] A. Rényi and G. Szekeres, On the height of trees, *J. Austral. Math. Soc.*, 7 (1967) 497-507.

[2] J. Riordan, The enumeration of trees by height and diameter, *IBM Journal of Research and Development*, 4 (1960) 473-478.

ORTHOGONAL LATIN SQUARES WITH SMALL SUBSQUARES

W.D. WALLIS AND L. ZHU

We prove that there exist a pair of orthogonal Latin squares of side v with orthogonal subsquares of side n for all $v \geq 3n$, in the cases $n = 3, 4, 5$.

1. INTRODUCTION

Suppose S is a set of v symbols. A *Latin square* (of side v) based on S is a $v \times v$ array each of whose rows and columns are permutations of S. A *subsquare* of side r is an $r \times r$ subarray which is itself a Latin square based on some r-subset of S.

A *transversal* in a Latin square is a set of positions, one per row and one per column, among which the symbols occur precisely once each. A transversal square is a Latin square whose main diagonal is a transversal, and we usually relabel symbols so that the (i,i) entry is the symbol i.

We say Latin squares A and B of the same side are *orthogonal* if, for each symbol x in A, the set of positions where A has an x form a transversal in B. Alternatively if A and B are based on the same v-set S, they are orthogonal if and only if the v^2 pairs (a_{ij}, b_{ij}) contain each ordered pair on S precisely once.

If two orthogonal Latin squares have subsquares occupying the same positions in each, the subsquares must themselves be orthogonal. We refer to them as *orthogonal subsquares*. A pair of orthogonal Latin squares of side v with orthogonal subsquares of side n will be denoted on LS(v,n); LS(v,1) is abbreviated to LS(v). The set of all orders v such that there exists an LS(v,n) is denoted P(n).

Our interest here is in finding the set P(n) for some small values of n. It is easy to prove that an LS(v,n) can exist only if $v \geq 3n$ (see, for example, [11]). We prove that this condition is sufficient for $n = 3$, 4 and 5.

2. SELF-ORTHOGONAL LATIN SQUARES

There has been some study of Latin squares which are orthogonal to their own transposes. Such squares are called *self-orthogonal*. If a self-orthogonal Latin square contains a subsquare which is symmetrically placed about the diagonal, this subsquare and its transpose constitute a pair of orthogonal subsquares. Such an arrangement is called a "self-orthogonal Latin square with self-orthogonal subsquare"; if the main square has side v and the subsquare has side n, we denote it SOLS(v,n). It is clear that an SOLS(v,n) is a special kind of LS(v,n) which explains our interest in them (although they are of no help in the determination of P(3), since there is no self-orthogonal Latin square of side 3).

All the best known existence results concerning LS(v,n) actually deal with SOLS(v,n). There can be no SOLS(v,n) unless $v \geq 3n+1$ [10]. Drake and Lenz [6] prove the following sufficient conditions for SOLS(v,n) (and for LS(v,n)):

Theorem 1. *The following designs always exist:*

(i) $SOLS(v,4)$, $v > 66$;
(ii) $SOLS(v,5)$, $v > 67$;
(iii) $SOLS(v,7)$, $v > 62$;
(iv) $SOLS(v,8)$, $v > 70$;
(v) $SOLS(v,9)$, $v > 71$;
(vi) $SOLS(v,n)$, $v \geq 8n$; $10 \leq n \leq 63$;
(vii) $SOLS(v,n)$, $v > 8n + 62$, $63 \leq n \leq 303$;
(viii) $SOLS(v,n)$, $v > 4n + 2$, $n \geq 304$.

Bennett and Mendelsohn [1,2] have investigated the existence of SOLS for small n, and eliminated some cases. They prove, among other things:

Theorem 2. *If $n = 4$ or 5 and $v \geq 3n + 1$, a $SOLS(v,n)$ exists except possibly in the following cases:*

$n = 4$, $v = 14, 18, 19, 23, 26, 27, 30, 38$;
$n = 5$, $v = 14, 22, 23, 26, 27, 28, 30, 34, 38$.

We know a number of "smallest possible" embeddings of self-orthogonal Latin squares:

Theorem 3 [10].

(i) $SOLS(3n+1,n)$ *exists when* $n \neq 2,3,6$;
(ii) $SOLS(3n+2,n)$ *exists when* $n \neq 3$, n *odd*;
(iii) $SOLS(3n+3,n)$ *exists when* $4 \leq n \leq 21$, $n \neq 6$.

We shall use the following well-known constructions.

Theorem 4. *(Chain Rule). If $LS(k,n)$ exists and $LS(v,k)$ exists, then $LS(v,n)$ exists. If $SOLS(k,n)$ exists and $SOLS(v,k)$ exists, then $SOLS(v,n)$ exists.*

Theorem 5. *(Product Rule). If $LS(v)$ and $LS(m)$ exist then $LS(mv,m)$ and $LS(mv,v)$ exist. If $LS(v,n)$ and $LS(m)$ exist then $LS(mv,mn)$ and $LS(mv,n)$ exist; and similarly for self-orthogonal squares.*

Theorem 6. *(Singular Direct Product)* [3,pp428-432]. *Suppose there are orthogonal transversal squares of order v_1. Suppose also that $LS(v_2-v_3)$ and $LS(v_2,v_3)$ exist. Then $LS(v_1(v_1(v_2-v_3)+v_3, n)$ exists for $n = v_1, v_2, v_3$ and v_2-v_3.*

This theorem also remains true when "LS" is replaced by "SOLS", except that an $SOLS(v_1)$ is also needed in order to produce $SOLS(v_1(v_2-v_3)+v_3,v_1)$.

Finally, we state two recent results.

Theorem 7. [*13, Theorem 1*]. *If n is even, m is prime to 6 and m > 2n, and if LS(n) exists, then LS(m+n,n) exists.*

Theorem 8. [*15, 9*]. *If p is a prime greater than 7 then there exists LS(p+3,3).*

3. CONSTRUCTIONS USING PAIRWISE BALANCED DESIGNS

Given positive integers v and λ and a set of positive integers K, a *pairwise balanced design* PB(v;K;λ) is a way of choosing a collection of *blocks*, or subsets, from a v-set, such that any two members of the v-set are found together in precisely λ subsets, and the size of every subset is a number of K. (Observe that we allow elements in K which are the size of no block.)

Suppose B is a PB(v;K;1) with v-set $\{1,2,\ldots,v\}$. For each $k \in K$, select a transversal square A_k of side k, with diagonal $(1,2,\ldots,k)$; write $A = \{A_k : k \in K\}$. Now select a way of ordering each block of B. A Latin square $L(B,A)$ is defined as follows: given the block $B = (b_1,b_2,\ldots,b_k)$ of B, which has size k, examine A_k, and if A_k has (i,j) entry H then $L(B,A)$ has (b_i,b_j) entry b_h, for $1 \le i \le k$, $1 \le j \le k$, do this for every block of B. Then $L(B,A)$ is a transversal square of side v.

This construction has been widely used in the construction of orthogonal Latin squares by many authors, in the following way. For each k in K, one selects m transversal squares $A_k^1, A_k^2, \ldots, A_k^m$ of side k, and writes $A^i = \{A_k^i : k \in K\}$. Then $L(B,A^i)$ is orthogonal to $L(B,A^j)$ provided A_k^i is orthogonal to A_k^j. So this construction gives m orthogonal Latin squares of side v *provided* there are m orthogonal transversal squares of side k for every k in K. The squares have orthogonal subsquares of every K which is the order of a block. The obvious shortcoming is that one cannot construct a pair of orthogonal Latin squares by using a pairwise balanced design with blocks of size 3, because there is no pair of orthogonal transversal squares of that order.

However, suppose B contains just one block of size 3 : {1,2,3} say. One can carry out the construction with

$$A_3^1 = A_3^2 = \begin{bmatrix} 1 & 3 & 2 \\ 3 & 2 & 1 \\ 2 & 1 & 3 \end{bmatrix},$$

and obtain two transversal squares which would be orthogonal except that the pairs on {1,2,3} are not properly represented. Now make the following substitution: both Latin squares have top left 3×3 block

$$\left[\begin{array}{ccc}1 & 3 & 2\\ 3 & 2 & 1\\ 2 & 1 & 3\end{array}\right. .$$

In one of them, replace the block by

$$\left[\begin{array}{ccc}1 & 2 & 3\\ 3 & 1 & 2\\ 2 & 3 & 1\end{array}\right. .$$

We no longer have transversal squares, but we have orthogonal Latin squares with subsquares of all sizes of blocks.

Similarly a pairwise balanced design with just two 3-blocks can be handled. If they are disjoint, there is no problem; if they are {1,2,3} and {1,4,5}, then

$$\text{replace}\quad \left[\begin{array}{ccccc}1 & 3 & 2 & 5 & 4\\ 3 & 2 & 1 & & \\ 2 & 1 & 3 & & \\ 5 & & & 4 & 1\\ 4 & & & 1 & 5\end{array}\right. \qquad \text{by}\quad \left[\begin{array}{ccccc}1 & 2 & 3 & 4 & 5\\ 3 & 1 & 2 & & \\ 2 & 3 & 1 & & \\ 5 & & & 1 & 4\\ 4 & & & 5 & 1\end{array}\right. .$$

To generalize this we define a *brush* (or k-brush) with centre x to be a set of k 3-sets which all contain the common element x but are otherwise disjoint. We say two brushes are *disjoint* if and only if their sets of *elements* are disjoint. We have:

Theorem 9. *Suppose there exists a PB(v;K;1) in which $2 \notin K$, $6 \notin K$ and the set of all 3-blocks comprises a union of disjoint brushes. Then there exists a LS(v,n) for every n which is the size of a block which contains no non-central element of a brush, and also a LS(v,3).*

If all 3-blocks are disjoint, we can take any member of a 3-block as centre of its brush (which is a 1-brush). As no other block can meet a 3-block in more than one point (since $\lambda = 1$), the last condition then becomes vacuous.

Corollary 9.1. *Suppose $v \equiv 2,3,18$ or $19 \pmod{20}$, $v > 3$. Then LS(v,3), LS(v,4) and LS(v,5) exist.*

Proof. It is well-known [7] that a (v,5,1) - balanced incomplete block design, on PB(v:{5};1), exists whenever $v \equiv 1$ or $5 \pmod{20}$. Consider any such design with more than 5 points. If any two points are deleted, we obtain a design with v-2 points which contains one block of size 3, and several blocks of size 4 and 5. If three non-collinear points are deleted, one obtains a (v-3) - point design with three disjoint 3-blocks, together with blocks of size 4 and 5. Theorem 9 is satisfied, and we have all the required values of v.

Drake and Larson [5] have investigated the spectrum of "proper" pairwise balanced designs - ones with no blocks of size 2,3 or 6. If PB(k) denotes the set of all sizes of proper pairwise balanced designs which actually have at least one block of size k, they prove:

Theorem 10. *There is a proper pairwise balanced design of size v if and only if v belongs to PB(4).*

PB(4) = {13,16,17,20,21,22,24,25,28,29} ∪ A ∪ {v:v ≥ 31};

PB(5) = {17,20,21,24,25,29,31,32,33} ∪ B ∪ {v:v ≥ 35};

where A is empty or is {30} and B is some subset of {28,30,34}.

Of course, if v lies in PB(k), there exists not only LS(v,k) but also SOLS(v,k).

4. CONSTRUCTION FROM SPOUSE-AVOIDING MIXED DOUBLES.

A *sharply resolvable spouse-avoiding mixed doubles round robin tournament* of order n, or SR(n), is a schedule whereby n couples may play mixed doubles tennis in such a way that:

(i) no player appears on court with his or her spouse;

(ii) each player has every other player an opponent exactly once, and every other player of the opposite sex as a partner exactly once, except for his or her spouse;

(iii) the games are organized into rounds such that if n is even every player is in one game in every round, and if n is odd one couple sits out of each round and every other player is in one game in that round.

Suppose such a tournament exists. Write r[a,b|c,d] to mean that Mr a and Mrs b play against Mr c and Mrs d in round r. The two men meet in precisely one match, so their names determine the names of their partners and the round number. Let us define ℓ_{ij} to be the name of Mr i's partner when he plays Mr j, and r_{ij} to be the round number - in other words, the typical match is $r_{ij}[i,\ell_{ij}|j,\ell_{ji}]$. Then define $\ell_{ii} = i$ for all i; if n is even, $r_{ii} = n$ for all i, while if n is odd, $r_{ii} = i$, and we ensure that the pair who sits out in round i is Mr and Mrs i. Then, as is pointed out in [12], $L = (\ell_{ij})$ is a self-orthogonal Latin square of order n and $R = (r_{ij})$ is a symmetric Latin square orthogonal to it.

Wang and Wilson (see [14, p.23]) construct several examples of SR(n) with subtournaments. Their technique is as follows. Let n = g+k, where n,g and k are even. G is the cyclic group of order g. Write f for ½g and h for ½n-2k. The subtournament is based on symbols $A_1,A_2,\dots,A_k$, and its rounds are labelled $A_2,A_3,\dots,A_k$. The whole tournament is based on the symbols $A_1,A_2,\dots,A_k$ and the g elements of G. To indicate its generation we use the following notation:

if [P] is the match [p,q|r,s] then [P+x] is [p+x,q+x|r+x,s+x] where addition is in G except that $A_i+x = A_i$ for $x \in G$. Then the construction uses "starter" matches of the following kind:

f matches $[H_2],[H_3],\dots,[H_{f+1}]$ with all four symbols from G;

h matches $[F_1],[F_2],\dots,[F_h]$ with all four symbols from G;

SR(4) in SR(18):

$H_2 = [\ 0,\ 1 \mid 7,\ 8]$
$H_3 = [\ 0,\ 7 \mid 1,\ 4]$
$F_1 = [10,12 \mid 2,11]$
$I_1 = [\ 1,\ 0 \mid 3,A_1]$
$I_2 = [\ 8,\ 6 \mid 11,A_2]$
$I_3 = [\ 5,\ 2 \mid 9,A_3]$
$I_4 = [\ 7,\ 3 \mid 12,A_4]$
$J_1 = [\ 0,\ 8 \mid A_1,13]$
$J_2 = [\ 4,\ 9 \mid A_2,\ 7]$
$J_3 = [\ 6,10 \mid A_3,\ 4]$
$J_4 = [13,\ 5 \mid A_4,\ 1]$

SR(4) in SR(22):

$H_2 = [\ 0,\ 1 \mid 9,10]$
$H_3 = [\ 1,\ 6 \mid 0,\ 9]$
$F_1 = [\ 9,17 \mid 15,10]$
$F_2 = [\ 6,12 \mid 13,\ 2]$
$F_3 = [14,\ 7 \mid 4,\ 8]$
$I_1 = [\ 1,\ 0 \mid 3,A_1]$
$I_2 = [\ 8,\ 6 \mid 11,A_2]$
$I_3 = [16,13 \mid 2,A_3]$
$I_4 = [\ 5,\ 1 \mid 10,A_4]$
$J_1 = [\ 7,\ 9 \mid A_1,14]$
$J_2 = [\ 0,\ 3 \mid A_2,\ 5]$
$J_3 = [12,\ 4 \mid A_3,16]$
$J_4 = [17,11 \mid A_4,15]$

SR(4) in SR(26):

$H_2 = [\ 0,\ 1 \mid 11,12]$
$H_3 = [\ 0,\ 9 \mid 1,18]$
$F_1 = [15,\ 8 \mid 21,\ 4]$
$F_2 = [\ 5,17 \mid 12,19]$
$F_3 = [\ 0,10 \mid 8,16]$
$F_4 = [11,\ 2 \mid 20,12]$
$F_5 = [19,13 \mid 7,18]$
$I_1 = [\ 1,\ 0 \mid 3,A_1]$
$I_2 = [13,11 \mid 16,A_2]$
$I_3 = [\ 6,\ 3 \mid 10,A_3]$
$I_4 = [\ 9,\ 5 \mid 14,A_4]$
$J_1 = [\ 4,\ 6 \mid A_1,17]$
$J_2 = [19,22 \mid A_2,\ 1]$
$J_3 = [23,\ 3 \mid A_3,12]$
$J_4 = [18,\ 8 \mid A_4,10]$

SR(4) in SR(30):

$H_2 = [\ 0,\ 1 \mid 13,14]$
$H_3 = [\ 0,\ 7 \mid 1,10]$
$F_1 = [\ 8,23 \mid 16,\ 7]$
$F_2 = [13,\ 5 \mid 22,\ 9]$
$F_3 = [25,11 \mid 10,\ 4]$
$F_4 = [14,19 \mid 7,18]$
$F_5 = [\ 1,16 \mid 5,24]$
$F_6 = [\ 0,14 \mid 12,20]$
$F_7 = [15,25 \mid 9,13]$
$I_1 = [\ 1,\ 0 \mid 3,A_1]$
$I_2 = [17,15 \mid 20,A_2]$
$I_3 = [24,21 \mid 2,A_3]$
$I_4 = [\ 6,\ 2 \mid 11,A_4]$
$J_1 = [\ 4,\ 6 \mid A_1,17]$
$J_2 = [19,22 \mid A_2,\ 1]$
$J_3 = [23,\ 3 \mid A_3,12]$
$J_4 = [18,\ 8 \mid A_4,10]$

SR(8) in SR(34):

$H_2 = [\ 0,\ 1 \mid 13,14]$
$H_3 = [\ 0,15 \mid 7,16]$
$H_4 = [\ 0,\ 5 \mid 9,20]$
$H_5 = [\ 0,12 \mid 11,\ 9]$
$F_1 = [22,15 \mid 10,24]$
$I_1 = [\ 1,\ 0 \mid 2,A_1]$
$I_2 = [\ 4,\ 1 \mid 6,A_2]$
$I_3 = [\ 8,\ 4 \mid 11,A_3]$
$I_4 = [17,12 \mid 21,A_4]$
$I_5 = [12,\ 6 \mid 20,A_5]$
$I_6 = [18,10 \mid 23,A_6]$
$I_7 = [\ 3,20 \mid 9,A_7]$
$I_8 = [\ 5,21 \mid 15,A_8]$
$J_1 = [\ 0,13 \mid A_1,\ 3]$
$J_2 = [\ 7,14 \mid A_2,22]$
$J_3 = [13,17 \mid A_3,\ 5]$
$J_4 = [14,16 \mid A_4,11]$
$J_5 = [16,19 \mid A_5,23]$
$J_6 = [19,25 \mid A_6,18]$
$J_7 = [24,\ 8 \mid A_7,\ 2]$
$J_8 = [25,\ 7 \mid A_8,\ 9]$

SR(8) in SR(38):

$H_2 = [\ 0,\ 1 \mid 15,16]$
$H_3 = [\ 0,\ 5 \mid 9,24]$
$H_4 = [\ 0,13 \mid 11,20]$
$H_5 = [\ 0,12 \mid 23,\ 9]$
$F_1 = [29,27 \mid 13,\ 6]$
$F_2 = [\ 7,21 \mid 24,11]$
$F_3 = [\ 3,22 \mid 21,\ 9]$
$I_1 = [\ 1,\ 0 \mid 2,A_1]$
$I_2 = [\ 4,\ 1 \mid 6,A_2]$
$I_3 = [\ 9,\ 5 \mid 12,A_3]$
$I_4 = [19,14 \mid 23,A_4]$
$I_5 = [10,\ 4 \mid 16,A_5]$
$I_6 = [20,12 \mid 25,A_6]$
$I_7 = [22,13 \mid 0,A_7]$
$I_8 = [28,18 \mid 8,A_8]$
$J_1 = [26,28 \mid A_1,29]$
$J_2 = [17,20 \mid A_2,\ 2]$
$J_3 = [11,15 \mid A_3,19]$
$J_4 = [27,\ 3 \mid A_4,\ 8]$
$J_5 = [18,25 \mid A_5,17]$
$J_6 = [15,23 \mid A_6,\ 7]$
$J_7 = [14,24 \mid A_7,26]$
$J_8 = [\ 5,16 \mid A_8,10]$

SR(8) in SR(42):

$H_2 = [\ 0,\ 1 \mid 17,18]$
$H_3 = [\ 0,23 \mid 15,10]$
$H_4 = [\ 0,\ 4 \mid 21,13]$
$H_5 = [\ 0,16 \mid 25,11]$
$F_1 = [\ 4,16 \mid 15,24]$
$F_2 = [25,\ 9 \mid 7,29]$
$F_3 = [13,32 \mid 33,20]$
$F_4 = [32,13 \mid 20,\ 3]$
$F_5 = [27,\ 7 \mid 17,23]$
$I_1 = [\ 1,\ 0 \mid 2,A_1]$
$I_2 = [\ 3,\ 1 \mid 6,A_2]$
$I_3 = [\ 5,\ 2 \mid 10,A_3]$
$I_4 = [\ 9,\ 5 \mid 11,A_4]$
$I_5 = [21,15 \mid 28,A_5]$
$I_6 = [26,19 \mid 30,A_6]$
$I_7 = [23,14 \mid 29,A_7]$
$I_8 = [14,\ 4 \mid 22,A_8]$
$J_1 = [16,21 \mid A_1,28]$
$J_2 = [\ 0,\ 8 \mid A_2,31]$
$J_3 = [18,25 \mid A_3,\ 6]$
$J_4 = [24,26 \mid A_4,27]$
$J_5 = [12,22 \mid A_5,18]$
$J_6 = [19,30 \mid A_6,33]$
$J_7 = [31,10 \mid A_7,12]$
$J_8 = [\ 8,11 \mid A_8,17]$

FIGURE 1

k matches $[I_1],[I_2],\dots,[I_k]$ of form $[u,v \mid w,A_j]$, for some $u,v,w \in G$;

k matches $[J_1],[J_2],\dots,[J_k]$ of form $[u,v \mid A_j,w]$, for some $u,v,w \in G$.

A_1 is used as the diagonal element of S. The rounds of the SR(n) are:

Round A_2: the f matches $[H_2]+x$, $x = 0,1,\dots,f-1$ together with round A_2 of the subtournament;

Round A_i, $3 \le i \le f+1$: the f matches $[H_i]+2x$, $x = 0,1,\dots,f-1$, together with round A_i of the subtournament;

Round A_i, $f+2 \le i \le g$: the f matches $[H_{1+i-f}]+2x+1$, $x = 0,1,\dots,f-1$ together with round A_i of the subtournament;

Round x, $x \in G$: $[F_1]+x, [F_2]+x, \dots, [F_h]+x, [I_1]+x, [I_2]+x, \dots, [I_k]+x, [J_1]+x, [J_2]+x, \dots, [J_k]+x$.

Since h must be non-negative, n must be at least 4k.

Wang has constructed several examples by computer. Those listed in [13] - which yield SR(4) in SR(18), SR(22), SR(26) and SR(30) and also SR(8) in SR(34), SR(38) and SR(42) - are tabulated in Figure 1.

The subtournaments obviously give rise to self-orthogonal subsquares in the self-orthogonal Latin squares. So as a consequence we have:

Theorem 11. *The following designs exist:*

SOLS(18,4), SOLS(22,4), SOLS(26,4), SOLS(30,4), SOLS(34,8), SOLS(38,8), SOLS(42,8).

In fact, each design yields a set of three pairwise orthogonal Latin squares with pairwise orthogonal subsquares of the given size.

5. EXISTENCE OF LS(v,3).

We know that an LS(v,3) cannot exist for $v < 9$. We prove that examples exist for all other orders v.

Lemma 12.1. *There is an LS(v,3) for v = 11, 17, 18, 19, 20, 23, 32.*

Proof. For v = 11 we present a pair of squares in Figure 2. The cases v = 20 and v = 32 follow from Theorem 8 : 20 = 17+3, 32 = 29+3. The other four values come from pairwise balanced designs via Theorem 9 : orders 18, 19 and 23 come from Corollary 9.1, and a suitable PB(17;{5,4,3};1) is shown in Figure 3.

The next result is essentially Theorem 3 of [15]. We present it here so that the proof will be available in English.

A	C	B	4	5	6	7	0	1	2	3
C	B	A	5	6	7	0	1	2	3	4
B	A	C	7	0	1	2	3	4	5	6
1	2	3	0	4	A	B	6	C	7	5
2	3	4	6	1	5	A	B	7	C	0
3	4	5	1	7	2	6	A	B	0	C
4	5	6	C	2	0	3	7	A	B	1
5	6	7	2	C	3	1	4	0	A	B
6	7	0	B	3	C	4	2	5	1	A
7	0	1	A	B	4	C	5	3	6	2
0	1	2	3	A	B	5	C	6	4	7

A	B	C	2	3	4	5	6	7	0	1
C	A	B	7	0	1	2	3	4	5	6
B	C	A	6	7	0	1	2	3	4	5
5	3	6	0	A	7	4	B	1	C	2
6	4	7	3	1	A	0	5	B	2	C
7	5	0	C	4	2	A	1	6	B	3
0	6	1	4	C	5	3	A	2	7	B
1	7	2	B	5	C	6	4	A	3	0
2	0	3	1	B	6	C	7	5	A	4
3	1	4	5	2	B	7	C	0	6	A
4	2	5	A	6	3	B	0	C	1	7

FIGURE 2

0 1 2 3 4	3 5 B G	4 8 A F
0 5 6 7 8	3 6 C F	0 D E
0 9 A B C	3 7 A D	0 F G
1 5 9 D F	3 8 9 E	1 7 C
1 6 A E G	4 5 C E	1 8 B
2 7 B E F	4 6 B D	2 5 A
2 8 C D G	4 7 9 G	2 6 9

FIGURE 3

Lemma 12.2. *If $v \equiv 2 \pmod 4$ then there is an LS(v,3) for all v except 2 and 6.*

Proof. If 3 divides v then the Lemma is easy: if $v = 3.2 = 6$ there is nothing to prove; if $v = 3.6 = 18$ then the result is given by Lemma 12.1; if $v > 18$ then $v = 3q$ where orthogonal Latin squares of side q exist, and the Lemma follows from the product rule.

Now suppose $v \equiv 2 \pmod 4$ and v is prime to 3. Write $v = n + 3$. Now n is odd, $n \equiv 3 \pmod 4$ and n is prime to 3. Since $n \equiv 3 \pmod 4$ it has a prime divisor, p say, congruent to 3 modulo 4. As p cannot be 3, $p \geq 7$, so there is an LS(p+3,3) by Theorem 8. If we write $n = pm$, then m is odd, and $m > 3$, so there exist orthogonal transversal squares of side m. So by singular direct product there exists an LS(r,3) when $r = m[(p+3)-3] + 3 = mp + 3 = v$.

Theorem 12. *There is an LS(v,3) for all $v \geq 9$.*

Proof. First consider v modulo 3. If $v = 3k$ for some k, the product rule provides an LS(v,3) for all v in the range except $v = 18$. If $v = 3k + 1$ then $v = k(4-1) + 1$; using singular direct product we can construct an LS(v,3) for all v except 10 and 19. The Lemmas cover 10, 18 and 19. So, using Lemma 12.2, we see that the only cases to be considered are when $v \equiv 2 \pmod 3$ but $v \not\equiv 2 \pmod 4$: that is, $v \equiv 5,8,11 \pmod{12}$.

If $v = 12h + 5$, then $v = (12h-5) + 10$, and LS(v,10) exists when $12h - 5 > 20$ from Theorem 7. If $v = 12h + 11$, then $v = (12h+1) + 10$, and LS(v,10) exists when $12h + 1 > 20$ by Theorem 7. So we have an LS(v,10) - and by consequence of the chain rule an LS(v,3), since we already know LS(10,3) - when $v \equiv 5$ or $11 \pmod{12}$, except when $v = 17,29$, 11 or 23. In three of these four exceptional cases of v, LS(v,3) was constructed in Lemma 12.1. To construct LS(29,3) we observe that LS(29,9) exists by Theorem 3 and LS(9,3) exists by the product rule, so LS(29,3) exists by the chain rule.

If $v \equiv 8 \pmod{12}$ we again subdivide. If $v = 24t + 8$ then $v = 8(3t+1)$. If LS(3t+1,3) exists then LS(v,3) is given by the product rule. But any LS(3t+1,3) exists for $t \geq 3$. So the only exceptions are $t = 1$, $t = 2$, i.e., $v = 32$, $v = 56$. Now $56 = 4.14$, and LS(14,3) has already been constructed, so LS(56,3) exists; LS(32,3) is in Lemma 12.1.

Finally, if $v = 24t+20$, $v = 4.(6t+5)$. Now we have found all possible odd orders, so LS(6t+5,3) exists for $t > 0$, and LS(v,3) exists when $t > 0$. The only remaining case, $v = 20$, is treated in Lemma 12.1.

By an entirely different method, Dinitz and Stinson [4] have proven Theorem 12 for *odd* v; but there seems to be no hope of applying their techniques to the even case.

6. EXISTENCE OF LS(v,4)

From Theorem 2 we know of an LS(v,4) for all possible orders except 12,14,18, 19,23,26,27,30 and 38. Orders 18,19,23 and 38 are covered by Corollary 9.1. Using Theorem 7 with $n = 4$, $m = 23$ we get order 27 (and order 23 could also be handled in this way). Order 12 is done by the Product Rule. Orders 26 and 30 are given by Theorem 11. Finally, Zhu [16] has recently constructed an LS(14,4); it is shown in Figure 4. We have:

Theorem 13. *There is an LS(v,4) for all v ≥ 12.*

Observe that we have also shown that there is an SOLS(v,4) for every possible order except perhaps 14,19,23 and 27. (Order 38 is provided by Theorem 10.)

1	C	0	D	A	6	7	9	8	3	4	2	B	5
4	2	C	1	D	7	8	0	9	5	3	B	6	A
6	5	3	C	2	8	9	1	0	4	B	7	A	D
5	7	6	4	C	9	0	2	1	B	8	A	D	3
B	6	8	7	5	0	1	3	2	9	A	D	4	C
9	0	1	2	3	A	B	C	D	8	7	6	5	4
2	3	4	5	6	D	C	B	A	1	0	9	8	7
3	4	5	6	7	B	A	D	C	2	1	0	9	8
7	8	9	0	1	C	D	A	B	6	5	4	3	2
C	9	D	A	4	5	6	8	7	0	2	3	1	B
8	D	A	3	B	4	5	7	6	C	9	1	2	0
D	A	2	B	9	3	4	6	5	7	C	8	0	1
A	1	B	8	0	2	3	5	4	D	6	C	7	9
0	B	7	9	8	1	2	4	3	A	D	5	C	6

1	4	A	5	8	2	9	3	7	D	B	C	6	0
D	2	5	A	6	3	0	4	8	B	C	7	1	9
B	D	3	6	A	4	1	5	9	C	8	2	0	7
C	B	D	4	7	5	2	6	0	9	3	1	8	A
0	C	B	D	5	6	3	7	1	4	2	9	A	8
7	8	9	0	1	A	D	B	C	6	5	4	3	2
9	0	1	2	3	B	C	A	D	8	7	6	5	4
6	7	8	9	0	C	B	D	A	5	4	3	2	1
8	9	0	1	2	D	A	C	B	7	6	5	4	3
3	A	4	7	9	1	8	2	6	0	D	B	C	5
A	3	6	8	4	0	7	1	5	2	9	D	B	C
2	5	7	3	C	9	6	0	4	A	1	8	D	B
4	6	2	C	B	8	5	9	3	1	A	0	7	D
5	1	C	B	D	7	4	8	2	3	0	A	9	6

FIGURE 4

7. EXISTENCE OF LS(v,5).

From Theorem 2, we need only discuss LS(v,5) for v = 15,19,22,23,26,27,28,30, 34 and 38. Orders 19,22,23 and 38 follow from Corollary 9.1; order 15 comes from the Product Rule.

For $v = 28$, we know there exists a resolvable balanced incomplete block design with $v = 28$, $k = 4$ and $\lambda = 1$ (see, for example, [8]). Add a new point x to each line in one parallel class; the result is a PB(29;{5,4};1). Now delete one point y, other than x. We have a PB(28;{5,4,3};1). The 3-blocks are all disjoint, since they come from lines which contained y in the 29-point design. So Theorem 9 applies.

In Figure 5 we present the blocks of a PB(30;{8,7,5,4,3};1) in which there are blocks of size 8,5 and 4 which satisfy Theorem 9: the blocks ABC56789, 12345 and ALQ4 are satisfactory. So LS(30,5) exists. This design also proves the existence of LS(30,8), and gives an alternative construction for LS(30,4); however it cannot be used for LS(30,7), since all the blocks of size 7 intersect non-central points of brushes.

A J T	A P R 3	D J Q 5	H J S 8	E P U 2 8	I J R 2 9
A N U	B D T 1	D L U 6	H K U 5	F J V 3 6	I K T 4 7
B E S	B G Q 2	D P V 7	H M R 7	F L S 2 7	I L V 1 8
B F R	B H V 4	E L T 9	I M Q 6	F M U 4 9	1 2 3 4 5
C G K	B I U 3	E M V 5	I N S 5	G J U 1 7	A D E F G H I
C I P	C D M 2	F K Q 8	D K S 3 9	G M T 3 8	B J K L M N P
A K V 2	C E J 4	F P T 5	D N R 4 8	G P S 4 6	C Q R S T U V
A L Q 4	C F N 1	G L R 5	E K R 1 6	H N T 2 6	A B C 5 6 7 8 9
A M S 1	C H L 3	G N V 9	E N Q 3 7	H P Q 1 9	

FIGURE 5

We cover v = 34 by starting with the affine plane on 64 points, which is a PB(64;8;1) in which the blocks fall into nine parallel classes. Delete three parallel lines, and delete five points from some line ℓ which was parallel to the first three. Finally, we delete one further point x, not on ℓ. This produces some more blocks of size 3; but the new 3-blocks made derive from blocks which contained x and also one of the points which were deleted from ℓ, so they are all disjoint from each other and from ℓ. There are still many blocks of size 5, and three of size 8. So LS(34,5) and LS(34,8) exist.

0	4	A	V	W	7	B	E	X	Y	Z	6	I	C	5	J	H	D	F	9	8	G	2	1	3	K
9	1	5	B	V	W	8	C	F	X	Y	Z	7	J	D	6	K	I	E	G	A	H	3	2	4	0
B	A	2	6	C	V	W	9	D	G	X	Y	Z	8	K	E	7	0	J	F	H	I	4	3	5	1
I	C	B	3	7	D	V	W	A	E	H	X	Y	Z	9	0	F	8	1	K	G	J	5	4	6	2
H	J	D	C	4	8	E	V	W	B	F	I	X	Y	Z	A	1	G	9	2	0	K	6	5	7	3
1	I	K	E	D	5	9	F	V	W	C	G	J	X	Y	Z	B	2	H	A	3	0	7	6	8	4
4	2	J	0	F	E	6	A	G	V	W	D	H	K	X	Y	Z	C	3	I	B	1	8	7	9	5
C	5	3	K	1	G	F	7	B	H	V	W	E	I	0	X	Y	Z	D	4	J	2	9	8	A	6
K	D	6	4	0	2	H	G	8	C	I	V	W	F	J	1	X	Y	Z	E	5	3	A	9	B	7
6	0	E	7	5	1	3	I	H	9	D	J	V	W	G	K	2	X	Y	Z	F	4	B	A	C	8
G	7	1	F	8	6	2	4	J	I	A	E	K	V	W	H	0	3	X	Y	Z	5	C	B	D	9
Z	H	8	2	G	9	7	3	5	K	J	B	F	0	V	W	I	1	4	X	Y	6	D	C	E	A
Y	Z	I	9	3	H	A	8	4	6	0	K	C	G	1	V	W	J	2	5	X	7	E	D	F	B
X	Y	Z	J	A	4	I	B	9	5	7	1	0	D	H	2	V	W	K	3	6	8	F	E	G	C
7	X	Y	Z	K	B	5	J	C	A	6	8	2	1	E	I	3	V	W	0	4	9	G	F	H	D
5	8	X	Y	Z	0	C	6	K	D	B	7	9	3	2	F	J	4	V	W	1	A	H	G	I	E
2	6	9	X	Y	Z	1	D	7	0	E	C	8	A	4	3	G	K	5	V	W	B	I	H	J	F
W	3	7	A	X	Y	Z	2	E	8	1	F	D	9	B	5	4	H	0	6	V	C	J	I	K	G
V	W	4	8	B	X	Y	Z	3	F	9	2	G	E	A	C	6	5	I	1	7	D	K	J	0	H
8	V	W	5	9	C	X	Y	Z	4	G	A	3	H	F	B	D	7	6	J	2	E	0	K	1	I
3	9	V	W	6	A	D	X	Y	Z	5	H	B	4	I	G	C	E	8	7	K	F	1	0	2	J
D	E	F	G	H	I	J	K	0	1	2	3	4	5	6	7	8	9	A	B	C	V	Z	Y	X	W
J	K	0	1	2	3	4	5	6	7	8	9	A	B	C	D	E	F	G	H	I	X	W	V	Z	Y
E	F	G	H	I	J	K	0	1	2	3	4	5	6	7	8	9	A	B	C	D	Z	Y	X	W	V
F	G	H	I	J	K	0	1	2	3	4	5	6	7	8	9	A	B	C	D	E	W	V	Z	Y	X
A	B	C	D	E	F	G	H	I	J	K	0	1	2	3	4	5	6	7	8	9	Y	X	W	V	Z

FIGURE 6

0	2	Y	1	X	A	C	E	B	D	K	M	P	L	N	5	7	9	6	8	F	H	J	G	I	3	4
X	3	0	Y	4	B	D	A	C	E	L	N	K	M	P	6	8	5	7	9	G	I	F	H	J	1	2
2	X	1	3	Y	C	E	B	D	A	M	P	L	N	K	7	9	6	8	5	H	J	G	I	F	4	0
Y	0	X	4	1	D	A	C	E	B	N	K	M	P	L	8	5	7	9	6	I	F	H	J	G	2	3
4	Y	3	X	2	E	B	D	A	C	P	L	N	K	M	9	6	8	5	7	J	G	I	F	H	0	1
5	6	7	8	9	F	H	Y	G	X	0	2	4	1	3	A	C	E	B	D	K	M	P	L	N	I	J
7	8	9	5	6	X	I	F	Y	J	1	3	0	2	4	B	D	A	C	E	L	N	K	M	P	G	H
9	5	6	7	8	H	X	G	I	Y	2	4	1	3	0	C	E	B	D	A	M	P	L	N	K	J	F
6	7	8	9	5	Y	F	X	J	G	3	0	2	4	1	D	A	C	E	B	N	K	M	P	L	H	I
8	9	5	6	7	J	Y	I	X	H	4	1	3	0	2	E	B	D	A	C	P	L	N	K	M	F	G
A	B	C	D	E	K	L	M	N	P	5	7	Y	6	X	F	H	J	G	I	0	2	4	1	3	8	9
C	D	E	A	B	M	N	P	K	L	X	8	5	Y	9	G	I	F	H	J	1	3	0	2	4	6	7
E	A	B	C	D	P	K	L	M	N	7	X	6	8	Y	H	J	G	I	F	2	4	1	3	0	9	5
B	C	D	E	A	L	M	N	P	K	Y	5	X	9	6	I	F	H	J	G	3	0	2	4	1	7	8
D	E	A	B	C	N	P	K	L	M	9	Y	8	X	7	J	G	I	F	H	4	1	3	0	2	5	6
F	G	H	I	J	0	1	2	3	4	A	B	C	D	E	K	M	Y	L	X	5	7	9	6	8	N	P
H	I	J	F	G	2	3	4	0	1	C	D	E	A	B	X	N	K	Y	P	6	8	5	7	9	L	M
J	F	G	H	I	4	0	1	2	3	E	A	B	C	D	M	X	L	N	Y	7	9	6	8	5	P	K
G	H	I	J	F	1	2	3	4	0	B	C	D	E	A	Y	K	X	P	L	8	5	7	9	6	M	N
I	J	F	G	H	3	4	0	1	2	D	E	A	B	C	P	Y	N	X	M	9	6	8	5	7	K	L
K	L	M	N	P	5	6	7	8	9	F	G	H	I	J	0	1	2	3	4	A	C	E	Y	B	D	X
M	N	P	K	L	7	8	9	5	6	H	I	J	F	G	2	3	4	0	1	B	D	X	A	C	E	Y
P	K	L	M	N	9	5	6	7	8	J	F	G	H	I	4	0	1	2	3	C	E	Y	B	D	X	A
L	M	N	P	K	6	7	8	9	5	G	H	I	J	F	1	2	3	4	0	D	X	A	C	E	Y	B
N	P	K	L	M	8	9	5	6	7	I	J	F	G	H	3	4	0	1	2	E	Y	B	D	X	A	C
1	4	2	0	3	G	J	H	F	I	6	9	7	5	8	L	P	M	K	N	X	A	C	E	Y	B	D
3	1	4	2	0	I	G	J	H	F	8	6	9	7	5	N	L	P	M	K	Y	B	D	X	A	C	E

FIGURE 7

Finally, we exhibit an LS(26,5) and an LS(27,5). In fact, both are self-orthogonal squares with self-orthogonal subsquares. The square of side 26, shown in Figure 6, has a subsquare of side 5 in its lower right corner. The square of side 27 is shown in Figure 7. It has many subsquares of side 5. However, the subsquare based on $\{0,5,A,F,K\}$, made up of the rows and columns which have these symbols on the diagonal, is a self-orthogonal subsquare and is symmetrically situated. Notice further that the bottom right corner is a self-orthogonal subsquare of side 7. So the array is both an SOLS(27,5) and an SOLS(27,7).

We have proven:

Theorem 14. *There is an LS(v,5) for all $v \geq 15$.*

REFERENCES

[1] F.E. Bennett and N.S. Mendelsohn, Conjugate orthogonal Latin square graphs. *Congressus Num.* 23 (1979), 179-192.

[2] F.E. Bennett and N.S. Mendelsohn, On the spectrum of Stein quasigroups. *Bull. Austral. Math. Soc.* 21 (1980), 47-63.

[3] J. Denés and A.D. Keedwell, *Latin Squares and Their Applications*. Academic Press, London, 1974.

[4] J.H. Dinitz and D.R. Stinson, MOLS with holes. *Discrete Math.* (to appear).

[5] D.A. Drake and J.A. Larson, Pairwise balanced designs whose line sizes do not divide six. *J. Combinatorial Theory* (Series A) (to appear).

[6] D.A. Drake and H. Lenz, Orthogonal Latin squares with orthogonal subsquares. *Archiv. der Math.* 34 (1980), 565-576.

[7] H. Hanani, On balanced incomplete block designs with blocks having five elements. *J. Combinatorial Theory* (Series A) 12 (1972), 184-201.

[8] H. Hanani, D.K. Ray-Chaudhuri and R.M. Wilson, On resolvable designs. *Discrete Math.* 3 (1972), 343-357.

[9] A. Hedayat and E. Seiden, On the theory and application of sum composition of Latin squares and orthogonal Latin squares. *Pacific J. Math.* 54 (1974), 85-113.

[10] K. Heinrich, Self-orthogonal Latin squares with self-orthogonal subsquares. *Ars Combinatoria* 3 (1977), 251-266.

[11] E.T. Parker, Nonextendibility conditions on mutually orthogonal Latin squares. *Proc. Amer. Math. Soc.* 13 (1962), 219-221.

[12] W.D. Wallis, Spouse-avoiding mixed doubles tournaments. *Ann. NY Acad. Sci.* 319 (1979), 549-554.

[13] W.D. Wallis and L. Zhu, Existence of orthogonal diagonal Latin squares. *Ars Combinatoria* 12 (1981), 51-68.

[14] S.P. Wang, *On self-orthogonal Latin squares and partial transversals of Latin squares.* Ph.D. Thesis, Ohio State University, 1978.

[15] Zhu Lie, On a method of sum composition to construct orthogonal Latin squares. *Acta Math. Appl. Sinica* 3 (1977), 56-61. (Chinese.)

[16] L. Zhu, Orthogonal diagonal Latin squares of order 14. *J. Austral. Math. Soc.* Series A (to appear).

k-SETS OF $(n-1)$-DIMENSIONAL SUBSPACES OF $PG(3n-1,q)$

L.R.A. CASSE AND P.R. WILD

A k-set of $PG(3n-1,q)$ is a collection of $(n-1)$-dimensional subspaces any three of which generate the whole space. We give a representation of k-sets with certain properties as a collection of k-arcs in a translation plane. We give a characterization of the (q^n+1)-sets, q odd, of Thas [3].

1. INTRODUCTION

A k-set of $PG(3n-1,q)$ is a collection of k $(n-1)$-dimensional subpsaces such that any three generate the whole space.

Thas [3] shows that if K is a k-set of $PG(3n-1,q)$ then

(i) $k \le q^n+1$ if q is odd

and (ii) $k \le q^n+2$ if q is even.

Also, if K is a (q^n+1)-set of $PG(3n-1,q)$, then each member X of K is contained in a unique $(2n-1)$-dimensional subspace Y which is skew to the remaining members of K; Y is called the tangent space to K at X. If q is even, the q^n+1 tangent spaces to a (q^k+1)-set K meet in an $(n-1)$-dimensional subspace N, called the nucleus of K; $K \cup \{N\}$ is a (q^n+2)-set. If q is odd, the q^n+1 tangent spaces to the (q^n+1)-set K form a dual (q^n+1)-set.

Thas [3] has constructed (q^n+1)-sets using ovals of $PG(2,q^n)$ in the following way. Let $GF(q^n)$ be an extension of $GF(q)$ and let $PG(3n-1,q^n)$ be the corresponding extension of $PG(3n-1,q)$. Let π be a plane of $PG(3n-1,q^n)$ such that π and its conjugates generate the whole space. Each point of π with its conjugates generate an $(n-1)$-dimensional subspace of $PG(3n-1,q^n)$ which contains an $(n-1)$-dimensional subspace of $PG(3n-1,q)$. Thas [3] shows that the points of a (q^n+1)-arc in π determine in this way the subspaces of a (q^n+1)-set in $PG(3n-1,q)$.

In this paper we give a construction for k-sets of $PG(3n-1,q)$ from k-arcs in a translation plane representable as a spread in $PG(2n-1,q)$. The k-sets we construct have the property that there is a $(2n-1)$-dimensional subspace S which either contains or is skew to each member of the k-set, with the property that the projections onto

S of the other members from each member of the k-set skew to S all belong to a fixed spread in S. We show that they are characterised by this property. It is easy to verify that the (q^n+1)-sets of Thas [3] have this property, and so arise from our method when the translation plane is desarguesian. In the last section we show that, when q is odd, the example of Thas [3] is characterised by the property that the projection from just one member of the (q^n+1)-set yields a regular spread.

2. SPREADS AND TRANSLATION PLANES

The reader is referred to Bruck and Bose [1] for details of the following remarks on spreads and translation planes.

A spread of $PG(2n-1,q)$ is a collection W of q^n+1 mutually skew $(n-1)$-dimensional subspaces.

A translation plane is a projective plane which has a line ℓ with the property that for any pair of points not on ℓ there is an elation with axis ℓ mapping one of the points to the other.

A spread W of $PG(2n-1,q)$ yields a translation plane of order q^n in the following way. Embed $PG(2n-1,q)$ as a hyperplane in $PG(2n,q)$. The structure whose points are the points of $PG(2n,q)\setminus PG(2n-1,q)$ and whose lines are the n-dimensional subspaces of $PG(2n,q)$ whose intersection with $PG(2n-1,q)$ is a member of W, with the natural incidence, is an affine translation plane. This affine plane may be completed to a projective plane π by adjoining $PG(2n-1,q)$ as the line at infinity and the members of W as the points on it. Conversely any translation plane of order q^n whose kernel contains $GF(q)$ may be represented in this way. π may be coordinatized in the following way. The points of $PG(2n,q)$ correspond to 1-dimensional vector subspaces of $(2n+1)$-dimensional vector space V. $PG(2n-1,q)$ corresponds to a 2n-dimensional vector subspace of V. Each $X \in W$ corresponds to an n-dimensional vector subspace of V. Denote by X_∞, X_0, and X_I, three subspaces in W, corresponding to vector subspaces $J(\infty)$, $J(O)$, and $J(I)$. Bases $e_1,\ldots,e_n$ for $J(\infty)$ and $e_1',\ldots,e_n'$ for $J(O)$ may be chosen so that $e_1+e_1',\ldots,e_n+e_n'$ is a basis for $J(I)$. Then corresponding to each member X_U of $W\setminus\{X_\infty\}$ is a vector subspace of the form $J(U) = \{xU+x' \mid x \in J(\infty)\}$ where U is a linear transformation of $J(\infty)$, and if $x = \sum_{i=1}^{n} x_ie_i$ then $x' = \sum_{i=1}^{n} x_ie_i'$. The set S of linear transformations of $J(\infty)$ obtained in this way is called a spread set. $O,I \in S$ and for any $X,Y \in S$, $X-Y$ is non-singular and for any $x,y \in J(\infty)$ there is a unique $Z \in S$ with $y = xZ$.

Extend $e_1,\ldots,e_n,e_1',\ldots,e_n'$ to a basis for V by adjoining e^*. Then any point of $PG(2n,q)\backslash PG(2n-1,q)$ corresponds to a 1-dimensional vector subspace containing a vector $y+x'+e^*$ where $x,y \in J(\infty)$. This point of π is given the coordinates (x,y). Choose a non-zero unit vector $1 \in J(\infty)$ (or a unit point $(1,1)$ of π). The point on the line at infinity corresponding to $X_U \in W\backslash\{X_\infty\}$ is given the coordinate $(1U)$. The point corresponding to X_∞ is given the coordinate (∞). Thus π is coordinatized by $J(\infty)$ which becomes a quasifield by defining multiplication so that $ab = aB$ where $b = 1B$ for $B \in S$.

A (q^n+1)-set K of $PG(3n-1,q)$ gives rise to spreads (and hence translation planes) in the following way. Let $X \in K$ and let S_{2n-1} be a $(2n-1)$-dimensional subspace of $PG(3n-1,q)$ skew to X. Any $Y \in K\backslash\{X\}$ generates with X a $(2n-1)$-dimensional subspace which intersects S_{2n-1} in an $(n-1)$-dimensional subspace Y'. The tangent space to K at X meets S_{2n-1} in an $(n-1)$-dimensional subspace X'. $W = \{Y' | Y \in K\backslash\{X\}\} \cup \{X'\}$ is a set of q^n+1 $(n-1)$-dimensional subspaces of S_{2n-1} which are mutually skew and so form a spread. We say the projection of K from X onto S_{2n-1} yields the spread W.

If S_{2n} is any $2n$-dimensional subspace of $PG(3n-1,q)$ containing S_{2n-1} then W determines a translation plane π whose affine points are the points of $S_{2n}\backslash S_{2n-1}$. Suppose S_{2n-1} has the property that every member of K is either skew to it or contained in it. Then every $Y \in K$ either lies in W and so determines a point on the line at infinity of π or meets S_{2n} in exactly one point and so determines an affine point of π. Thus for a suitable choice of S_{2n-1}, the (q^n+1)-set K determines a set of q^n+1 points of π.

3. k-SETS FROM TRANSLATION PLANES

In this section we give our construction for k-sets of $PG(3n-1,q)$. Let π be a translation plane of order q^n with translation line ℓ, representable as a spread W in $PG(2n-1,q)$ and coordinatized by the quasifield $J(\infty)$ (an n-dimensional vector space over $GF(q)$).

Consider n k-arcs $K_1,\ldots,K_n$ of π with the following properties:

(i) any point of ℓ belonging to one of the k-arcs belongs to all of them;

(ii) $K_1,\ldots,K_n$ has a common affine point 0 with coordinates $(0,0)$;

(iii) $K_1,\ldots,K_n$ are in perspective from 0 with axis ℓ (i.e.

for all i,j if $X_i \in K_i\setminus\{0\}$ there is a $X_j \in K_j\setminus\{0\}$ with $0, X_i, X_j$ collinear and if $X_i, Y_i \in K_i$ and $X_j, Y_j \in K_j$ with $0, X_i, X_j$ collinear and $0, Y_i, Y_j$ collinear then the lines X_iY_i and X_jY_j meet in a point on ℓ);

(iv) if a line through 0 meets $K_1,\dots,K_n$ again in affine points with coordinates $(x_1,y_1),\dots,(x_n,y_n)$ then the vectors $x_1,\dots,x_n$ form a basis for $J(\infty)$.

Theorem 1. *Let* π, $K_1,\dots,K_n$ *have the properties described above. Then there exists a* k-*set* K *in* $PG(3n-1,q)$.

Proof. Let S_{2n-1} be a $(2n-1)$-dimensional subspace of $PG(3n-1,q)$. Let W be a spread of S_{2n-1}. Let $S^1_{2n},\dots,S^n_{2n}$ be n $2n$-dimensional subspaces of $PG(3n-1,q)$ containing S_{2n-1} which generate $PG(3n-1,q)$. π may be represented by W in each of $S^1_{2n},\dots,S^n_{2n}$. The point 0 of π determines n points of $PG(3n-1,q)$, one in each of $S^1_{2n},\dots,S^n_{2n}$. These n points generate an $(n-1)$-dimensional subspace $0'$ of $PG(3n-1,q)$. By properties (i) and (iii) a line of π through 0 meets $K_1,\dots,K_n$ again in a point of ℓ common to them, in n affine points, or in no point at all. Suppose a line of π through 0 meets $K_1,\dots,K_n$ again in n affine points with coordinates $(x_1,y_1),\dots,(x_n,y_n)$. Let A_i be the point of S^i_{2n} determined by (x_i,y_i). The n points $A_1,\dots,A_n$ generate an $(n-1)$-dimensional subspace of $PG(3n-1,q)$. Any point of ℓ common to $K_1,\dots,K_n$ corresponds to an $(n-1)$-dimensional subspace belonging to W. Hence it is clear that $K_1,\dots,K_n$ determine a set K of k $(n-1)$-dimensional subspaces. We show that K is a k set.

By (iv) it follows that the subspace S generated by $0'$ and a subspace in $K\setminus\{0'\}$ meets S_{2n-1} in an $(n-1)$-dimensional subspace. Hence S has dimension $2n-1$ and so $0'$ is skew to each member of $K\setminus\{0'\}$. Further by (iii) the subspaces in $K\setminus\{0'\}$ are mutually skew since they are projected from $0'$ into distinct members of W in S_{2n-1}. Now by (iii) the $(2n-1)$-dimensional subspace generated by two subspaces of K intersects S_{2n-1} in a member of W and since $K_1,\dots,K_n$ are k-arcs it follows that no three subspaces of K pairwise determine the same subspace of W. Hence any three members of K generate $PG(3n-1,q)$ and K is a k-set. □

The k-sets we have constructed in Theorem 1 have the property that there exists a $(2n-1)$-dimensional subspace S_{2n-1} with the property that each element of the k-set either is contained in S_{2n-1} or is skew to it and is such that the projection of the k-set from each member skew to S_{2n-1} onto S_{2n-1} yields (n-1)-dimensional subspaces belonging to a given spread. We show that they are characterized by this property.

<u>Theorem 2</u>. *Let* K *be a* k-*set in* $PG(3n-1,q)$. *Suppose* S_{2n-1} *is* $(2n-1)$-*dimensional subspace of* $PG(3n-1,q)$ *with the property that each* $X \in K$ *either is contained in* S_{2n-1} *or is skew to it. Suppose the projection of* K *from* $X \in K$ *onto* S_{2n-1} *yields* $(n-1)$-*dimensional subspaces belonging to a given spread* W *for each* X *skew to* S_{2n-1}. *Then* K *determines* n k-*arcs in the translation plane* π *determined by* W *and these arcs have properties (i), (ii), (iii) and (iv).*

<u>Proof</u>. Let $S_{2n}^1,\ldots,S_{2n}^n$, be n 2n-dimensional subspaces of $PG(3n-1,q)$ containing S_{2n-1} which generate $PG(3n-1,q)$. Each subspace in K contained in S_{2n-1} belongs to W. Any other subspace in K meets each of $S_{2n}^1,\ldots,S_{2n}$ in a point (not belonging to S_{2n-1}). Since the points of $S_{2n}^i \backslash S_{2n-1}$ and the subspaces of W correspond to the points of π, K and S_{2n}^i determine a set K_i of k points in π for each i. π may be coordinatized as before by an n-dimensional vector space over $GF(q)$, and with a suitable choice of bases for the vector spaces corresponding to $S_{2n}^1,\ldots,S_{2n}^n$ there exists $0 \in K$ giving rise to the point $(0,0)$ of π in each S_{2n}^i. Thus (ii) holds. Any $X \in K$ belonging to W determines the same point on the line at infinity in each S_{2n}^i, and any other $X \in K\backslash\{0\}$ gives rise to n affine points $(x_1,y_1),\ldots,(x_n,y_n)$ of π on a line through $(0,0)$. Thus (i) holds. Further since 0 is skew to each $X \in K\backslash\{0\}$ the n vectors $x_1,\ldots,x_n$ are linearly independent so that (iv) holds. Now the (2n-1)-dimensional subspace of $PG(3n-1,q)$ containing two members of K meets S_{2n-1} in an (n-1)-dimensional subspace belonging to W. Since K is a k-set it follows that each set K_i in π is a k-arc and further they are in perspective from $(0,0)$ with axis the line at infinity. Hence (iii) holds and the proof is complete. □

The examples of (q^n+1)-sets given by Thas [3] satisfy the hypothesis of Theorem 2. Indeed any line of the plane π of $PG(3n-1,q^n)$, with its conjugates generates a suitable S_{2n-1} for which the spread W is the regular spread consisting of the q^n+1 subspaces generated by each point of that line with its conjugates. Hence the examples of Thas may be represented by n (q^n+1)-arcs of the desarguesian plane satisfying properties (i), (ii), (iii) and (iv). It follows that these (q^n+1)-arcs are images of each other under homologies with centre 0 and axis ℓ.

4. (q^n+1)-SETS WITH REGULAR PROJECTIONS

We have seen that the (q^n+1)-sets of Thas [3] have the property that the projection from each member yields a regular spread. We show that when q is odd they are characterized by the property that the projection from just one member yields a regular spread.

Theorem 3. *Let K be a (q^n+1)-set of $PG(3n-1,q)$ with q odd. Suppose that the projection of K from some $X \in K$ onto a $(2n-1)$-dimensional subspace S_{2n-1} skew to X yields a regular spread W. Then K results from the construction of Thas [3].*

Proof. Let $GF(q^n)$ be a field extension of $GF(q)$. Let $PG(3n-1,q^n)$ be the corresponding extension of $PG(3n-1,q)$. Let S'_{2n-1} be the subspace of $PG(3n-1,q^n)$ containing S_{2n-1}. Since W is a regular spread there is a line T of S'_{2n-1} which meets the extension of each member of W in a point of $PG(3n-1,q^n)$ (Dembowski, [2], p. 133). Let $S(X)$ be the tangent space to K at X and let S'_{2n} be the $2n$-dimensional subspace of $PG(3n-1,q^n)$ which contains $S(X)$ and T. We may coordinatize $PG(3n-1,q^n)$ by homogeneous $3n$-tuples $(x_0,\dots,x_{3n-1})$ so that X has equations $x_n = x_{n+1} = \dots = x_{3n-1} = 0$, T has equations $x_0 = \dots = x_{n-1} = 0$, $\lambda^{n-1}x_n = \lambda^{n-2}x_{n+1} = \dots = x_{2n-1}$, $\lambda^{n-1}x_{2n} = \dots = x_{3n-1}$, and $S(X)$ has equations $x_{2n} = x_{2n+1} = \dots = x_{3n-1} = 0$, where λ is a primitive element of $GF(q^n)$.

Let $K = \{X, Y_1, \dots, Y_{q^n}\}$ and let P_i be the point of intersection of S'_{2n} with the extension of Y_i. P_i has coordinates $(a_{i0},\dots,a_{in-1},c_i,c_i\lambda,\dots,c_i\lambda^{n-1},1,\lambda,\dots,\lambda^{n-1})$ for some $a_{i0},\dots,a_{in-1},c_i \in GF(q^n)$. The line joining P_i and P_j, $i \neq j$, meets the extension of $S(X)$ in the point P_{ij} with coordinates $(a_{i0}-a_{j0},\dots,a_{in-1}-a_{jn-1},c_i-c_j,(c_i-c_j)\lambda,\dots,(c_i-c_j)\lambda^{n-1},0,\dots,0)$. The plane determined by three points P_i,P_j,P_k, $i \neq j \neq k \neq i$, meets the extension of $S(X)$ in the line joining P_{ij} and P_{ik} which meets the extension of X in the point with coordinates

$$\left(\frac{a_{i0}-a_{j0}}{c_i-c_j} - \frac{a_{i0}-a_{k0}}{c_i-c_k},\dots,\frac{a_{in-1}-a_{jn-1}}{c_i-c_j} - \frac{a_{in-1}-a_{kn-1}}{c_i-c_j},0,\dots,0\right).$$

Since K is a (q^n+1)-set the conjugates in $GF(q^n)$ over $GF(q)$ of this point must generate X. Hence for all $\ell=0,\dots,n-1$

$$\frac{a_{i\ell}-a_{j\ell}}{c_i-c_j} - \frac{a_{i\ell}-a_{k\ell}}{c_i-c_j} \neq 0 .$$

Hence the points $(1,c_i,a_{i\ell})$, $i=1,\dots,q^n$, of $PG(2,q^n)$ form a q^n-arc for each ℓ. Since q is odd each of these q^n-arcs is contained in a unique conic. All these conics contain the point $(0,0,1)$. It follows that for $\ell=0,\ldots,n-1$, $a_{i\ell} = \alpha_\ell c_i^2+\beta_\ell c_i+\gamma_\ell$ for some $\alpha_\ell, \beta_\ell, \gamma_\ell \in GF(q^n)$, $i=1,\ldots,q^n$ and so $a_{i\ell} = \alpha'_\ell a_{i1}+\beta'_\ell c_i+\gamma'_\ell$ for some $\alpha'_\ell, \beta'_\ell, \gamma'_\ell \in GF(q^n)$, $i=1,\ldots,q^n$.

It follows that the points P_i, $i=1,\ldots,q^n$ lie on a plane of $PG(3n-1,q^n)$ and K may be constructed as in Thas [3]. □

REFERENCES

[1] R.H. Bruck and R.C. Bose, The construction of translation planes from projective spaces, *J. of Algebra*, 1, (1964), 85-102.

[2] P. Dembowski, *Finite Geometries*, Springer-Verlag, New York, 1968.

[3] J.A. Thas, The m-dimensional projective space $S_m(M_n(GF(q)))$ over the total matrix algebra $M_n(GF(q))$ on $n\times n$ matrices with elements in the Galois field $GF(q)$, *Rend. Mat.*, 4, (1971), 459-532.

SUBTREES OF LARGE TOURNAMENTS

NICHOLAS C. WORMALD

By an oriented graph we mean a graph in which each edge has been directed. A tournament is an oriented complete graph. Let f(n) *by the least integer for which every tournament on* f(n) *vertices contains every oriented tree on* n *vertices. D. Sumner has conjectured that* $f(n) = 2n - 2$. *F.R.K. Chung has shown that* $f(n) \leqslant cn \exp((\log n)^{\varepsilon+\frac{1}{2}})$ *for any* $\varepsilon > 0$ *and for some* c *depending on* ε. *In this note this bound is improved to* $f(n) \leqslant n \log_2(2n/e)$, *for* $n \geqslant 4$.

It was conjectured by Sumner [5] that for $n \geqslant 2$ every tournament on $2n - 2$ vertices contains every orientation of every tree on n vertices (or n-*tree*). Burr [1] probably intended to make the more general conjecture that every orientation of every (2n-2)-chromatic graph contains every orientation of every n-tree. (He actually conjectured 2n-chromatic instead.) But even Sumner's conjecture seems to be extremely difficult. Consideration of the orientation of $K_{1,n-1}$ in which the vertex of degree $n - 1$ is a sink shows why $2n - 2$ cannot be replaced by a smaller integer in either conjecture.

In the positive direction, Sumner's conjecture was verified in [4] for certain classes of trees and tournaments, in particular for all trees in case the tournament is near-regular and for all tournaments in case the tree is an oriented path or an oriented caterpillar whose "spine" is a directed path. In [1], Burr verified the weakened version of his conjecture, that for $n \geqslant 3$ every $(n-1)^2$-chromatic oriented graph contains every oriented n-tree. Similarly, let f(n) denote the least integer such that every tournament on f(n) vertices contains every oriented n-tree. As already observed, $f(n) \geqslant 2n - 2$, so Sumner's conjecture becomes that $f(n) = 2n - 2$ for $n \geqslant 2$. Chung [2] established the upper bound $f(n) \leqslant cn \exp((\log n)^{\frac{1}{2}+\varepsilon})$ for any $\varepsilon > 0$ and some constant c depending on ε. In the theorem below, this bound is sharpened to $f(n) \leqslant n \log_2(2n/e)$ for $n \geqslant 4$. Terminology not defined here is explained by Harary [3].

We first isolate a property of all tournaments in general, which is probably well-known. By a *dominating set* of a tournament T we mean a subset Q of the vertices of T with the property that each vertex in $T - Q$ is adjacent from some vertex in Q.

Lemma. *Each tournament* T *on* k *vertices has a dominating set of cardinality at most* $\lfloor \log_2(k+1) \rfloor$.

Proof. This is by induction on k. The case $k = 1$ is immediate, so take

$k > 1$. Since the mean indegree of vertices of T is $(k-1)/2$ there is a vertex v of indegree at most $\lfloor (k-1)/2 \rfloor$. By induction, the tournament T' induced by the vertices adjacent to v has a dominating set Q' of cardinality at most $\lfloor \log_2 \lfloor (k+1)/2 \rfloor \rfloor$. Thus $Q' \cup \{v\}$ is a dominating set of T, of cardinality at most $\lfloor \log_2 (k+1) \rfloor$, which completes the inductive step.

Theorem. *For* $n \geq 4$,

$$f(n) \leq n \log_2 (2n/e).$$

Proof. Let $m = f(n-1) + \lfloor \log_2 (n-1) \rfloor + 1$, with $f(n)$ defined as above, and let T be a tournament on m vertices. Let D denote the subtournament of T induced by the vertices of indegree at most $n-2$. There are at most $2n-3$ such vertices, since if D had more vertices than this, at least one would have indegree at least $n-1$ in D and hence indegree at least $n-1$ in T. Hence by the Lemma, D has a dominating set Q of cardinality at most $\lfloor \log_2 (n-1) \rfloor + 1$.

If t is an oriented n-tree with a vertex v of indegree 0 and outdegree 1, then $t - v$ is contained in $T - Q$ as $|V(T-Q)| \geq f(n-1)$. Denote the vertex set of a copy of $t - v$ in $T - Q$ by S, and let u denote the vertex in S which is a copy of the vertex adjacent from v in t. Then either u has indegree at least $n-1$, or u is in D in which case it is adjacent from a vertex in Q. In either case, u is adjacent from some vertex v' in $T - S$. Thus v' together with the copy of $t - v$ with vertex set S yields a copy of t in T. Similarly, if t is an oriented n-tree with a vertex of indegree 1 and outdegree 0, the converse argument with indegrees replaced by outdegrees, etc., shows that t is again in T. Thus T contains every oriented n-tree. So $m \geq f(n)$, that is,

$$f(n) \leq f(n-1) + \lfloor \log_2 (n-1) \rfloor + 1. \qquad (1)$$

We proceed by induction on n. The theorem can be verified for $n = 4$ by checking $f(4) = 6$ by brute force. In any case, the results of [4] verify $f(n) = 2n - 2$ for $2 \leq n \leq 8$, so we may assume

$$f(n-1) \leq (n-1) \log_2 (2(n-1)/e).$$

Hence the upper bound (1) on $f(n)$ yields

$$\begin{aligned} f(n) &\leq n \log_2 (n-1) + n \log_2 (2/e) - \log_2 (2/e) + 1 \\ &\leq (n \log n - n(\log n - \log(n-1)) + n \log(2/e) + 1)/\log 2 \\ &\leq (n \log n + n \log(2/e))/\log 2 \\ &\leq n \log_2 (2n/e) \end{aligned}$$

as required.

ACKNOWLEDGEMENT

The author is grateful for support by the Australian Department of Science and Technology, under the Queen Elizabeth II Fellowship Scheme.

REFERENCES

[1] S.A. Burr, Subtrees of directed graphs and hypergraphs, *Congressus Numerantium* 28 (1980), 227-239.

[2] F.R.K. Chung, A note on subtrees in tournaments, Internal Memorandum of Bell Laboratories (1981).

[3] F. Harary, *Graph Theory*. Addison-Wesley, Reading, Mass., (1969).

[4] K.B. Reid and N.C. Wormald, Embedding oriented n-trees in tournaments, to appear.

[5] D. Sumner, Private communication with K.B. Reid.